P9-CAD-932

DATA HANDLING IN SCIENCE AND TECHNOLOGY

Volume 1 Microprocessor Programming and Applications for Scientists and Engineers
by R.R. Smardzewski

Volume 2 Chemometrics, a textbook
by D.L. Massart, B.G.M. Vandeginste, S.N. Deming, Y. Michotte and L. Kaufman

Volume 3 Experimental design: a chemometric approach
by S.N. Deming and S.L. Morgan

DATA HANDLING IN SCIENCE AND TECHNOLOGY — VOLUME 3

Experimental design: a chemometric approach

STANLEY N. DEMING

Department of Chemistry, University of Houston, Houston, TX 77004, U.S.A.

and

STEPHEN L. MORGAN

Department of Chemistry, University of South Carolina, Columbia, SC 29208, U.S.A.

ELSEVIER

Amsterdam — Oxford — New York — Tokyo 1987

ELSEVIER SCIENCE PUBLISHERS B.V.
Sara Burgerhartstraat 25
P.O. Box 211, 1000 AE Amsterdam, The Netherlands

Distributors for the United States and Canada:

ELSEVIER SCIENCE PUBLISHING COMPANY INC.
655, Avenue of the Americas
New York, NY 10010, U.S.A.

First edition 1987
Second impression 1988
Third impression 1990

Library of Congress Cataloging-in-Publication Data

Deming, Stanley N., 1944-
Experimental design.

(Data handling science and technology ; v. 3)
Bibliography: p.
Includes index.
1. Chemistry, Analytic--Statistical methods.
2. Experimental design. I. Morgan, Stephen L.,
1949- . II. Title. III. Series.
QD75.4.S8D46 1987 543'.00724 86-32825

ISBN 0-444-42734-1 (Hardbound)
ISBN 0-444-43032-6 (Paperback)

Printed in The Netherlands

Contents

To
Bonnie, Stephanie, and Michael,
and to
Linda

Preface

As analytical chemists, we are often called upon to participate in studies that require the measurement of chemical or physical properties of materials. In many cases, it is evident that the measurements to be made will not provide the type of information that is required for the successful completion of the project. Thus, we find ourselves involved in more than just the measurement aspect of the investigation – we become involved in carefully (re)formulating the questions to be answered by the study, identifying the type of information required to answer those questions, making appropriate measurements, and interpreting the results of those measurements. In short, we find ourselves involved in the areas of experimental design, data acquisition, data treatment, and data interpretation.

These four areas are not separate and distinct, but instead blend together in practice. For example, data interpretation must be done in the context of the original experimental design, within the limitations of the measurement process used and the type of data treatment employed. Similarly, data treatment is limited by the experimental design and measurement process, and should not obscure any information that would be useful in interpreting the experimental results. The experimental design itself is influenced by the data treatment that will be used, the limitations of the chosen measurement process, and the purpose of the data interpretation.

Data acquisition and *data treatment* are today highly developed areas. Fifty years ago, measuring the concentration of fluoride ion in water at the parts-per-million level was quite difficult; today it is routine. Fifty years ago, experimenters dreamed about being able to fit models to large sets of data; today it is often trivial.

Experimental design is also today a highly developed area, but it is not easily or correctly applied. We believe that one of the reasons "experimental design" is not used more frequently (and correctly) by scientists is because the subject is usually taught from the point of view of the statistician rather than from the point of view of the researcher. For example, one experimenter might have heard about factorial designs at some point in her education, and applies them to a system she is currently investigating; she finds it interesting that there is a "highly significant interaction" between factors A and B, but she is disappointed that all this work has not revealed to her the particular combination of A and B that will give her optimal results from her system. Another experimenter might be familiar with the least squares fitting of straight lines to data; the only problems he chooses to investigate are those that can be reduced to straight-line relationships. A third experimenter might be asked to do a screening study using Plackett-Burman designs; instead, he transfers out of the research division.

We do not believe that a course on the design of experiments must necessarily be preceded by a course on statistics. Instead, we have taken the approach that both subjects can be developed simultaneously, complementing each other as needed, in a course that presents the fundamentals of experimental design.

It is our intent that the book can be used in a number of fields by advanced undergraduate students, by beginning graduate students, and (perhaps more important) by workers who have already completed their formal education. The material in this book has been presented to all three groups, either through regular one-semester courses, or through intensive two- or three-day short courses. We have been pleased by the confidence these students have gained from the courses, and by their enthusiasm as they study further in the areas of statistics and experimental design.

The text is intended to be studied in one way only – from the beginning of Chapter 1 to the end of Chapter 12. The chapters are highly integrated and build on each other: there are frequent references to material that has been covered in previous chapters, and there are many sections that hint at material that will be developed more fully in later chapters.

The text can be read "straight through" without working any of the exercises at the ends of the chapters; however, the exercises serve to reinforce the material presented in each chapter, and also serve to expand the concepts into areas not covered by the main text. Relevant literature references are often given with this latter type of exercise.

The book has been written around a framework of linear models and matrix least squares. Because we authors are so often involved in the measurement aspects of investigations, we have a special fondness for the estimation of purely experimental uncertainty. The text reflects this prejudice. We also prefer the term "purely experimental uncertainty" rather than the traditional "pure error", for reasons we as analytical chemists believe should be obvious.

One of the important features of the book is the *sums of squares and degrees of freedom tree* that is used in place of the usual ANOVA tables. We have found the "tree" presentation to be a more effective teaching tool than ANOVA tables by themselves.

A second feature of the book is its emphasis on degrees of freedom. We have tried to remove the "magic" associated with knowing the source of these numbers by using the symbols n (the total number of experiments in a set), p (the number of parameters in the model), and f (the number of distinctly different factor combinations in the experimental design). Combinations of these symbols appear on the "tree" to show the degrees of freedom associated with various sums of squares (e.g., $n-f$ for SS_{pe}).

A third feature is the use of the $\boldsymbol{J}$ matrix (a matrix of mean replicate response) in the least squares treatment. We have found it to be a useful tool for teaching the effects (and usefulness) of replication.

We are grateful to a number of friends for help in many ways. Grant Wernimont and L. B. Rogers first told us why statistics and experimental design should be important to us as analytical chemists. Ad Olansky and Lloyd Parker first told us why a *clear presentation* of statistics and experimental design should be important to us; they and Larry Bottomley aided greatly in the early drafts of the manuscript. Kent Linville provided many helpful comments on the early drafts of the first two chapters. A large portion of the initial typing was done by Alice Ross; typing of the final manuscript was done by Lillie Gramann. Their precise work is greatly appreciated.

We are grateful also to the Literary Executor of the late Sir Ronald A. Fisher, F. R. S., to Dr. Frank Yates, F. R. S., and to Longman Group Ltd., London, for permission to partially reprint Tables III and V from their book *Statistical Tables for Biological, Agricultural and Medical Research*, 6th ed., 1974.

Finally, we would like to acknowledge our students who provided criticism as we developed the material presented here.

S. N. Deming
Houston, Texas
August 1986

S. L. Morgan
Columbia, South Carolina

CHAPTER 1

System Theory

General system theory is an organized thought process to be followed in relating cause and effect. The system is treated as a bounded whole with inputs and outputs external to the boundaries, and transformations occurring within the boundaries. Inputs, outputs, and transformations can be important or trivial – the trick is to determine which. The system of determination involves a study and understanding of existing theory in the chosen field; a study and understanding of past observations and experiments more specific to the chosen problem; and new experiments specific to the problem being studied.

General system theory is a highly versatile tool that provides a useful means of investigating many research and development projects. Although other approaches to research and development often focus on the detailed internal structure and organization of the system, our approach here will be to treat the system as a whole and to be concerned with its overall behavior.

1.1. Systems

A *system* is defined as a regularly interacting or interdependent group of items forming a unified whole. A system is described by its borders, by what crosses the borders, and what goes on inside. Thus, we often speak of a solar system when referring to a sun, its planets, their moons, etc.; a thermodynamic system when we are describing compounds in equilibrium with each other; and a digestive system if we are discussing certain parts of the body. Other examples of systems are ecological systems, data processing systems, and economic systems. We even speak of *the* system when we mean some part of the established order around us, some regularly interacting or interdependent group of items forming a unified whole of which we are a part.

General system theory views a system as possessing three basic elements – *inputs*, *transforms*, and *outputs* (see Figure 1.1). An example of a simple system is the mathematical relationship

$$y = x + 2. \tag{1.1}$$

This algebraic system is diagrammed in Figure 1.2. The *input* to the system is the independent variable x. The *output* from the system is the dependent variable y.

Figure 1.1. General system theory view of relationships among inputs, transforms, and outputs.

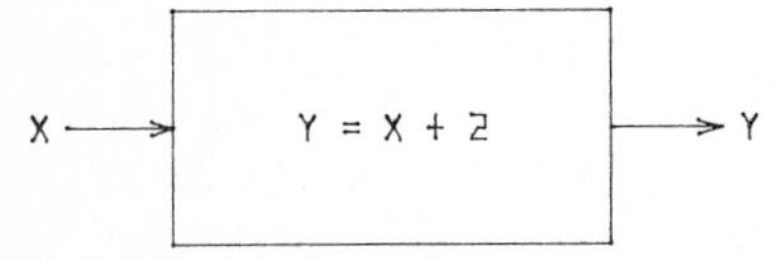

Figure 1.2. General system theory view of the algebraic relationship $y = x + 2$.

The *transform* that relates the output to the input is the well defined mathematical relationship given in Equation 1.1. The mathematical equation transforms a given value of the input, x, into an output value, y. If $x = 0$, then $y = 2$. If $x = 5$, then $y = 7$, and so on. In this simple system, the transform is known with certainty.

Figure 1.3 is a system view of a wine-making process. In this system, there are

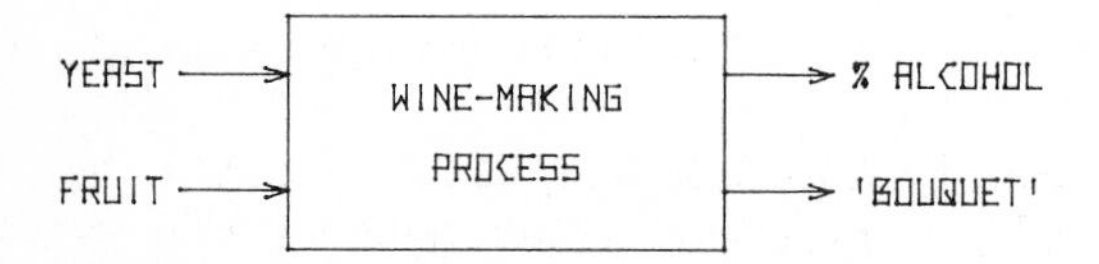

Figure 1.3. General system theory view of a wine-making process.

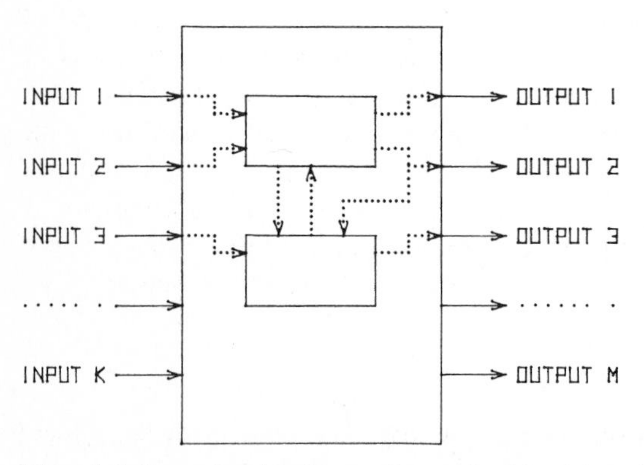

Figure 1.4. General system theory view emphasizing internal structures and relationships within a system.

two inputs (yeast and fruit), *two* outputs (percent alcohol and "bouquet"), and a transform that is probably *not* known with certainty.

Most systems are much more complex than the simple examples shown here. In general, there will be many inputs, many outputs, many transforms, and considerable subsystem structure. A more realistic view of most systems is probably similar to that shown in Figure 1.4.

1.2. Inputs

We will define a system *input* as a quantity or quality that might have an influence upon the system.

The definition of a system input is purposefully broad. It could have been made narrower to include only those quantities and qualities that *do* have an influence upon the system. However, because a large portion of the early stages of much research and development is concerned with determining which inputs do have an influence and which do not, such a narrow definition would assume that a considerable amount of work had already been carried out. The broader definition used here allows the inclusion of quantities and qualities that might eventually be shown to have no influence upon the system, and is a more useful definition for the early and speculative stages of most research.

We will use the symbol x with a subscript to represent a given input. For example, x_1 means "input number one", x_2 means "input number two", x_i means "the ith input", and so on.

The intensity setting of an input is called a *level*. It is possible for an input to be at different levels at different times. Thus, x_1 might have had the value 25 when we were first interested in the system; now we might want x_1 to have the value 17. To designate these different sets of conditions, a second subscript is added. Thus, $x_{11} = 25$ means that in the first instance, $x_1 = 25$; and $x_{12} = 17$ means that in the second instance, $x_1 = 17$.

Ambiguity is possible with this notation: e.g., x_{137} might refer to the level of input number one under the 37th set of conditions, or it might refer to the level of input 13 under the seventh set of conditions, or it might refer to input number 137. To avoid this ambiguity, subscripts greater than nine can be written in parentheses or separated with commas. Thus, $x_{1(37)}$ and $x_{1,37}$ refer to the level of input number one under the 37th set of conditions, $x_{(13)7}$ and $x_{13,7}$ refer to the level of input 13 under the seventh set of conditions, and $x_{(137)}$ and x_{137} refer to input number 137.

Input variables and factors

A system *variable* is defined as a quantity or quality associated with the system that may assume any value from a set containing more than one value. In the algebraic system described previously, "x" is an *input variable*: it can assume any one of an infinite set of values.

"Yeast" and "fruit" are input variables in the wine-making process. In the case of yeast, the amount of a given strain could be varied, or the particular type of yeast could be varied. If the variation is of extent or quantity (e.g., the use of one ounce of yeast, or two ounces of yeast, or more) the variable is said to be a *quantitative variable*. If the variation is of type or quality (e.g., the use of *Saccharomyces cerevisiae*, or *Saccharomyces ellipsoideus*, or some other species) the variable is said to be a *qualitative variable*. Thus, "yeast" could be a qualitative variable (if the amount added is always the same, but the type of yeast is varied) or it could be a quantitative variable (if the type of yeast added is always the same, but the amount is varied). Similarly, "fruit" added in the wine-making process could be a qualitative variable or a quantitative variable. In the algebraic system, x is a quantitative variable.

A *factor* is defined as one of the elements contributing to a particular result or situation. It is an input that *does* have an influence upon the system.

In the algebraic system discussed previously, x is a factor; its value determines what the particular result y will be. "Yeast" and "fruit" are factors in the wine-making process; the type and amount of each contributes to the alcohol content and flavor of the final product.

In the next several sections we will further consider factors under several categorizations.

Known and unknown factors

In most research and development projects it is important that as many factors as possible be known. Unknown factors can be the witches and goblins of many projects – unknown factors are often uncontrolled, and as a result such systems appear to behave excessively randomly and erratically. Because of this, the initial phase of many research and development projects consists of screening a large number of input variables to see if they are factors of the system; that is, to see if they have an *effect* upon the system.

The proper identification of factors is clearly important (see Table 1.1). If an input variable *is* a factor and it *is* identified as a factor, the probability is increased for the success of the project. If an input variable truly *is* a factor but it *is not*

TABLE 1.1
Possible outcomes in the identification of factors.

Type of input variable	Identified as a factor	Not identified as a factor
A factor	Desirable for research and development	Random and erratic behavior
Not a factor	Unnecessary complexity	Desirable for research and development

included as an input variable and/or *is not* identified as a factor, random and erratic behavior might result. If an input variable *is not* a factor but it is falsely identified as a factor, an unnecessary input variable will be included in the remaining phases of the project and the work will be unnecessarily complex. Finally, if an input variable *is not* a factor and *is not* identified as a factor, ignoring it will be of no consequence to the project.

The first and last of the above four possibilities are the desired outcomes. The second and third are undesirable outcomes, *but undesirable for different reasons and with different consequences*. The third possibility, falsely identifying an input variable as a factor, is unfortunate but the consequences are not very serious: it might be expensive, in one way or another, to carry the variable through the project, but its presence will not affect the ultimate outcome. However, the second possibility, *not* identifying a factor, can be very serious: omitting a factor can often cause the remaining results of the project to be worthless.

In most research and development, the usual approach to identifying important factors uses a statistical test that is concerned with the risk (α) of stating that an input variable is a factor when, in fact, it is not – a risk that is of relatively little consequence (see Table 1.1). Ideally, the identification of important factors should also be concerned with the potentially much more serious risk (β) of stating that an input variable is not a factor when, in fact, it is a factor (see Table 1.1). This subject is discussed further in Chapter 6.

In our representations of systems, a known factor will be shown as a solid arrow pointing toward the system; an unknown factor will be shown as a dotted arrow pointing toward the system (see Figure 1.5).

Controlled and uncontrolled factors

The word "control" is used here in the sense of exercising restraint or direction over a factor – that is, the experimental setting of a factor to a certain quantitative or qualitative value.

Controllable factors are desirable in experimental situations because their effects can usually be relatively easily and unambiguously detected and evaluated. Examples of individual controllable factors include x, yeast, fruit, temperature, concentration, time, amount, number, and size.

Uncontrollable factors are undesirable in experimental situations because their effects cannot always be easily or unambiguously detected or evaluated. Attempts

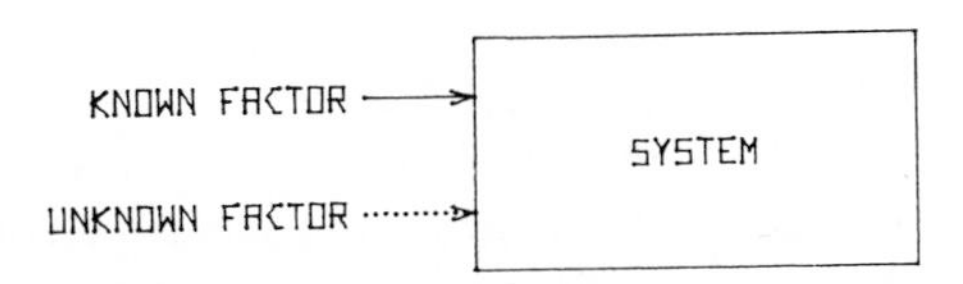

Figure 1.5. Symbols for a known factor (solid arrow) and an unknown factor (dotted arrow).

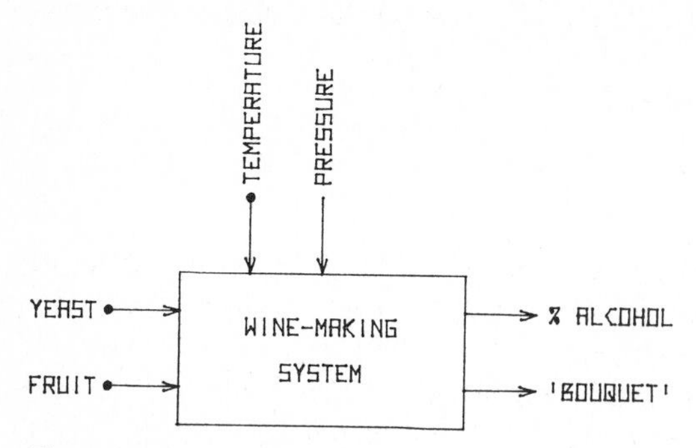

Figure 1.6. Symbols for controlled factors (arrows with dot at tail) and an uncontrolled factor (arrow without dot).

are often made to minimize their effects statistically (e.g., through randomization of experiment order – see Section 12.2) or to separate their effects, if known, from those of other factors (e.g., by measuring the level of the uncontrolled factor during each experiment and applying a "known correction factor" to the experimental results). Examples of individual uncontrollable factors include incident gamma ray background intensity, fluctuations in the levels of the oceans, barometric pressure, and the much maligned "phase of the moon".

A factor that is uncontrollable by the experimenter might nevertheless be controlled by some other forces. Incident gamma ray background intensity, fluctuations in the levels of the oceans, barometric pressure, and phase of the moon cannot be controlled by the experimenter, but they are "controlled" by the "Laws of Nature". Such factors are usually relatively constant with time (e.g., barometric pressure over a short term), or vary in some predictable way (e.g., phase of the moon).

A controlled factor will be identified as an arrow with a dot at its tail; an uncontrolled factor will not have a dot. In Figure 1.6, temperature is shown as a controlled known factor; pressure is shown as an uncontrolled known factor. "Yeast" and "fruit" are presumably controlled known factors.

Intensive and extensive factors

Another categorization of factors is based upon their dependence on the size of a system. The value of an *intensive factor* is not a function of the size of the system. The value of an *extensive factor* is a function of the size of the system.

The temperature of a system is an intensive factor. If the system is, say, 72°C, then it is 72°C independent of how large the system is. Other examples of intensive factors are pressure, concentration, and time.

The mass of a system, on the other hand, does depend upon the size of the system and is therefore an extensive factor. Other examples of extensive factors are volume and heat content.

Masquerading factors

A *true factor* exerts its effect directly upon the system and is correctly identified as doing so. A *masquerading factor* also exerts its effect directly upon the system but is incorrectly assigned some other identity.

"Time" is probably the most popular costume of masquerading factors. Consider again the wine-making process shown in Figure 1.3, and imagine the effect of using fruit picked at different times in the season. In general, the later it is in the season, the more sugar the fruit will contain. Wine begun early in the season will probably be "drier" than wine started later in the season. Thus, when considering variables that have an effect upon an output called "dryness", "time of year" might be identified as a factor – it has been correctly observed that wines are drier when made earlier in the season.

But time is not the true factor; sugar content of the fruit is the true factor, and it is masquerading as time. The masquerade is successful because of the high correlation between time of year and sugar content.

A slightly different type of masquerade takes place when a single, unknown factor influences two outputs, and one of the outputs is mistaken as a factor. G.E.P. Box has given such masquerading factors the more dramatic names of either "latent variables" or "lurking variables".

Suppose it is observed that as foaming in the wine-making system increases, there is also an increase in the alcohol content. The process might be envisioned as shown in Figure 1.7. Our enologists seek to increase the percent alcohol by introducing foaming agents. Imagine their surprise and disappointment when the intentionally increased foaming does not increase the alcohol content. The historical evidence is clear that increased foaming is accompanied by increased alcohol content! What is wrong?

What might be wrong is that foaming and percent alcohol are both *outputs* related to the lurking factor "sugar content" (see Figure 1.8). The behavior of the system might be such that as sugar content is increased, foaming increases and alcohol content increases; as sugar content is decreased, foaming decreases and

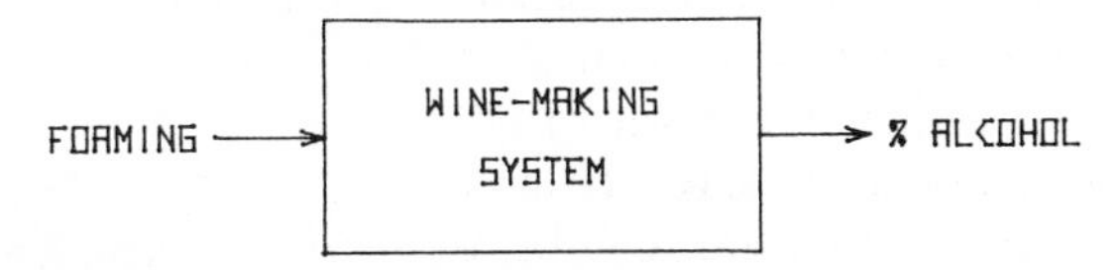

Figure 1.7. One view of a wine-making process suggesting a relationship between foaming and percent alcohol.

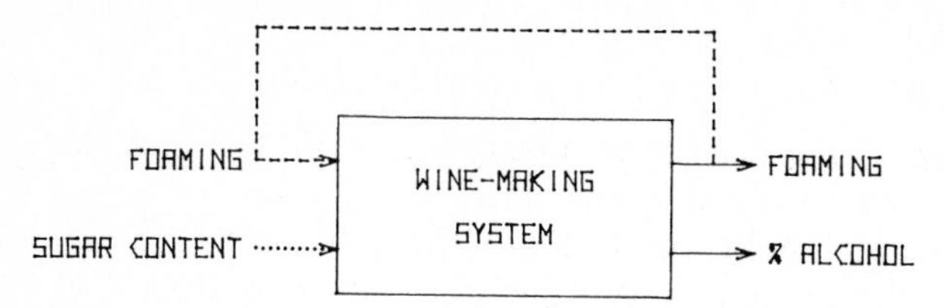

Figure 1.8. Alternate view of a wine-making process suggesting sugar content as a masquerading factor.

alcohol content decreases. "Sugar content", the true factor, is thus able to masquerade as "foaming" because of the high correlation between them. A corrected view of the system is given in Figure 1.9.

Experiment vs. observation

Think for a moment why it is possible for factors to masquerade as other variables. Why are we sometimes fooled about the true identity of a factor? Why was "time of year" said to be a factor in the dryness of wine? Why was "foaming" thought to be a factor in the alcohol content of wine?

One very common reason for the confusion is that the factor is identified on the basis of *observation* rather than *experiment*. In the wine-making process, it was observed that dryness is related to the time of the season; the causal relationship between time of year and dryness was *assumed*. In a second example, it was observed that percent alcohol is related to foaming; the causal relationship between foaming and alcohol content was assumed.

But consider what happened in the foaming example when additional foaming agents were introduced. The single factor "foaming" was *deliberately* changed. An *experiment* was performed. The result of the experiment clearly *disproved* any hypothesized causal relationship between alcohol content and foaming.

An *observation* involves a measurement on a system as it is found, unperturbed and undisturbed. An *experiment* involves a measurement on a system *after it has been deliberately perturbed and disturbed by the experimenter.*

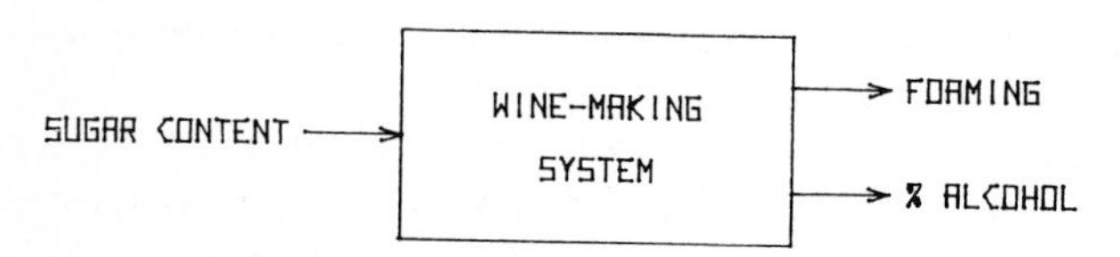

Figure 1.9. Corrected view of a wine-making process showing relationships between sugar content and both foaming and percent alcohol.

1.3. Outputs

We will define a system *output* as a quantity or quality that might be influenced by the system.

Again, the definition of system output is purposefully broad. It could have been made narrower to include only those quantities and qualities that *are* influenced by the system. However, to do so presupposes a more complete knowledge of the system than is usually possessed at the beginning of a research and development project. System outputs that are influenced by the system are called *responses*.

System outputs can include such quantities and qualities as yield, color, cost of raw materials, and public acceptance of a product.

We will use the symbol y to represent a given output. For example, y_1 means "output number one", y_2 means "output number two", and y_i means the ith output.

The intensity value of an output is called its level. It is possible for an output to be at different levels at different times; in keeping with the notation used for inputs, a second subscript is used to designate these different sets of conditions. Thus, y_{25} refers to the level of the output y_2 under the fifth set of conditions, and $y_{(11)(17)}$ or $y_{11,17}$ refers to the level of the 11th input under the 17th set of conditions.

The symbol $\bar{y}$ will be used to represent the mean or average level of output from a system (see Section 3.1 and Equation 3.2). Thus, $\bar{y}_1$ refers to the average level of the first output, $\bar{y}_5$ refers to the average level of the fifth output, etc.

Occasionally, a second subscript will be used in symbols representing the average level of output. In these cases, a group of individual values of output were obtained under identical conditions (see Section 5.6). Thus, $\bar{y}_{1i}$ would refer to the average of all values of the output y_1 obtained at the particular *set of experimental conditions under which* y_{1i} *was obtained.*

Important and unimportant responses

Most systems have more than one response. The wine-making process introduced in Section 1.1 is an example. Percent alcohol and bouquet are two responses, but there are many additional responses associated with this system. Examples are the amount of carbon dioxide evolved, the extent of foaming, the heat produced during fermentation, the turbidity of the new wine, and the concentration of ketones in the final product. Just as factors can be classified into many dichotomous sets, so too can responses. One natural division is into *important responses* and *unimportant responses*, although the classification is not always straightforward.

The criteria for classifying responses as important or unimportant are seldom based solely upon the system itself, but rather are usually based upon elements external to the system. For example, in the wine-making process, is percent alcohol an important or unimportant response? To a down-and-outer on skid row, alcohol content could well be a very important response; any flavor would probably be ignored. Yet to an enophile of greater discrimination, percent alcohol might be of no consequence whatever: "bouquet" would be supremely important.

Almost all responses have the potential of being important. Events or circumstances external to the system usually reveal the category – important or unimportant – into which a given response should be placed.

Responses as factors

Most of the responses discussed so far have an effect on the universe outside the system; it is the importance of this effect that determines the importance of the response.

Logically, if a response has an effect on some other system, then it must be a factor of that other system. It is not at all unusual for variables to have this dual identity as response and factor. In fact, most systems are seen to have a rather complicated internal subsystem structure in which there are a number of such factor-response elements (see Figure 1.4). The essence of responses as factors is illustrated in the drawings of Rube Goldberg in which an initial cause triggers a series of intermediate factor-response elements until the final result is achieved.

Occasionally, a response from a system will act as a true factor to the same system, a phenomenon that is generally referred to as *feedback*. (This is not the same as the situation of masquerading factors.) Feedback is often classified as *positive* if it enhances the response being returned as a factor, or *negative* if it diminishes the response.

Heat produced during a reaction in a chemical process is an example of positive feedback that is often of concern to chemical engineers (see Figure 1.10). Chemical reactions generally proceed more rapidly at higher temperatures than they do at lower temperatures. Many chemical reactions are exothermic, giving off heat as the reaction proceeds. If this heat is not removed, it increases the temperature of the system. If the temperature of the system increases, the reaction will proceed faster. If it proceeds faster, it produces more heat. If it produces more heat, the temperature goes still higher until at some temperature either the rate of heat removed equals the rate of heat produced (an equilibrium), or the reaction "goes out of control" (temperature can no longer be controlled), often with disastrous consequences.

Examples of negative feedback are common. A rather simple example involves pain. If an animal extends an extremity (system factor = length of extension) toward

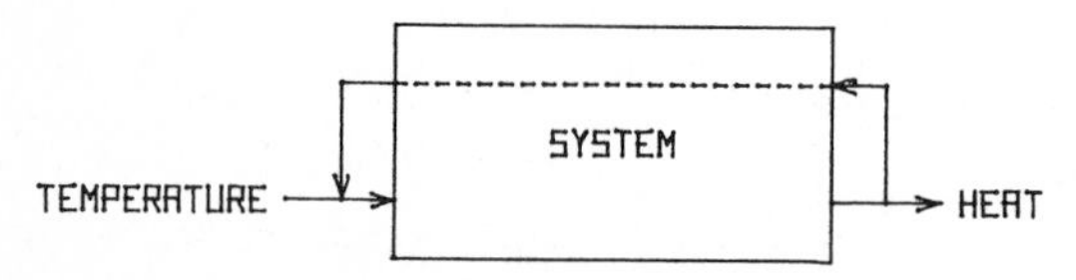

Figure 1.10. Example of positive feedback: as heat is produced, the temperature of the system increases.

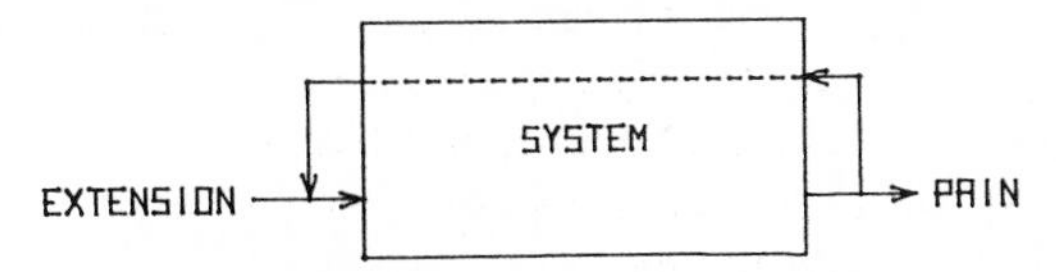

Figure 1.11. Example of negative feedback: as pain increases, extension decreases.

an unpleasant object such as a cactus, and the extremity encounters pain (system response = pain), the pain is transformed into a signal that causes the extremity to contract (i.e., negative extension), thereby decreasing the amount of pain (see Figure 1.11). Negative feedback often results in a stable equilibrium. This can be immensely useful for the control of factors within given tolerances.

Known and unknown responses

There are many examples of systems for which one or more important responses were unknown. One of the most tragic involved the drug thalidomide. It was known that one of the responses thalidomide produced when administered to humans was that of a tranquilizer; it was not known that when taken during pregnancy it would also affect normal growth of the fetus and result in abnormally shortened limbs of the newborn.

Another example of an unknown or unsuspected response would be the astonished look of a young child when it first discovers that turning a crank on a box produces not only music but also the surprising "jack-in-the-box".

Unknown important responses are destructive in many systems: chemical plant explosions caused by impurity built up in reactors; Minimata disease, the result of microorganisms metabolizing inorganic mercury and passing it up the food chain; the dust bowl of the 1930's – all are examples of important system responses that were initially unknown and unsuspected.

As Table 1.2 shows, it is desirable to know as many of the important responses from a system as possible. If *all* of the responses from a system are known, then the *important* ones can be correctly assessed according to *any* set of external criteria (see the section on important and unimportant responses). In addition, because

TABLE 1.2
Possible outcomes in the identification of responses.

Type of output variable	Identified as a response	Not identified as a response
Important response	Desirable for research and development	Possibility of unexpected serious consequences
Unimportant response	Unnecessary data acquired (but available if response becomes important later)	Desirable for research and development

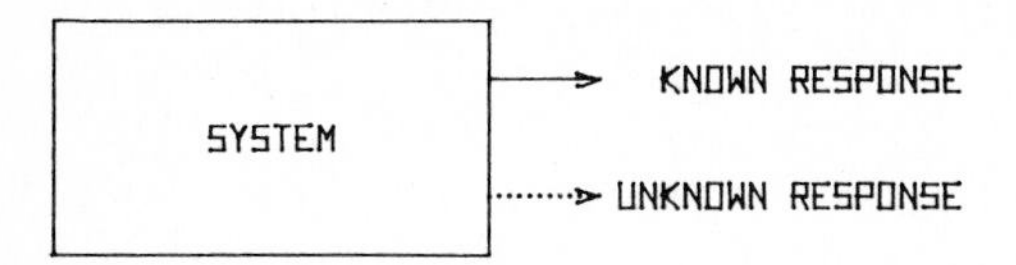

Figure 1.12. Symbols for a known response (solid arrow) and an unknown response (dotted arrow).

most decision-making is based upon the responses from a system, bad decisions will not be made because of *incomplete* information if all of the responses are known.

A known response will be represented as a solid arrow pointing away from the system; an unknown response will be represented as a dotted arrow pointing away from the system (see Figure 1.12).

Controlled and uncontrolled responses

A *controlled response* is defined as a response capable of being set to a specified level by the experimenter; an *uncontrolled response* is a response not capable of being set to a specified level by the experimenter.

It is important to realize that an experimenter has no direct control over the responses. The experimenter can, however, have direct control over the *factors*, and this control will be transformed by the system into an indirect control over the responses. If the behavior of a system is known, then the response from a system can be set to a specified level by setting the inputs to appropriate levels. If the behavior of a system is not known, then this type of control of the response is not possible.

Let us reconsider two of the systems discussed previously, looking first at the algebraic system, $y = x + 2$. In this system, the experimenter does not have any direct control over the response y, but does have an indirect control on y through the system by direct control of the factor x. If the experimenter wants y to have the value 7, the factor x can be set to the value 5. If y is to have the value -3, then by setting $x = -5$, the desired response can be obtained.

This appears to be a rather trivial example. However, the example appears trivial only because the behavior of the system is known. If we represent Equation 1.1 as

$$y = A(x) \tag{1.2}$$

where A is the transform that describes the behavior of the system (see Figure 1.13), then the inverse of this transform, A^{-1}, relates x to y

$$x = A^{-1}(y). \tag{1.3}$$

In this example, the inverse transform may be obtained easily and is expressed as

$$x = y - 2 \tag{1.4}$$

(see Figure 1.14). Because this inverse transform is easily known, the experimenter can cause y to have whatever value is desired.

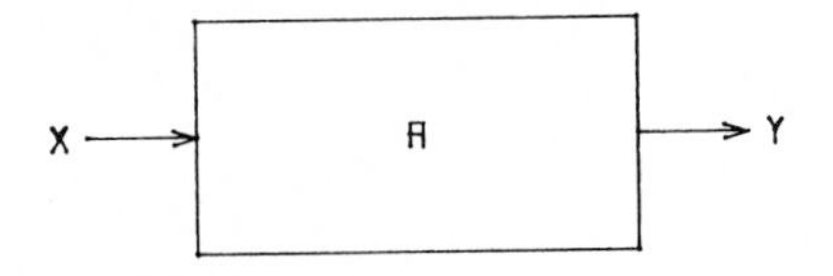

Figure 1.13. Representation of system transform A: $y = Ax$.

Now let us consider the wine-making example and ask, "Can we control temperature to produce a product of any exactly specified alcohol content (up to, say, 12%)"? The answer is that we probably cannot, and the reason is that the system behavior is not known with certainty. We know that *some* transform relating the alcohol content to the important factors must exist – that is,

$$\text{percent alcohol} = W(\text{temperature}, ...). \qquad (1.5)$$

However, the transform W is probably unknown or not known with certainty, and therefore the inverse of the transform (W^{-1}) cannot be known, or cannot be known with certainty. Thus,

$$\text{temperature} = W^{-1}(\text{percent alcohol}, ...) \qquad (1.6)$$

is not available: we cannot specify an alcohol content and calculate what temperature to use.

If the response from a system is to be set to a specified level by setting the system factors to certain levels, then the behavior of the system must be known. This statement is at the heart of most research and development projects.

Intensive and extensive responses

Another categorization of responses is based upon their dependence on the size or throughput of a system.

The value of an *intensive response* is not a function of the size or throughput of the system. Product purity is usually an example of an intensive response. If a manufacturing facility can produce 95% pure material, then it will be 95% pure, whether we look at a pound or a ton of the material. Other examples of intensive responses might be color, density, percent yield, alcohol content, and flavor.

The value of an *extensive response* is a function of the size of the system. Examples of extensive responses are total production in tons, cost, and profit.

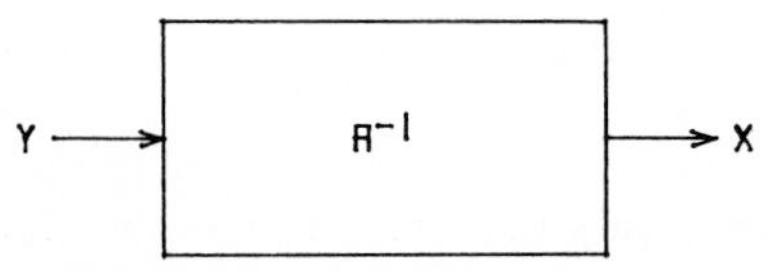

Figure 1.14. Representation of system inverse transform A^{-1}: $x = A^{-1}y$.

1.4. Transforms

The third basic element of general system theory is the *transform*. As we have seen, it is the link between the factors and the responses of a system, transforming levels of the system's factors into levels of the system's responses.

Just as there are many different types of systems, there are many different types of transforms. In the algebraic system pictured in Figure 1.2, the system transform is the algebraic relationship $y = x + 2$. In the wine-making system shown in Figure 1.3, the transform is the microbial colony that converts raw materials into a finished wine. Transforms in the chemical process industry are usually sets of chemical reactions that transform raw materials into finished products.

We will take the broad view that the system transform is that part of the system that actively converts system factors into system responses. A system transform *is not a description* of how the system behaves; a description of how the system behaves (or is thought to behave) is called a *model*. Only in rare instances are the system transform and the description of the system's behavior the same – the algebraic system of Figure 1.2 is an example. In most systems, a complete description of the system transform is not possible – approximations of it (incomplete models) must suffice. Because much of the remainder of this book discusses models, their formulation, their uses, and their limitations, only one categorization of models will be given here.

Mechanistic and empirical models

If a detailed theoretical knowledge of the system is available, it is often possible to construct a *mechanistic model* which will describe the general behavior of the system. For example, if a biochemist is dealing with an enzyme system and is interested in the rate of the enzyme catalyzed reaction as a function of substrate concentration (see Figure 1.15), the Michaelis-Menten equation might be expected to provide a general description of the system's behavior.

$$\text{rate} = \text{rate}_{\max} \times [\text{substrate}]/(K_m + [\text{substrate}]). \tag{1.7}$$

The parameter K_m would have to be determined experimentally.

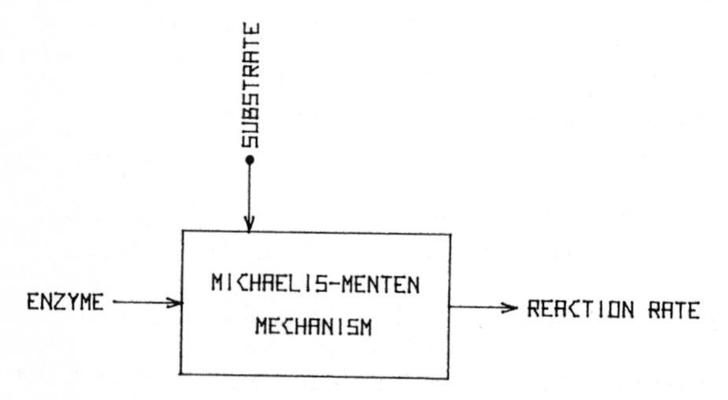

Figure 1.15. General system theory view of a mechanistic model of enzyme activity.

The Michaelis-Menten equation represents a mechanistic model because it is based upon an assumed chemical reaction mechanism of how the system behaves. If the system does indeed behave in the assumed manner, then the mechanistic model is adequate for describing the system. If, however, the system does not behave in the assumed manner, then the mechanistic model is inadequate. The only way to determine the adequacy of a model is to carry out experiments to see if the system does behave as the model predicts it will. (The design of such experiments will be discussed in later chapters.) In the present example, if "substrate inhibition" occurs, the Michaelis-Menten model would probably be found to be inadequate; a different mechanistic model would better describe the behavior of the system.

If a detailed theoretical knowledge of the system is either not available or is too complex to be useful, it is often possible to construct an *empirical model* which will approximately describe the behavior of the system over some limited set of factor levels. In the enzyme system example, at relatively low levels of substrate concentration, the rate of the enzyme catalyzed reaction is found to be described rather well by the simple expression

$$\text{rate} = k[\text{substrate}] \tag{1.8}$$

The adequacy of empirical models should also be tested by carrying out experiments at many of the sets of factor levels for which the model is to be used.

Most of the models in this book are empirical models that provide an approximate description of the true behavior of a system. However, the techniques presented for use with empirical models are applicable to many mechanistic models as well.

References

Bertalanffy, L. (1968). *General System Theory. Foundations, Development, Applications*. George Braziller, New York.

Box, G.E.P., Hunter, W.G., and Hunter, J.S. (1978). *Statistics for Experimenters. An Introduction to Design, Data Analysis, and Model Building*. Wiley, New York.

Gold, H.J. (1977). *Mathematical Modelling of Biological Systems – an Introductory Guidebook*. Wiley, New York.

Hofstadter, D.R. (1979). *Gödel, Escher, Bach: an Eternal Golden Braid*. Basic Books, New York.

Jeffers, J.R. (1978). *An Introduction to Systems Analysis: with Ecological Applications*. University Park Press, Baltimore, Maryland.

Laszlo, E. (1972). *Introduction to Systems Philosophy*. Harper and Row, New York.

Mandel, J. (1964). *The Statistical Analysis of Experimental Data*. Wiley, New York.

Shewhart, W.A. (1939). *Statistical Method from the Viewpoint of Quality Control*. Lancaster Press, Pennsylvania.

Suckling, C.J., Suckling, K.E., and Suckling, C.W. (1978). *Chemistry through Models. Concepts and Applications of Modelling in Chemical Science, Technology and Industry*. Cambridge University Press, Cambridge.

Vemuri, V. (1978). *Modeling of Complex Systems: an Introduction*. Academic Press, New York.

Weinberg, G.M. (1975). *An Introduction to General Systems Thinking*. Wiley, New York.

Williams, H.P. (1978). *Model Building in Mathematical Programming*. Wiley, New York.

Wilson, E.B., Jr. (1952). *An Introduction to Scientific Research*. McGraw-Hill, New York.

Exercises

1.1. General system theory

Complete the following table by listing other possible inputs to and outputs from the wine-making process shown in Figure 1.3. Categorize each of the inputs according to your estimate of its expected influence on each of the outputs – strong (S), weak (W), or none (N).

Inputs	Outputs			
	% alcohol	"bouquet"	clarity	...
Amount of yeast	W			
Type of yeast	S			
Amount of fruit		W		
Type of fruit		S		
Amount of sugar	S			
.				
.				
.				

1.2. General system theory

Discuss the usefulness of a table such as that in Exercise 1.1 for planning experiments. What are its relationships to Tables 1.1 and 1.2?

1.3. General system theory

Choose a research or development project with which you are familiar. Create a table of inputs and outputs for it such as that in Exercise 1.1. Draw a system diagram representing your project (see, for example, Figure 1.3).

1.4. General system theory

Suppose you are in charge of marketing the product produced in Exercise 1.1. Which output is most important? Why? Does its importance depend upon the market you wish to enter? What specifications would you place on the outputs to produce an excellent wine? A profitable wine?

1.5. Important and unimportant outputs

"The automobile has provided an unprecedented degree of personal freedom and mobility, but its side effects, such as air pollution, highway deaths, and a dependence on foreign oil supplies, are undesirable. The United States has tried to regulate the social cost of these side effects through a series of major federal laws... however, each law has been aimed at a single goal, either emission reduction, safety, or fuel efficiency, with little attention being given to the conflicts and trade-offs between goals". [Lave, L.B. (1981). "Conflicting Objectives in Regulating the Automobile", *Science*, 212, p. 893.] Comment, and generalize in view of Table 1.2.

1.6. Inputs

In the early stages of much research, it is not always known which of the system inputs actually affect the responses from the system; that is, it is not always known which inputs are factors, and which are not. One point of view describes all inputs as factors, and then seeks to discover which are *significant* factors, and which are not significant. How would you design an experiment (or set of experiments) to prove that a factor exerted a significant effect upon a response? How would you design an experiment (or set of experiments) to prove that a factor had absolutely no effect on the response? [See, for example, Fisher, R.A. (1966). *Design of Experiments*, 8th ed., p. 11. Hafner, New York; or Draper, N.R., and Smith, H. (1966). *Applied Regression Analysis*, p. 20. Wiley, New York.]

1.7. Symbology

Indicate the meaning of the following symbols: x, y, x_1, x_2, y_1, y_2, x_{11}, x_{13}, x_{31}, y_{11}, y_{12}, y_{21}, y_{31}, y_{1i}, $\bar{y}_{1i}$, A, A^{-1}.

1.8. Qualitative and quantitative factors

List five quantitative factors. List five qualitative factors. What do the key words "type" and "amount" suggest?

1.9. Known and unknown factors

In pharmaceutical and biochemical research, crude preparations of material will often exhibit significant desired biological activity. However, that activity will sometimes be lost when the material is purified. Comment.

1.10. Controlled and uncontrolled factors

Comment on the wisdom of trying to *control* the following in a study of factors influencing employee attitudes: size of the company, number of employees, salary, length of employment. Would such a study be more likely to use *observations* or *experiments*?

1.11. Intensive and extensive factors

List five extensive factors. List five intensive factors. How can the extensive factor *mass* be converted to the intensive factor *density*?

1.12. Masquerading factors

There is a strong positive correlation between how much persons smoke and the incidence of lung cancer. Does smoking cause cancer? [See, for example, Fisher, R.A. (1959). *Smoking: the Cancer Controversy*. Oliver and Boyd, London.]

1.13. Masquerading factors

Data exist that show a strong positive correlation between human population and the number of storks in a small village. Do storks cause an increase in population? Does an increase in population cause storks? [See, for example, Box, G.E.P.,

Hunter, W.G., and Hunter, J.S. (1978). *Statistics for Experimenters. An Introduction to Design, Data Analysis, and Model Building*, p. 8. Wiley, New York.]

1.14. Important and unimportant responses

Discuss three responses associated with any systems with which you are familiar that were originally thought to be unimportant but were later discovered to be important.

1.15. Responses as factors

List five examples of positive feedback. List five examples of negative feedback.

1.16. Responses as factors

What is a "Rube Goldberg device"? Give an example. [See, for example, *Colliers* (magazine), 21 June 1930, p. 14.]

1.17. Known and unknown responses

Give examples from your childhood of discovering previously unknown responses. Which were humorous? Which were hazardous? Which were discouraging?

1.18. Controlled and uncontrolled responses

Plot the algebraic relationship $y = 10x - x^2$. How can the response y be controlled to have the value 9? Plot the algebraic relationship $x = 10y - y^2$. Can the response y be controlled to have the value 9?

1.19. Transforms

What is the difference between a transform and a model? Are the two ever the same?

1.20. Mechanistic and empirical models

Is the model $y = kx$ an empirical model or a mechanistic model? Why?

1.21. Mechanistic and empirical models

Trace the evolution of a mechanistic model that started out as an empirical model. How was the original empirical model modified during the development of the model? Did experimental observations or theoretical reasoning force adjustments to the model?

1.22. Research and development

The distinction between "research" and "development" is not always clear. Create a table such as Table 1.1 or 1.2 in which one axis is "motivation" with subheadings of "fundamental" and "applied", and the other axis is "character of work" with subheadings of "science" and "technology". Into what quarter might "research" be placed? Into what quarter might "development" be placed? Should there be distinct boundaries among quarters? [See, for example, Mosbacher, C.J.

(1977). *Research / Development*, January, p. 23; and Finlay, W.L. (1977). *Research / Development*, May, p. 85.]

1.23. General system theory

Composition Law: "The whole is more than the sum of its parts". Decomposition Law: "The part is more than a fraction of the whole". [Weinberg, G.M. (1975). *An Introduction to General Systems Thinking*, p. 43. Wiley, New York.] Are these two statements contradictory?

CHAPTER 2

Response Surfaces

Response surface methodology is an area of experimental design that deals with the optimization and understanding of system performance. In this chapter, several general concepts that provide a foundation for response surface methodology are presented, usually for the single-factor, single-response case (see Figure 2.1). In later chapters these concepts are expanded, and additional ones are introduced, to allow the treatment of multifactor systems.

2.1. Elementary concepts

A *response surface* is the graph of a system response plotted against one or more of the system factors. We consider here the simplest case, that of a single response plotted against only one factor. It is assumed that all other controllable factors are held constant, each at a specified level. As will be seen later, it is important that this assumption be true; otherwise, the single-factor response surface might appear to change shape or to be excessively noisy.

Figure 2.2 is a response surface showing a system response, y_1, plotted against one of the system factors, x_1. If there is no uncertainty associated with the response, and if the response is known for all values of the factor, then the response surface might be described by some mathematical model M that relates the response y_1 to the factor x_1.

$$y_1 = M(x_1). \tag{2.1}$$

For Figure 2.2 the exact relationship is

$$y_1 = 0.80 + 1.20x_1 - 0.05x_1^2. \tag{2.2}$$

Such a relationship (and corresponding response surface) might accurately represent reaction yield (y_1) as a function of reaction temperature (x_1) for a chemical process.

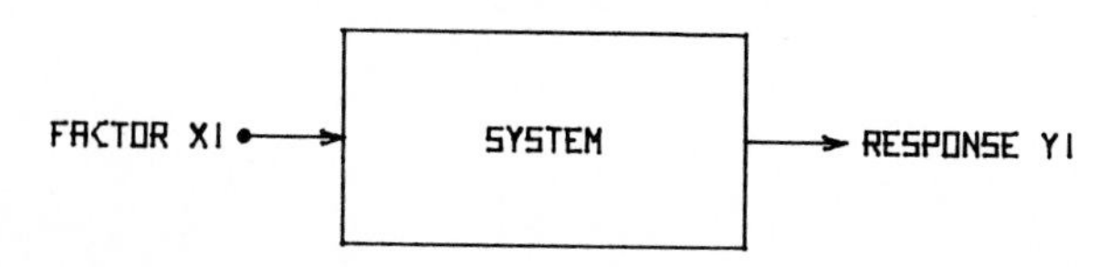

Figure 2.1. Single-factor, single-response system.

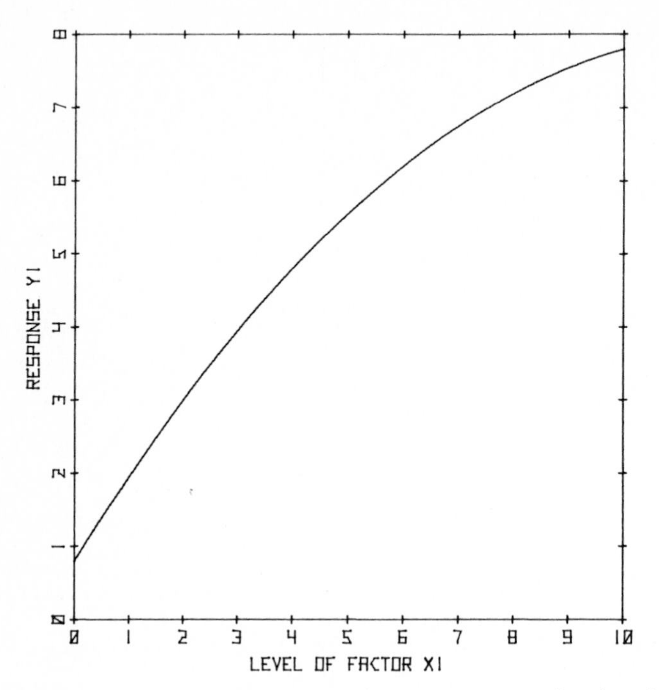

Figure 2.2. Response surface showing the response, y_1, as a function of the single factor, x_1.

It is more commonly the case that all points on the response surface are *not* known; instead, only a few values will have been obtained and the information will give an incomplete picture of the response surface, such as that shown in Figure 2.3. Common practice is to *assume* a functional relationship between the response and the factor (that is, to assume a model, either mechanistic or empirical) and find the values of the model parameters that fit the data. If a model of the form

$$y_1 = \beta_0 + \beta_1 x_1 + \beta_{11} x_1^2 \tag{2.3}$$

is assumed, model fitting methods discussed in later chapters would give the estimated equation

$$y_1 = 8.00 - 1.20 x_1 + 0.04 x_1^2 \tag{2.4}$$

for the relationship between y_1 and x_1 in Figure 2.3. This assumed response surface and the data points are shown in Figure 2.4.

The location of a point on the response surface must be specified by (1) stating the level of the factor, and (2) stating the level of the response. Stating only the coordinate of the point with respect to the factor locates the point in *factor space*; stating only the coordinate of the point with respect to the response locates the point in *response space*; and stating both coordinates locates the point in *experiment space*. If the seventh experimental observation gave the third point from the left in

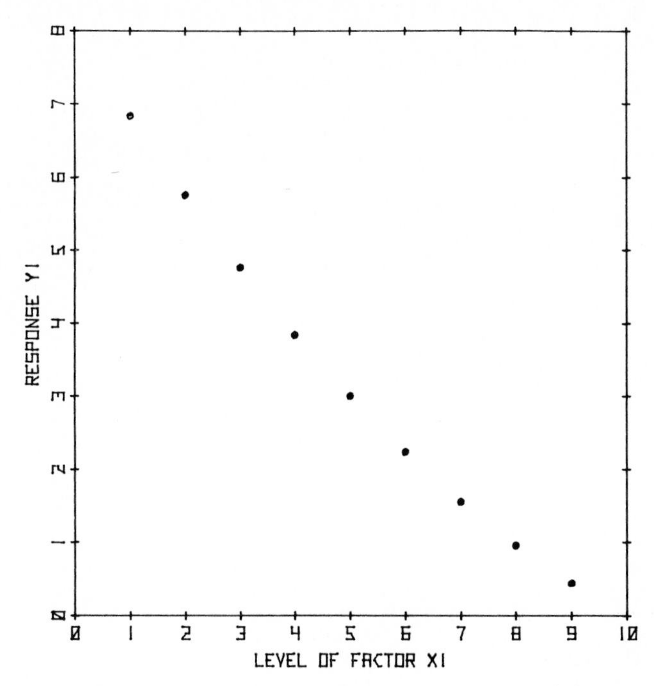

Figure 2.3. Response values obtained from experiments carried out at different levels of the factor x_1.

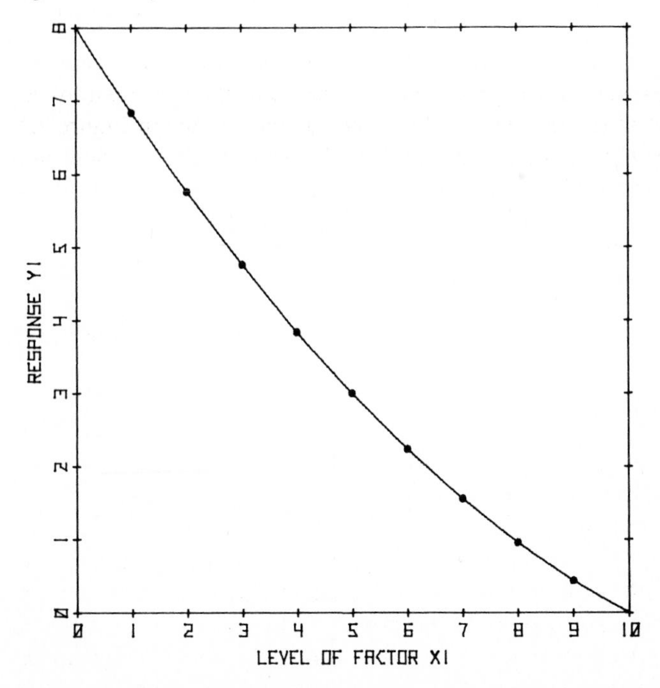

Figure 2.4. Response values obtained from experiments carried out at different levels of the factor x_1 and the assumed response surface.

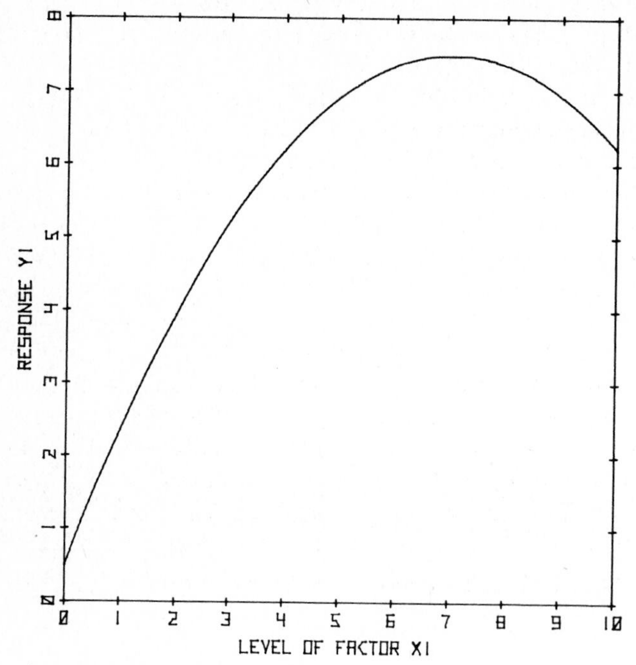

Figure 2.5. Response surface exhibiting a maximum at $x_1 = 7$.

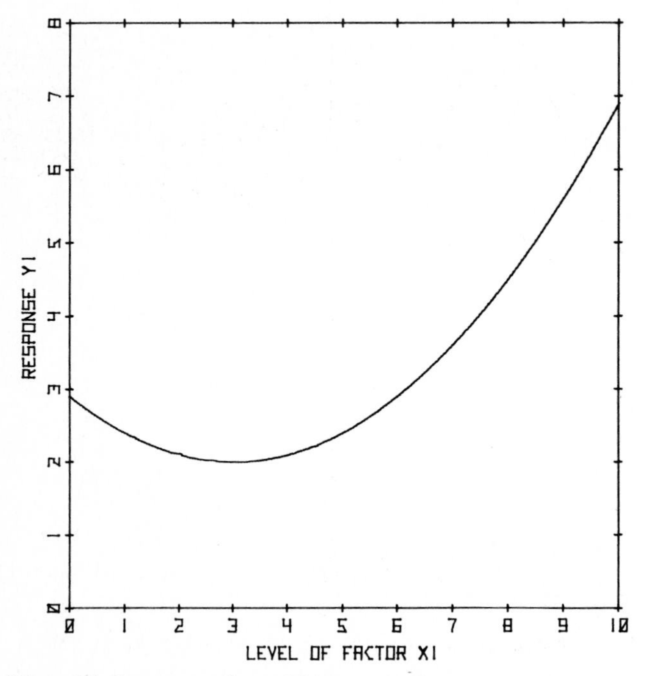

Figure 2.6. Response surface exhibiting a minimum at $x_1 = 3$.

Figure 2.3, the location of this point in experiment space would be $x_{17} = 3.00$, $y_{17} = 4.76$ (the first subscript on x and y refers to the factor or response number; the second subscript refers to the experiment number).

Figures 2.2, 2.3, and 2.4 show relationships between y_1 and x_1 that always increase or always decrease over the domains shown. The lowest and highest values of the response y_1 lie at the limits of the x_1 factor domain. Figure 2.2 is a response surface that is *monotonic increasing*; that is, the response always increases as the factor level increases. Figures 2.3 and 2.4 show response surfaces that are *monotonic decreasing*; the response always decreases as the factor level increases.

Figure 2.5 shows a relationship between y_1 and x_1 for which a *maximum* lies within the domain, specifically at $x_1 = 7$. Figure 2.6 is a response surface which exhibits a *minimum* at $x_1 = 3$. Each of these extreme points could be considered to be an *optimum*, depending upon what is actually represented by the response. For example, if the response surface shown in Figure 2.5 represents the yield of product in kilograms per hour vs. the feed rate of some reactant for an industrial chemical process, an optimum (maximum) yield can be obtained by operating at the point $x_1 = 7$. If the response surface shown in Figure 2.6 represents the percentage of impurity in the final product as a function of the concentration of reactant for the same industrial chemical process, there is clearly an optimum (minimum) at the point $x_1 = 3$.

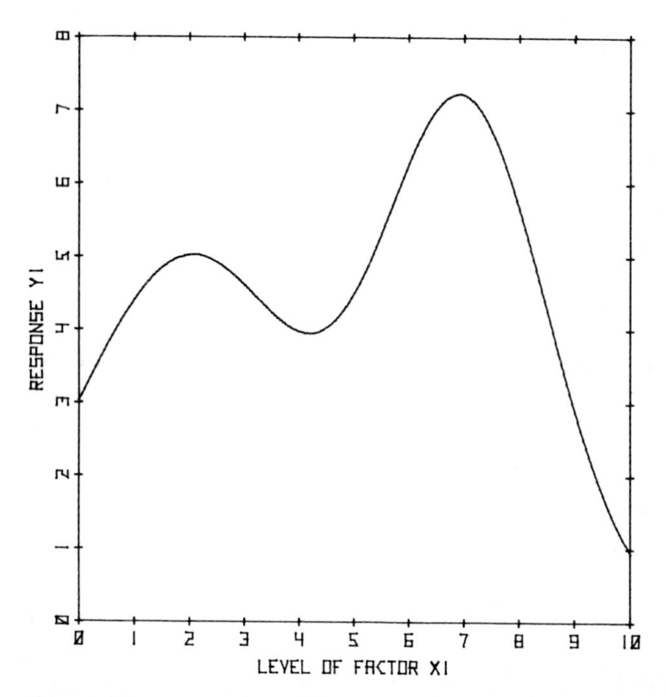

Figure 2.7. Response surface exhibiting two local maxima, one at $x_1 = 2$, the other (the global maximum) at $x_1 = 7$.

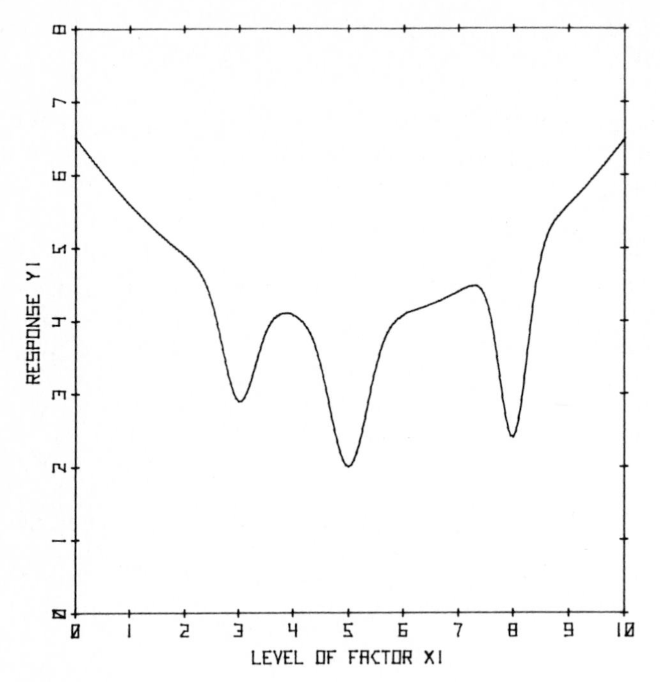

Figure 2.8. Response surface exhibiting three local minima, one at $x_1 = 3$, one at $x_1 = 8$, and the third (the global minimum) at $x_1 = 5$.

A *maximum* is the point in a region of experiment space giving the algebraically largest value of response. A *minimum* is the point in a region of experiment space giving the algebraically smallest value of response. An *optimum* is the point in a region of experiment space giving the *best* response.

The response surfaces shown in Figures 2.5 and 2.6 are said to be *unimodal* – they exhibit only one optimum (maximum or minimum) over the domain of the factor x_1. *Multimodal* response surfaces exhibit more than one optimum as shown in Figures 2.7 and 2.8. Each individual optimum in such response surfaces is called a *local optimum*; the best local optimum is called the *global optimum*. The response surface in Figure 2.7 has two *local maxima*, the rightmost of which is the *global maximum*. The response surface in Figure 2.8 has three *local minima*, the center of which is the *global minimum*.

2.2. Continuous and discrete factors and responses

A *continuous factor* is a factor that can take on any value within a given domain. Similarly, a *continuous response* is a response that can take on any value within a given range. Examples of continuous factors are pressure, volume, weight, distance, time, current, flow rate, and reagent concentration. Examples of continuous responses are yield, profit, efficiency, effectiveness, impurity concentration, sensitivity, selectivity, and rate.

Continuous factors and responses are seldom realized in practice because of the finite resolution associated with most control and measurement processes. For example, although temperature is a continuous factor and may in theory assume any value from −273.16°C upward, if the temperature of a small chemical reactor is adjusted with a common autotransformer, temperature must vary in a discontinuous manner (in "jumps") because the variable transformer can be changed only in steps as a sliding contact moves from one wire winding to the next. Another example arises in the digitization of a continuous voltage response: because analog-to-digital converters have fixed numbers of bits, the resulting digital values are limited to a finite set of possibilities. A voltage of 8.76397...V would be digitized on a ten-bit converter as either 8.76 or 8.77. Although practical limitations of resolution exist, it is important to realize that continuous factors and continuous responses can, in theory, assume any of the infinite number of values possible within a given set of bounds.

A *discrete factor* is a factor that can take on only a limited number of values within a given domain. A *discrete response* is a response that can take on only a

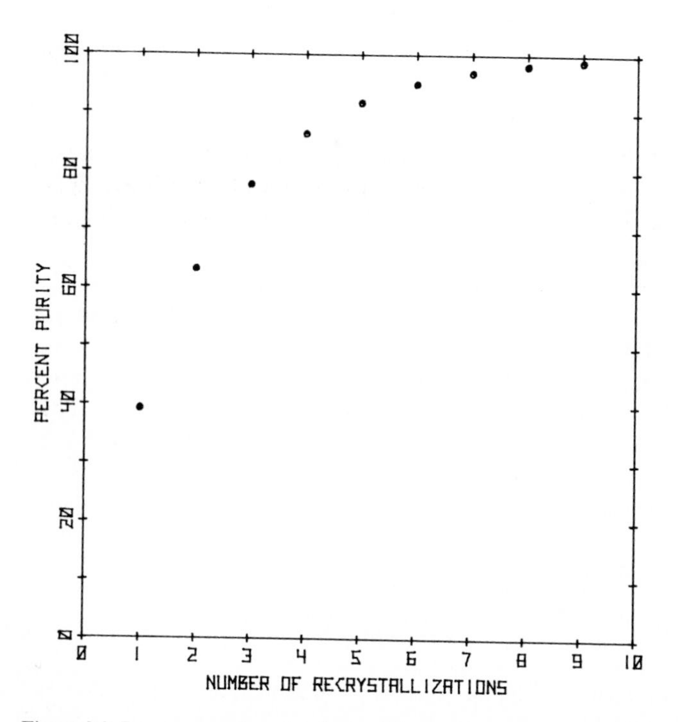

Figure 2.9. Response surface showing an inherently continuous response (percent purity of a protein) as a function of an inherently discrete factor (number of recrystallizations).

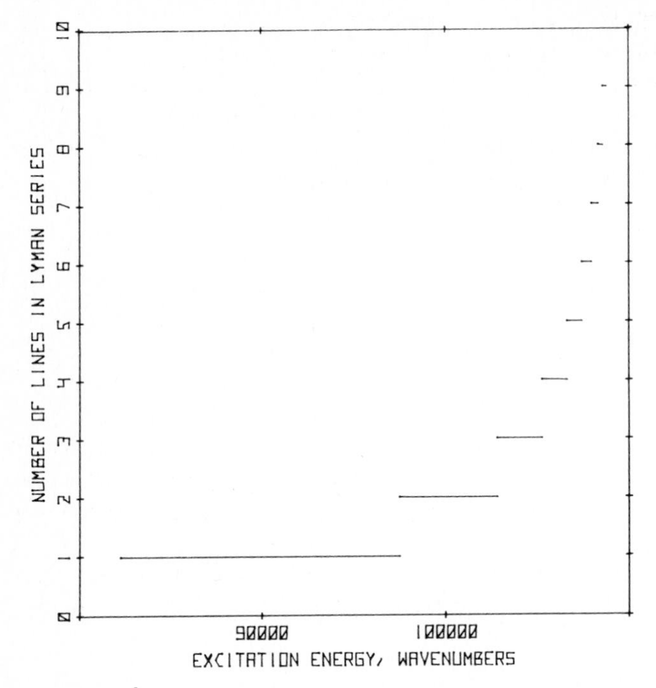

Figure 2.10. Response surface showing an inherently discrete response (number of lines in the Lyman series) as a function of an inherently continuous factor (excitation energy).

limited number of values within a given range. Examples of discrete factors are type of buffer, choice of solvent, number of extractions, type of catalyst, and real plates in a distillation column. Examples of discrete responses are the familiar postal rates, wavelengths of the hydrogen atom emission spectrum, number of radioactive decays in one second, number of peaks in a spectrum, and number of items that pass inspection in a sample of 100.

Figures 2.2–2.8 are response surfaces that show inherently continuous responses as functions of inherently continuous factors. Figure 2.9 illustrates a response surface for an inherently continuous response (the percent purity of a protein) plotted against an inherently discrete factor (number of recrystallizations); any percent purity from zero to 100 is possible, but only integer values are meaningful for the number of recrystallizations. Figure 2.10 is a response surface for an inherently discrete response (number of lines in the Lyman series of the hydrogen atom emission spectrum) as a function of an inherently continuous factor (energy of exciting radiation); only integer values are meaningful for the number of lines observed, but any excitation energy is possible.

The discrete factor "number of recrystallizations" (see Figure 2.9) is naturally ordered. It obviously makes sense to plot yield against the number of recrystallizations in the order 1, 2, 3, 4, 5, 6, 7, 8, 9; it would not make sense to plot the yield against a different ordering of the number of recrystallizations, say 3, 7, 5, 8, 2, 1, 4, 9, 6. Other discrete factors are not always as meaningfully ordered. Consider a chemical reaction for which the type of solvent is thought to be an important discrete factor. If we choose nine solvents at random, arbitrarily number them one through nine, and then evaluate the percent yield of the same chemical reaction carried out in each of the nine solvents, the results might be

Solvent	*Percent Yield*
1	54.8
2	88.7
3	34.8
4	83.7
5	68.7
6	89.9
7	74.7
8	90.0
9	83.6

If we now plot the percent yield vs. the solvent number, we will get the "response surface" shown in Figure 2.11. There seems to be a trend toward higher yield with

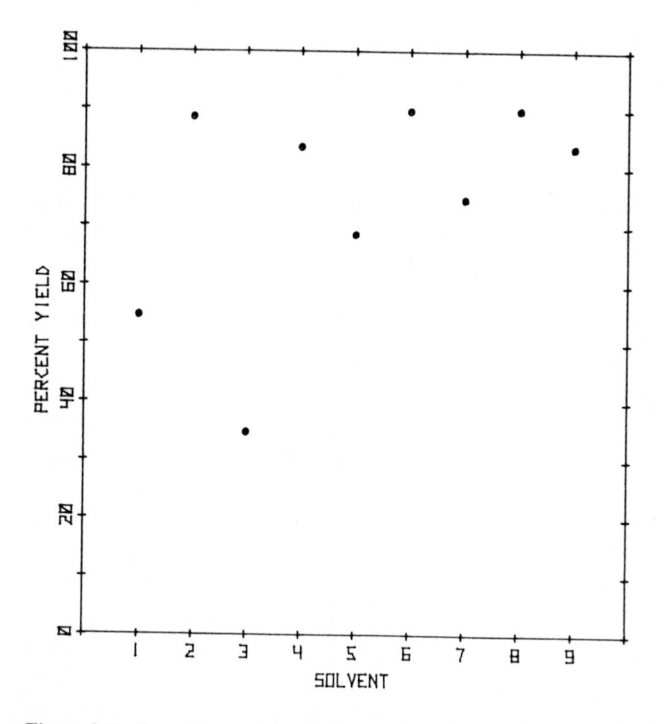

Figure 2.11. Response surface obtained from experiments using different solvents arbitrarily numbered 1-9.

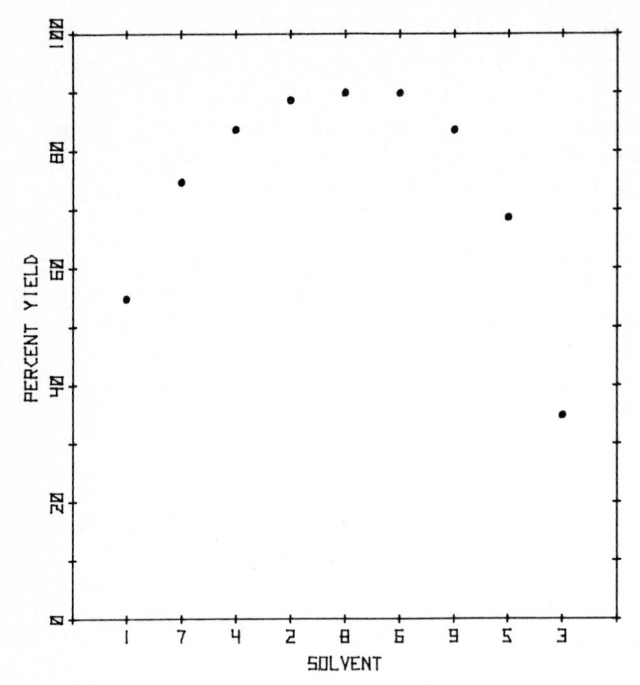

Figure 2.12. Response surface obtained from experiments using different solvents ranked in order of increasing dipole moment.

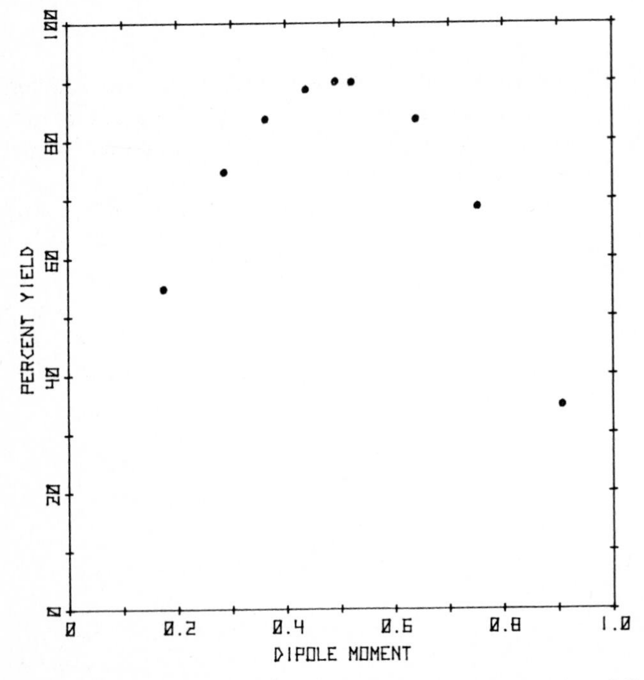

Figure 2.13. Response surface showing percent yield as a function of dipole moments of solvents.

increasing solvent number, but such a trend must be meaningless: it is difficult to imagine how an arbitrarily assigned number could possibly influence the yield of a chemical reaction. If a relationship does appear to exist between yield and solvent number, it must be entirely accidental – the variation in response is probably caused by some other property of the solvents.

When it is suspected that some particular property of the discrete factor is important, *reordering* or *ranking* the discrete factor in increasing or decreasing order of the suspected property might produce a smoother response surface. Many properties of solvents could have an influence on the yield of chemical reactions: boiling point, Lewis basicity, and dipole moment are three such properties. Perhaps in this case it is felt that dipole moment is an important factor for the particular chemical reaction under investigation. If we look up the dipole moments of solvents one through nine in a handbook, we might find the following:

Solvent	*Dipole Moment*
1	0.175
2	0.437
3	0.907
4	0.362
5	0.753
6	0.521
7	0.286
8	0.491
9	0.639

When arranged according to increasing dipole moment, the solvents are ordered 1, 7, 4, 2, 8, 6, 9, 5, 3. Figure 2.12 plots percent yield as a function of these reordered solvent numbers and suggests a discrete view of an underlying smooth functional relationship that might exist between percent yield and dipole moment.

The discrete factor "solvent number" is recognized as a simple bookkeeping designation. We can replace it with the continuous factor dipole moment and obtain, finally, the "response surface" shown in Figure 2.13.

A special note of caution is in order. Even when data such as that shown in Figure 2.13 is obtained, the suspected property might not be responsible for the observed effect; it may well be that a different, correlated property is the true cause (see Section 1.2 on masquerading factors).

2.3. Constraints and feasible regions

A factor or response that is not variable from infinitely negative values to infinitely positive values is said to be constrained.

Many factors and responses are *naturally constrained*. Temperature, for example, is usually considered to have a lower limit of $-273.16\,^{\circ}$C; temperatures at or below

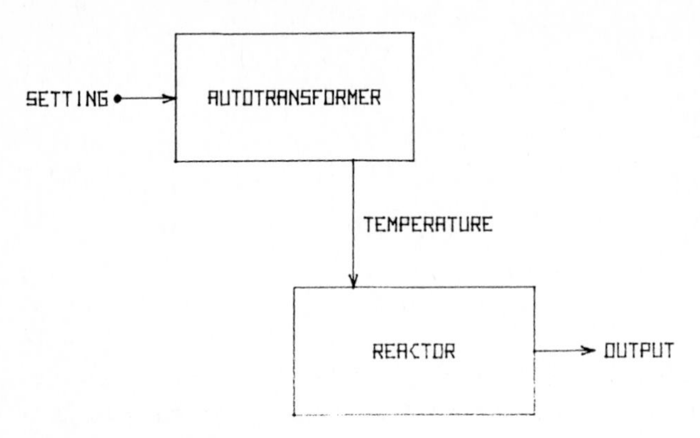

Figure 2.14. General system theory view showing how the use of an autotransformer imposes artificial constraints on the factor temperature.

that value are theoretically impossible. There is, however, no natural upper limit to temperature: temperature is said to have a natural *lower bound*, but it has no natural *upper bound*.

The voltage from an autotransformer has both a lower bound (0 V a.c.) and an upper bound (usually about 130 V a.c.) and is an example of a naturally constrained discrete factor. The upper constraint could be changed if the autotransformer were redesigned but, given a particular autotransformer, these constraints are "natural" in the sense that they are the limits available.

If the voltage from an autotransformer is to be used to adjust the temperature of a chemical reactor (see Figure 2.14), then the natural boundaries of the autotransformer voltage will impose *artificial constraints* on the temperature. The lower boundary of the autotransformer (0 V a.c.) would result in no heating of the chemical reactor. Its temperature would then be approximately ambient, say 25°C. The upper boundary of the autotransformer voltage would produce a constant amount of heat energy and might result in a reactor temperature of, say, 300°C. Thus, the use of an autotransformer to adjust temperature imposes artificial lower and upper boundaries on the factor of interest.

The natural constraint on temperature, the natural constraints on autotransformer voltage, and the artificial constraints on temperature are all examples of *inequality constraints*. If T is used to represent temperature and E represents voltage, these inequality constraints can be expressed, in order, as

$$-273.16°\text{C} < T, \quad (2.5)$$

$$0 \text{ V a.c.} \leqslant E \leqslant 130 \text{ V a.c.}, \quad (2.6)$$

$$25°\text{C} \leqslant T \leqslant 300°\text{C}. \quad (2.7)$$

Here the "less than or equal to" symbol is used to indicate that the boundary values themselves are included in the set of possible values these variables may assume. If the boundary values are not included, "less than" symbols ($<$) are used. The

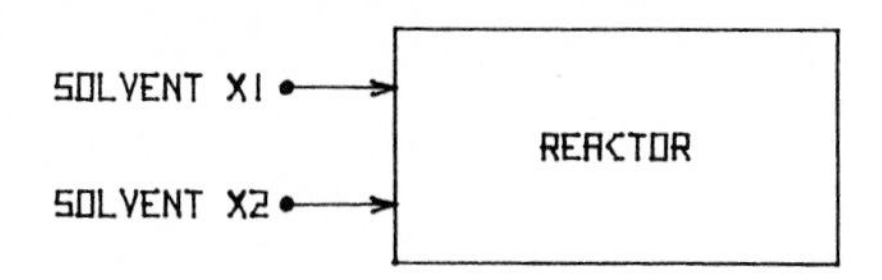

Figure 2.15. General system theory view of a process to which two different solvents are added.

presence of the additional "equal to" symbol in Equations 2.6 and 2.7 does not prevent the constraints from being inequalities.

It is often desirable to fix a given factor at some specified level. For example, if an enzymatic determination is always to be carried out at 37°C, we might specify

$$T = 37°\text{C}. \tag{2.8}$$

This is a simple example of an externally imposed *equality constraint*. Note that by imposing this constraint, temperature is no longer a variable in the enzymatic system.

Natural equality constraints exist in many real systems. For example, consider a chemical reaction in which a binary mixed solvent is to be used (see Figure 2.15). We might specify two continuous factors, the amount of one solvent (represented by x_1) and the amount of the other solvent (x_2). These are clearly continuous factors and each has only a natural lower bound. However, each of these factors probably should have an externally imposed upper bound, simply to avoid adding more total solvent than the reaction vessel can hold. If the reaction vessel is to contain 10 gallons, we might specify the inequality constraints

$$0 \text{ gal} \leqslant x_1 \leqslant 10 \text{ gal}, \tag{2.9}$$

$$0 \text{ gal} \leqslant x_2 \leqslant 10 \text{ gal}. \tag{2.10}$$

However, as Figure 2.16 shows, these inequality constraints are not sufficient to avoid overfilling the tank; any combination of x_1 and x_2 that lies to the upper right of the dashed line in Figure 2.16 will add a total of more than 10 gallons and will cause the reaction vessel to overflow. Inequality constraints of

$$0 \text{ gal} \leqslant x_1 \leqslant 5 \text{ gal} \tag{2.11}$$

$$0 \text{ gal} \leqslant x_2 \leqslant 5 \text{ gal} \tag{2.12}$$

will avoid overfilling the tank. The problem in this case, however, is that the reaction vessel will seldom be filled to capacity: in fact, it will be filled to capacity only when both x_1 and x_2 are 5 gallons (see Figure 2.16).

It is evident that an equality constraint is desirable. If we would like to vary both x_1 and x_2, and if the total volume of solvent must equal 10 gallons, then we want to choose x_1 and x_2 such that their location in factor space falls *on* the dashed line in Figure 2.16. An appropriate equality constraint is

$$x_1 + x_2 = 10 \text{ gal}. \tag{2.13}$$

Either (but not both) of the two inequality constraints given by Equations 2.9 and

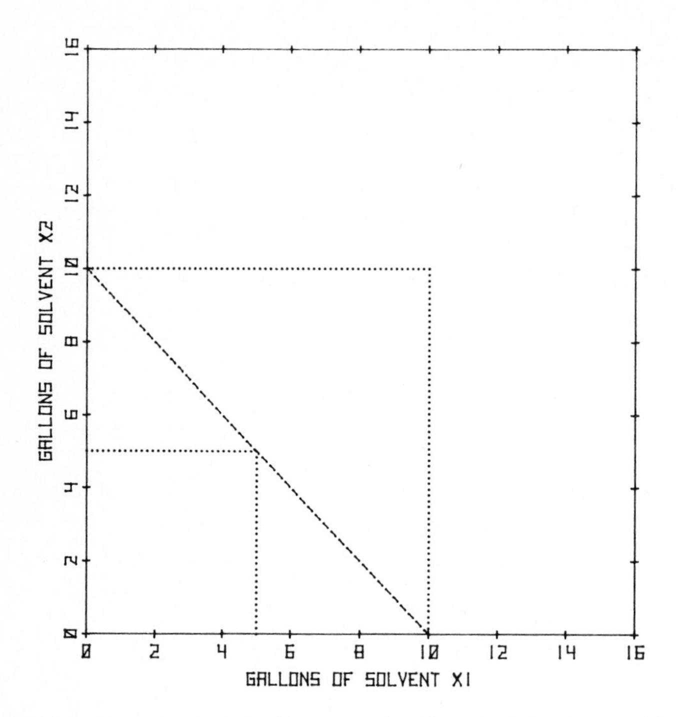

Figure 2.16. Graph of the factor space for Figure 2.15 showing possible constraints on the two different solvents.

2.10 also applies. *The specification of an equality constraint takes away a degree of freedom from the factors*. In this example, x_1 and x_2 cannot both be independently varied – mathematically, it is a single factor system.

Except in certain specialized areas, it is seldom practical to place an equality constraint on a continuous *response*: it is usually not possible to achieve an exact value for an output. For example, a purchasing agent for a restaurant might specify that the margarine it buys from a producer have a "spreadability index" of 0.50. When questioned further, however, the buyer will probably admit that if the spreadability index is between 0.45 and 0.55 the product will be considered acceptable. In general, if an equality constraint is placed on a continuous response, an inequality constraint is usually available.

When a system is constrained, the factor space is divided into feasible regions and nonfeasible regions. A *feasible region* contains permissible or desirable combinations of factor levels, or gives an acceptable response. A *nonfeasible region* contains prohibited or undesirable combinations of factor levels, or gives an unacceptable response.

2.4. Factor tolerances

Response surfaces provide a convenient means of discussing the subject of *factor tolerances*, limits within which a factor must be controlled to keep a system response within certain prescribed limits.

Consider the two response surfaces shown in Figures 2.17 and 2.18. Each response surface shows percent yield (y_1) as a function of reaction temperature (x_1), but for two different processes. In each process, the temperature is controlled at 85°C. The yield in each case is 93%. Let us suppose that the yield must be held between 92% and 94% for the process to be economically feasible. That is,

$$92\% \leqslant y_1 \leqslant 94\%. \tag{2.14}$$

The question of interest to us now is, "Within what limits must temperature (x_1) be controlled to keep the yield within the specified limits"?

Answering this question requires some knowledge of the response surface in the region around 85°C. We will assume that experiments have been carried out in this region for each process, and that the response surfaces shown in Figures 2.17 and 2.18 are good approximations of the true behavior of the system. Empirical models

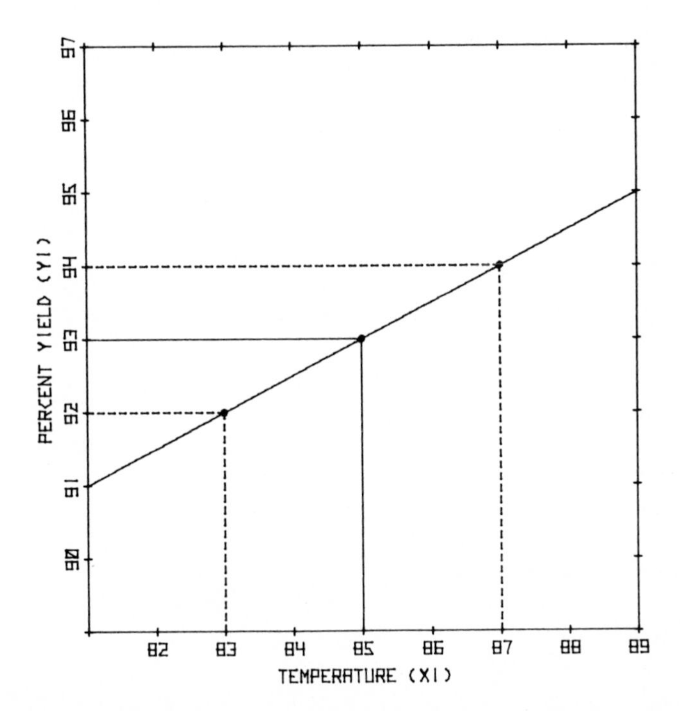

Figure 2.17. Relationship between percent yield and temperature for chemical process A. Note the relatively large variations in temperature that are possible while maintaining yield between 92% and 94%.

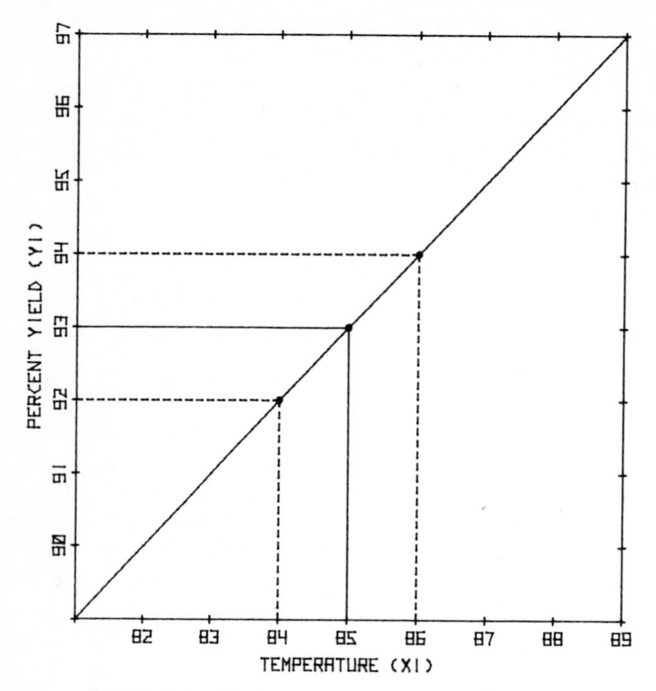

Figure 2.18. Relationship between percent yield and temperature for chemical process B. Note the relatively small variations in temperature that are required to maintained yield between 92% and 94%.

that adequately describe the dependence of yield upon temperature in the region of factor space near 85°C are

$$y_1 = 50.5 + 0.5x_1 \tag{2.15}$$

for the first process (Figure 2.17), and

$$y_1 = 8.0 + 1.0x_1 \tag{2.16}$$

for the second process (Figure 2.18), where x_1 is expressed in Celsius degrees. Because we know (approximately) the behavior of the system over the region of interest, we can use the inverse transform (see Section 1.4) to determine the limits within which the temperature must be controlled. For the first process,

$$x_1 = 2(y_1 - 50.5). \tag{2.17}$$

Substituting 92% and 94% for y_1, we find that temperature must be controlled between 83°C and 87°C. These limits on temperature, and the transformed limits on yield, are illustrated in Figure 2.17.

For the second process, the inverse transform is

$$x_1 = y_1 - 8.0. \tag{2.18}$$

Substituting 92% and 94% for y_1, we find that temperature must be controlled between 84°C and 86°C for this second process (see Figure 2.18). This is a tighter factor tolerance than was required for the first process.

The specification of factor tolerances is clearly important for the achievement of reproducible processes. It is especially important to specify factor tolerances if the developed process is to be implemented at several different facilities, with different types of equipment, at different times, with different operators, etc.

References

Bertalanffy, L. (1968). *General System Theory. Foundations, Development, Applications*. George Braziller, New York.

Box, G.E.P., Hunter, W.G., and Hunter, J.S. (1978). *Statistics for Experimenters. An Introduction to Design, Data Analysis, and Model Building*. Wiley, New York.

Hofstadter, D.R. (1979). *Gödel, Escher, Bach: an Eternal Golden Braid*. Basic Books, New York.

Mandel, J. (1964). *The Statistical Analysis of Experimental Data*. Wiley, New York.

Saunders, P.T. (1980). *An Introduction to Catastrophe Theory*. Cambridge University Press, Cambridge.

Weinberg, G.M. (1975). *An Introduction to General Systems Thinking*. Wiley, New York.

Wilde, D.J., and Beightler, C.S. (1967). *Foundations of Optimization*. Prentice-Hall, Englewood Cliffs, New Jersey.

Wilson, E.B., Jr. (1952). *An Introduction to Scientific Research*. McGraw-Hill, New York.

Exercises

2.1. Uncontrolled factors

Suppose the exact relationship for the response surface of Figure 2.2 is $y_1 = 0.80 + 1.20x_1 - 0.05x_1^2 + 0.50x_2$ (note the presence of the second factor, x_2) and that Figure 2.2 was obtained with x_2 held constant at the value zero. What would the response surface look like if x_2 were held constant at the value 4? If x_2 were not controlled, but were instead allowed to vary randomly between 0 and 1, what would repeated experiments at $x_1 = 5$ reveal? What would the entire response surface look like if it were investigated experimentally while x_2 varied randomly between 0 and 1?

2.2. Experiment space

Sketch the factor space, response space, and experiment space for Exercise 2.1.

2.3. Terminology

Give definitions for the following: maximum, minimum, optimum, unimodal, multimodal, local optimum, global optimum, continuous, discrete, constraint, equality constraint, inequality constraint, lower bound, upper bound, natural constraint, artificial constraint, degree of freedom, feasible region, nonfeasible region, factor tolerance.

2.4. Factor tolerances

Consider the response surface shown in Figure 2.7. Suppose it is desired to control the response to within plus or minus 0.5 units. Would the required factor tolerances on x_1 be larger around $x_1 = 2$ or around $x_1 = 6$?

2.5. Mutually exclusive feasible regions

Suppose that the response surface shown in Figure 2.5 represents the yield of product in kilograms per hour vs. the feed rate of a reactant x_1, and that the response surface shown in Figure 2.6 represents the percentage of impurity in the same product. Is it possible to adjust the feed rate of reactant x_1 to simultaneously achieve a yield greater than 7.25 kilograms per hour and a percentage impurity less than 2.1%? If not, are any other actions possible to achieve the goal of > 7.25 kg hr^{-1} and $< 2.1\%$ impurity? How could one or the other or both of the constraints be modified to achieve overlapping feasible regions?

2.6. Continuous and discrete factors

Specify which of the following factors are inherently continuous and which are inherently discrete: length, time, count, pressure, population, density, population density, energy, electrical charge, pages, temperature, and concentration.

2.7. Upper and lower bounds

Which of the factors listed in Exercise 2.6 have lower bounds? What are the values of these lower bounds? Which of the factors listed in Exercise 2.6 have upper bounds? What are the values of these upper bounds?

2.8. Maxima and minima

Locate the three local *minima* in Figure 2.7. Which is the global minimum? Locate the four local maxima in Figure 2.8. Which is the global maximum?

2.9. Degrees of freedom

A store has only three varieties of candy: A at 5 cents, B at 10 cents, and C at 25 cents. If a child intends to buy exactly $1.00 worth of candy in the store, how many degrees of freedom does the child have in choosing his selection of candy? Are the factors in this problem continuous or discrete? List all of the possible combinations of A, B, and C that would cost exactly $1.00.

2.10. Order arising from randomization

Suppose, by chance (one in 362,880) you happened to assign the letters H, C, I, D, G, B, F, A, E to the solvents numbered 1, 2, 3, 4, 5, 6, 7, 8, 9 in Figure 2.11. If these solvents and their percent yield responses were listed in alphabetical order A–I, and a plot of percent yield vs. solvent letter were made, what naive conclusion might be drawn?

2.11. Artificial constraints

List three factors for which artificial constraints are often imposed. What are the values of these constraints? If necessary, could the artificial constraints be removed easily, or with difficulty?

2.12. Equality constraints

Suppose you are given the task of preparing a *ternary* (three-component) solvent system such that the total volume be 1.00 liter. Write the equality constraint in terms of x_1, x_2, and x_3, the volumes of each of the three solvents. Sketch the three-dimensional factor space and clearly draw within it the planar, two-dimensional constrained feasible region. (Hint: try a cube and a triangle after examining Figure 2.16.)

2.13. Two-factor response surfaces

Obtain a copy of a geological survey topographic map showing contours of constant elevation. Identify the following features, if present: a local maximum, a local minimum, the global maximum, the global minimum, a broad ridge, a narrow ridge, a broad valley, a narrow valley, a straight valley or ridge, a curved valley or ridge, a plateau region, and a saddle region.

2.14. Catastrophic response surfaces

Sketch a response surface showing length (y_1) as a function of force exerted (x_1) on a rubber band that is stretched until it breaks. Give examples of other catastrophic response surfaces. [See, for example, Saunders, P.T. (1980). *An Introduction to Catastrophe Theory.* Cambridge University Press, Cambridge.]

2.15. Hysteresis

Suppose you leave your home, go to the grocery store, visit a friend, stop at the drug store, and then return home. Draw a respone surface of your round trip showing relative north-south distance (y_1) as a function of relative east-west distance (x_1). What does the term *hysteresis* mean and how does it apply to your round trip that you have described? Give examples of other systems and response surfaces that exhibit hysteresis. Why do there appear to be two different response surfaces in systems that exhibit hysteresis? Is there a factor that is not being considered? Is y_1 really a response? Is x_1 really a factor?

2.16. Equality and inequality constraints

Which would you sign, a \$100,000 contract to supply 1000 2.000-meter long metal rods, or a \$100,000 contract to supply 1000 metal rods that are 2.000 ± 0.001 meters in length?

2.17. Factor tolerances

Look up a standard method of measurement (e.g., the determination of iron in an ore, found in most quantitative analysis textbooks). Are factor tolerances specified? What might be the economic consequences of insufficiently controlling a factor that has narrow tolerances? What might be the economic consequences of controlling too closely a factor that actually has wide tolerance?

2.18. Factor tolerances

"The sponsoring laboratory may have all the fun it wants within its own walls by using nested factorials, components of variance, or anything else that the workers believe will help in the fashioning of a test procedure. At some time the chosen procedure should undergo the kind of mutilation that results from the departures from the specified procedure that occur in other laboratories". [Youden, W.J. (1971). "Experimental Design and ASTM Committees", Materials Research and Standards, 1, 862.] Comment.

2.19. Sampling theory

The equation $y = a \times \sin(bx)$ describes a sine wave of period $360/b$. If $a = 1$ and $b = 10$, evaluate y at $x = 0, 40, 80, \ldots, 320$, and 360. Plot the individual results. What is the apparent period of the plotted data? Do these discrete responses give an adequate representation of the true response from the system?

CHAPTER 3

Basic Statistics

It is usually not possible or practical to control a given system input *at* an exact level; in practice, most inputs are controlled *around* set levels within certain factor tolerances. Thus, controlled system inputs do exhibit some variation. Variation is also observed in the levels of otherwise constant system outputs, either because of instabilities within the system itself (e.g., a system involving an inherently random process, such as nuclear decay); or because of the transformation of variations in the system inputs into variations of the system output (see Figures 2.17 and 2.18); *or because of variations in an external measurement system* that is used to measure the levels of the outputs. This latter source of apparent system variation also applies to *measured* values of system inputs.

Thus, the single value 85°C might not completely characterize the level of input x_1. Similarly, the single value 93% yield might mislead us about the overall behavior of the output y_1. Will y_1 be 93% yield if we repeat the experiment using presumably identical conditions? Will it change? Over what limits? What might be expected as an "average" value?

To simplify the presentation of this chapter, we will look only at variation in the level of measured system outputs, but the same approach can also be applied to variation in the level of measured system inputs. We will assume that all controllable inputs (known and unknown) are fixed at some specified level (see Figure 3.1). Any variation in the output from the system will be assumed to be caused by variation in the uncontrolled inputs (known and unknown) or by small, unavoidable variations about the set levels of the controlled inputs.

The two questions to be answered in the chapter are, "What can be used to describe the *level* of output from the system"? and "What can be used to describe the *variation* in the level of output from the system"?

3.1. The mean

Suppose we carry out a single evaluation of response from the system shown in Figure 3.1. This number is designated y_{11} and is found to be 32.53. Does this number, by itself, represent the level of output from the system?

If we answer "yes", then we have done so on the basis of the assumption that the variation associated with the output is zero. In the absence of other information, we cannot be certain that there is no variation, but we can test this assumption by

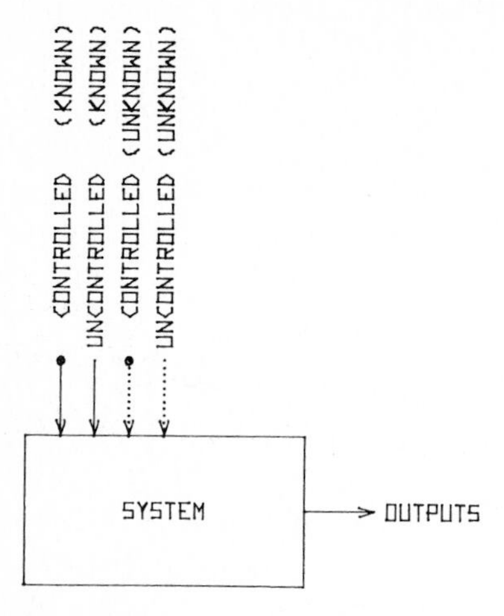

Figure 3.1 General system theory view showing known, unknown, controlled, and uncontrolled factors.

carrying out a second evaluation of response from the same system. If we did carry out another measurement and obtained the value $y_{12} = 32.53$, then we would feel more strongly that there is no measurable variation associated with the system output. But we have not proved that there is no uncertainty associated with the system; we have only increased our confidence in the assumption that there is no uncertainty. We will always be left with the question, "What if we evaluated the response one more time"? If, for example, we carried out our evaluation of response a third time and found $y_{13} = 32.55$, then we could conclude that the assumption about no measurable variation was false. (In general, it is easier to disprove an assumption or hypothesis than it is to prove it. The subject of hypothesis testing will be taken up again in Chapter 6.)

We can see that a single evaluation of output from a system does not, by itself, necessarily represent the true level of the output from a system. Consider the following data set obtained by repetitive evaluation of the output of the system shown in Figure 3.1.

Evaluation	*Response*
y_{11}	32.53
y_{12}	32.53
y_{13}	32.55
y_{14}	32.52
y_{15}	32.48
y_{16}	32.54
y_{17}	32.57
y_{18}	32.51
y_{19}	32.54

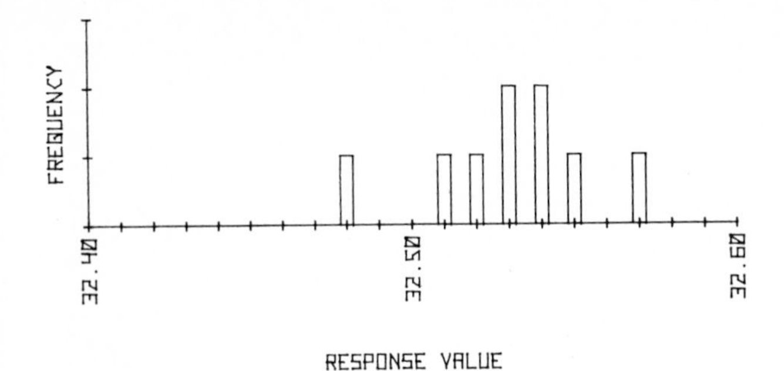

Figure 3.2 Plot showing frequency of obtaining a given response as a function of the response values themselves.

It is useful to reorder or rank the data set according to increasing (or decreasing) values of response. Such a reordering is

Evaluation	*Response*
y_{15}	32.48
y_{18}	32.51
y_{14}	32.52
y_{11}	32.53
y_{12}	32.53
y_{16}	32.54
y_{19}	32.54
y_{13}	32.55
y_{17}	32.57

The responses vary from a low value of 32.48 to a high value of 32.57, a response range of 0.09. Notice that the responses tend to cluster about a central value of approximately 32.53. Most of the values are very close to the central value; only a few are far away. Another representation of this ordered data set is given in Figure 3.2, where the vertical bars indicate the existence of a response with the corresponding value on the number line. Bars twice as high as the others indicate that there are two responses with this value. Again, notice that the data set tends to be clustered about a central value.

The measure of central tendency used throughout this book is the *mean*, sometimes called the *average*. It is defined as the sum (Σ) of all the response values divided by the number of response values. In this book, we will use $\bar{y}$ as the symbol for the mean (see Section 1.3).

$$\bar{y}_1 = \sum_{i=1}^{n} y_{1i}/n \tag{3.1}$$

where y_{1_i} is the value of the ith evaluation of response y_1, and n is the number of response evaluations. For the data illustrated in Figure 3.2,

$$\bar{y}_1 = (32.48 + 32.51 + 32.52 + 32.53 + 32.53 + 32.54 + 32.54 + 32.55 + 32.57)/9$$
$$= 32.53 \tag{3.2}$$

3.2. Degrees of freedom

Suppose a person asks you to choose two integer numbers. You might reply, "Five and nine". But you could just as easily have chosen one and six, 14 and 23, 398 and 437, etc. Because the person did not place any constraints on the integers, you had complete freedom in choosing the two numbers.

Suppose the same person again asks you to choose two integer numbers, this time with the restriction that the sum of the two numbers be ten. You might reply, "Four and six". But you could also have chosen five and five, -17 and 27, etc. A close look at the possible answers reveals that you don't have as much freedom to choose as you did before. In this case, you are free to choose the first number, but the second number is then fixed by the requirement that the sum of the numbers be ten. For example, you could choose three as the first number, but the second number must then be seven. In effect, the equality constraint that the sum of the two numbers be ten has taken away a degree of freedom (see Section 2.3).

Suppose the same person now asks you to choose two integer numbers, this time with the restrictions that the sum of the two numbers be ten and that the product of the two numbers be 24. Your only correct reply would be, "Four and six". The two equality constraints ($y_{11} + y_{12} = 10$ and $y_{11} \times y_{12} = 24$) have taken away two degrees of freedom and left you with no free choices in your answer.

The number of *degrees of freedom* (DF) states how many individual pieces of data in a set can be independently specified. In the first example, where there were no constraints, DF = 2. For each equality constraint upon the system, the number of degrees of freedom is decreased by one. Thus, in the second example, DF = 2 − 1 = 1, and in the third example, DF = 2 − 2 = 0.

In Section 3.1, the mean value of a set of data was calculated. If this mean value is used to characterize the data set, then the data set itself loses one degree of freedom. Thus, if there are n values of response, calculation of the mean value leaves only $n-1$ degrees of freedom in the data set. Only $n-1$ items of data are independent – the final item is fixed by the other $n-1$ items and the mean.

3.3. The variance

For any given set of data, the specification of a mean value allows each individual response to be viewed as consisting of two components – a part that is

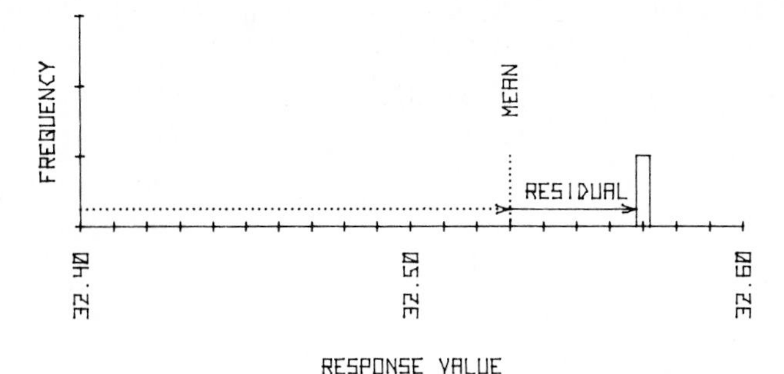

Figure 3.3. Illustration of a residual as the difference between a measured response and the mean of all responses.

described by the mean response ($\bar{y}_1$), and a *residual* part (r_{1i}) consisting of the difference or *deviation* that remains after the mean has been subtracted from the individual value of response.

$$r_{1i} = y_{1i} - \bar{y}_1 \tag{3.3}$$

The concept is illustrated in Figure 3.3 for data point seven of the example in Section 3.1. The total response (32.57) can be thought of as consisting of two parts – the contribution from the mean (32.53) and that part contributed by a deviation from the mean (0.04). Similarly, data point five of the same data set can be viewed as being made up of a mean response (32.53) and a deviation from the mean (−0.05), which add up to give the observed response (32.48). The residuals for the whole data set are shown in Figure 3.4. The set of residuals has $n-1$ degrees of freedom.

Information about the reproducibility of the data is contained in the residual parts of the responses. If the residuals are all small, then the dispersion is slight and the responses are said to cluster tightly – the reproducibility is good. On the other hand, if the residuals are all large, the variation in response is large, and the reproducibility is poor.

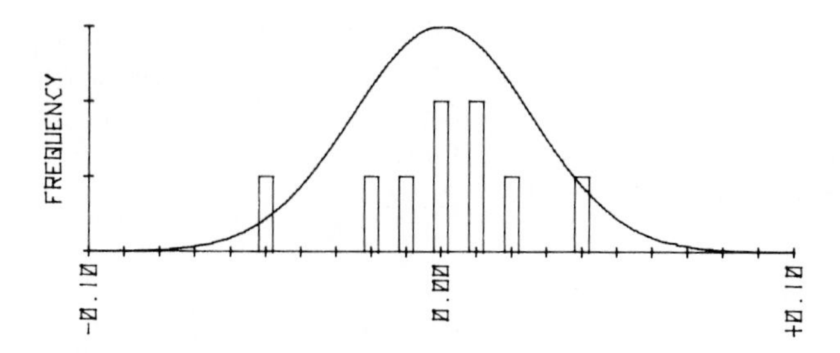

Figure 3.4. Plot showing frequency of obtaining a given residual as a function of the values of the residuals. The solid curve shows the estimated gaussian or normal distribution for the data.

One measure of variation about the mean of a set of data is the *variance* (s^2), defined as the sum of squares of residuals, divided by the number of degrees of freedom associated with the residuals.

$$s^2 = \left[\sum_{i=1}^{n} (y_{1i} - \bar{y}_1)^2\right] / (n-1) = \sum_{i=1}^{n} r_{1i}^2 / (n-1) \tag{3.4}$$

For the data in Section 3.1,

$$s^2 = \left[(-0.05)^2 + (-0.02)^2 + (-0.01)^2 + (0.01)^2 + (0.01)^2 + (0.02)^2 + (0.04)^2\right] / (9-1)$$

$$= 0.0052/8 = 0.00065 \tag{3.5}$$

The square root of the variance of a set of data is called the *standard deviation* and is given the symbol s.

$$s = (s^2)^{1/2} \tag{3.6}$$

For the data in Section 3.1,

$$s = (0.00065)^{1/2} = 0.025 \tag{3.7}$$

In much experimental work, the distribution of residuals (i.e., the frequency of

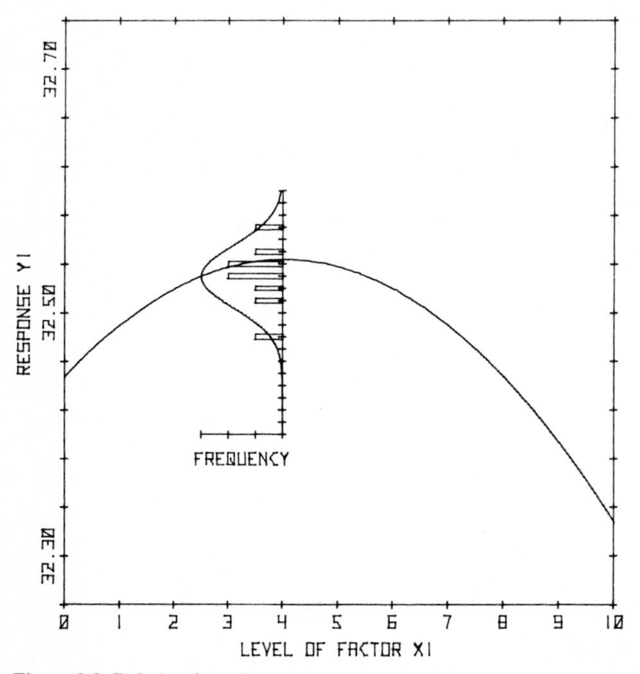

Figure 3.5. Relationship of response frequency plot to response surface.

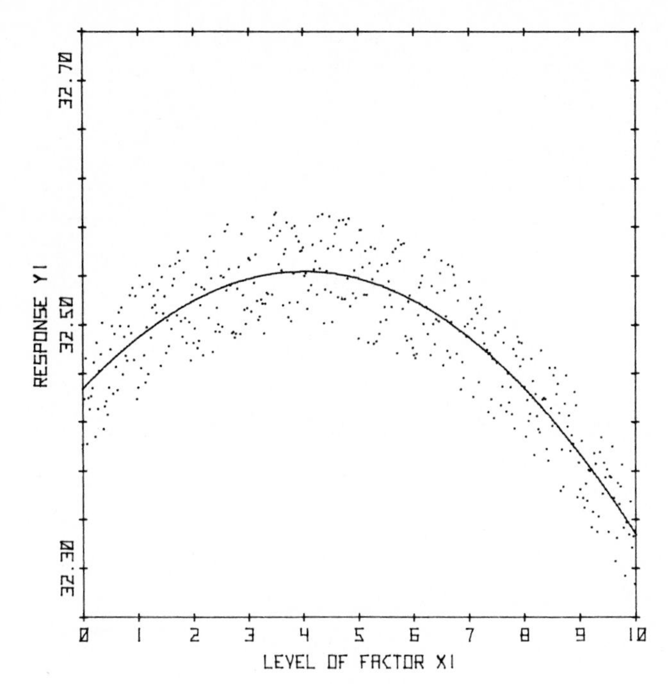

Figure 3.6. Example of homoscedastic noise.

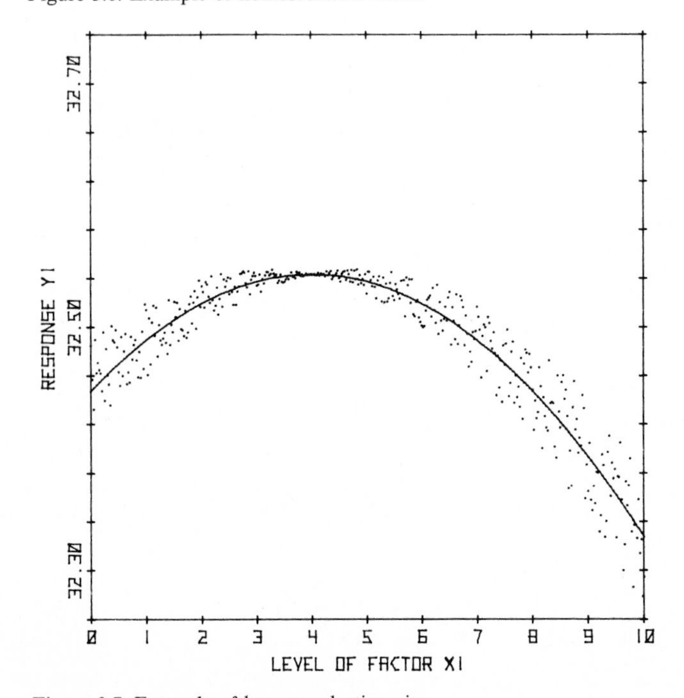

Figure 3.7. Example of heteroscedastic noise.

occurence of a particular value of residual vs. the value itself) follows a *normal* or *gaussian* curve described by the normalized equation

$$\text{frequency} = \left\{1/\left[s(2\pi)^{1/2}\right]\right\} \exp\left[-(y_1 - \bar{y}_1)^2/(2s^2)\right] \quad (3.8)$$

A scaled gaussian curve for the data of Section 3.1 is superimposed on Figure 3.4.

Figure 3.5 shows the current data set superimposed on an arbitrary response surface. The system has been fixed so that all factors are held constant, and only one point in factor space ($x_1 = 4$) has been investigated experimentally. It must be remembered that in practice we are not "all knowing" and would not have a knowledge of the complete response surface; we would know only the results of our experiments, and could only guess at what the remainder of the response surface looked like. If we can pretend to be all knowing for a moment, it is evident from Figure 3.5 that if a different region of factor space were to be explored, a different value for the mean would probably result. What is not evident from the figure, however, is what the variance would be in these other regions.

Two possibilities exist. If the variance of a response is constant throughout a region of factor space, the system is said to be *homoscedastic* over that region (see Figure 3.6). If the variance of a response is *not* constant throughout a region of factor space, the system is said to be *heteroscedastic* over that region (see Figure 3.7).

3.4. Sample statistics and population statistics

In the example we have been using (see Section 3.1), the response from the system was measured nine times. The resulting data is a *sample* of the conceptually infinite number of results that might be obtained if the system were continuously measured. This conceptually infinite number of responses constitutes what is called the *population* of measured values from the system. Clearly, in most real systems, it is not possible to tabulate all of the response values associated with this population. Instead, only a finite (and usually rather limited) number of values can be taken from it.

If it were possible to know the whole population of responses, then a mean and variance could be calculated for it. These descriptors are known as the *population mean* (μ) and the *population variance* (σ^2), respectively. For a given population, there can be only one value of μ and one value of σ^2. The population mean and the population variance are usually unknown ideal descriptors of a complete population about which only partial information is available.

Now consider a small sample ($n = 9$, say) drawn from an infinite population. The responses in this sample can be used to calculate the *sample mean*, $\bar{y}_1$, and the *sample variance*, s^2. It is highly improbable that the sample mean will equal exactly the population mean ($\bar{y}_1 = \mu$), or that the sample variance will equal exactly the population variance ($s^2 = \sigma^2$). It is true that the sample mean will be approximately equal to the population mean, and that the sample variance will be approximately equal to the population variance. It is also true (as would be expected) that as the number of responses in the sample increases, the closer the sample mean approximates the population mean, and the closer the sample variance approximates the population variance. The sample mean, $\bar{y}_1$, is said to be an *estimate* of the population mean, μ, and the sample variance, s^2, is said to be an estimate of the population variance, σ^2.

References

Arkin, H., and Colton, R.R. (1970). *Statistical Methods*, 5th ed. Barnes and Noble, New York.

Box, G.E.P., Hunter, W.G., and Hunter, J.S. (1978). *Statistics for Experimenters. An Introduction to Design, Data Analysis, and Model Building*. Wiley, New York.

Campbell, S.K. (1974). *Flaws and Fallacies in Statistical Thinking*. Prentice-Hall, Englewood Cliffs, New Jersey.

Davies, O.L., Ed. (1956). *The Design and Analysis of Industrial Experiments*. Hafner, New York.

Fisher, R.A. (1966). *The Design of Experiments*, 8th ed. Hafner, New York.

Fisher, R.A. (1970). *Statistical Methods for Research Workers*, 14th ed. Hafner, New York.

Himmelblau, D.M. (1970). *Process Analysis by Statistical Methods*. Wiley, New York.

Huff, D. (1954). *How to Lie with Statistics*. Norton, New York.

Kempthorne, O. (1952). *The Design and Analysis of Experiments*. Wiley, New York.

Mandel, J. (1964). *The Statistical Analysis of Experimental Data*. Wiley, New York.

Meyer, S.L. (1975). *Data Analysis for Scientists and Engineers*. Wiley, New York.

Moore, D.S. (1979). *Statistics. Concepts and Controversies*. Freeman, San Francisco, California.

Natrella, M.G. (1963). *Experimental Statistics* (Nat. Bur. of Stand. Handbook 91). US Govt. Printing Office, Washington, DC.

Shewhart, W.A. (1939). *Statistical Method from the Viewpoint of Quality Control*. Lancaster Press, Pennsylvania.

Youden, W.J. (1951). *Statistical Methods for Chemists*. Wiley, New York.

Exercises

3.1. Mean

Calculate the mean of the following set of responses: 12.61, 12.59, 12.64, 12.62, 12.60, 12.62, 12.65, 12.58, 12.62, 12.61.

3.2. Mean

Calculate the mean of the following set of responses: 12.83, 12.55, 12.80, 12.57, 12.84, 12.85, 12.58, 12.59, 12.84, 12.77.

3.3. Dispersion

Draw frequency histograms at intervals of 0.1 for the data in Exercises 3.1 and 3.2. Calculate the variance and standard deviation for each of the two sets of data.

3.4. Normal distribution

Use Equation 3.8 to draw a gaussian curve over each of the histograms in Exercise 3.3. Do the gaussian curves describe both sets of data adequately?

3.5. Degrees of freedom

A set of five numbers has a mean of 6; four of the numbers are 8, 5, 6, and 2. What is the value of the fifth number in the set?

3.6. Degrees of freedom

Given Equation 2.13, how many independent factors are there in Figure 2.15?

3.7. Degrees of freedom

It is possible to have two positive integer numbers a and b such that their mean is five, their product is 24, and $a/b=2$? Why or why not? Is it possible to have two positive integers c and d such that their mean is five, their product is 24, and $c/d=1.5$? Why or why not?

3.8. Sample and population statistics

A box contains only five slips of paper on which are found the numbers 3, 4, 3, 2, and 5. What is the mean of this data set, and what symbol should be used to represent it?

Another box contains 5,362 slips of paper. From this second box are drawn five slips of paper on which are found the numbers 10, 12, 11, 12, and 13. What is the mean of this sample, and what symbol should be used to represent it? What inference might be made about the mean of the numbers on all 5,362 slips of paper originally in the box? What inference might be made about the standard deviation of the numbers on all 5,362 slips of paper originally in the box?

3.9. Sample statistics

The five slips of paper drawn from the second box in Exercise 3.8 are replaced, and five more numbers are drawn from 5,362 slips of paper. The numbers are 5, −271, 84, 298, and 12. By itself, is this set of numbers surprising? In view of the five numbers previously drawn in Exercise 3.8, is this present set of numbers surprising? In view of the present set of numbers, is the previous set surprising?

3.10. Heteroscedastic noise

Poorly controlled factors are an important source of variation in response?. Consider the response surface of Figure 2.7 and assume that the factor x_1 can be controlled to ± 0.5 across its domain. What regions of the factor domain of x_1 will give large variations in response? What regions of x_1 will give small variations in response? Sketch a plot of *variation* in y_1 vs. x_1.

3.11. Reproducibility

"Reproducibility is desirable, but it should not be forgotten that it can be achieved just as easily by insensitivity as by an increase in precision. Example: All men are two meters tall give or take a meter". [Youden, W.J. (1961). "Experimental Design and ASTM Committees," *Materials Research and Standards*, 1, 862.] Comment.

3.12. Accuracy

"Accuracy" is related to the difference between a measured value (or the mean of a set of measured values) and the true value. It has been suggested that the term "accuracy" is impossible to define quantitatively, and that the term "inaccuracy" should be used instead. Comment.

3.13. Precision

"Precision" is related to the variation in a set of measured values. It has been suggested that the term "precision" is impossible to define quantitatively, and that the term "imprecision" should be used instead. Comment.

3.14. Accuracy

A student repeatedly measures the iron content in a mineral and finds that the mean of five measurements is 5.62%. The student repeats the measurements again in groups of five and obtains means of 5.61%, 5.62%, and 5.61%. On the basis of this information, what can be said about the student's accuracy (or inaccuracy)?

If the true value of iron in the mineral is 6.21%, what can be said about the student's inaccuracy?

3.15. Bias

The term "bias" is often used to express consistent inaccuracy of measurement. What is the bias of the measurement process in Exercise 3.14 (sign and magnitude?)

3.16. Accuracy and precision

Draw four representations of a dart board. Show dart patterns that might result if the thrower were a) inaccurate and imprecise; b) inaccurate but precise; c) accurate but imprecise; and d) accurate and precise.

3.17. Factor tolerances

Consider the response surface shown in Figure 2.17. A set of five experiments is to be carried out at $x_1 = 85\,°C$. If the actual temperatures were 84.2, 85.7, 83.2, 84.1, and 84.4, what would the corresponding responses be (assuming the exact relationship of Equation 2.15 and no other variations in the system)? What is the standard deviation of x_1? What is the standard deviation of y_1? What is the relationship between the standard deviation of x_1, the standard deviation of y_1, and the slope of the response surface in this case? In the general case?

3.18. Youden plots

The term "rugged" can be used to indicate insensitivity of a process to relatively small but common variations in factor levels. One method of evaluating the ruggedness of measurement methods is to use interlaboratory testing in which samples of the same material are sent to a number of laboratories. Presumably, different laboratories will use slightly different factor levels (because of differences in calibrating the equipment, the use of different equipment, different individuals carrying out the measurements, etc.) and the aggregate results reported to a central location will reveal the differences in response and thus the "ruggedness" of the method. W.J. Youden suggested that although the actual factor levels of a given laboratory would probably be slightly different from the factor levels specified by the method, the factor levels would nevertheless be relatively constant within that laboratory. Thus, if a laboratory reported a value that was higher than expected for one sample of material, it would probably report a value that was higher than expected for other samples of the same material; if a laboratory reported a value that was lower than expected for one sample, it would probably report other low sample values as well. Youden suggested sending to each of the laboratories participating in an interlaboratory study *two* samples of material having slightly different values of the property to be measured (e.g., sample No. 1 might contain 5.03% sodium; sample No. 2, 6.49% sodium). When all of the results are returned to the central location, the value obtained for sample No. 1 is plotted against the value of sample No. 2 for each laboratory.

What would the resulting "Youden plots" look like if the method was rugged? If the method was not rugged? If the laboratories were precise, but each was biased? If the laboratories were imprecise, but accurate? [See, for example, Youden, W.J. (1959). "Graphical Diagnosis of Interlaboratory Test Results", *Industrial Quality Control*, 24(May), p. 24.]

3.19. Sources of variation

Will variables that are true factors, but are not identified as factors *always* produce random and erratic behavior in a system (see Table 1.1)? Under what conditions might they not?

3.20. Homoscedastic and heteroscedastic response surfaces

In view of Exercise 3.17 and whatever insight you might have on the shapes of response surfaces in general, what fraction of response surfaces do you think might be homoscedastic? Why? Is it possible to consider any local region of the response surface shown in Figure 3.7 to be approximately homoscedastic?

3.21. Uncertainty

Note that in Figure 3.5 the mean $\bar{y}_1$ of the superimposed data set (32.53) does not coincide with the response surface (32.543). Assuming the drawing is correct, give two possible reasons for this discrepancy.

CHAPTER 4

One Experiment

The experiments that will be used to estimate the behavior of a system cannot be chosen in a whimsical or unplanned way, but rather, must be carefully designed with a view toward achieving a valid approximation to a region of the true response surface. In the next several chapters, many of the important concepts of the design and analysis of experiments are introduced at an elementary level for the single-factor single-response case. In later chapters, these concepts will be generalized to multifactor, multiresponse systems.

Let us imagine that we are presented with a system for which we know only the identities of the single factor x_1 and the single response y_1 (see Figure 4.1). Let us further suppose that we can carry out one experiment on the system – that is, we can fix the factor x_1 at a specified level and measure the associated response. We consider, now, the possible interpretations of this single observation of response. From this one piece of information, what can be learned about the effect of the factor x_1 on the behavior of the system?

4.1. A deterministic model

One possible model that might be used to describe the system pictured in Figure 4.1 is

$$y_{11} = \beta_0 \tag{4.1}$$

where y_{11} is the single measurement of response and β_0 is a *parameter* of the model. Figure 4.2 is a graphical representation. This model, an approximation to the system's true response surface, assumes that the response is constant (always the same) and does not depend upon the levels of any of the system's inputs. A model

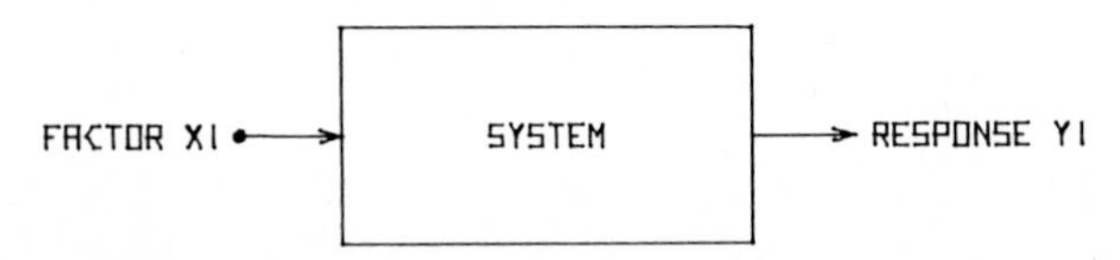

Figure 4.1. Single-factor, single-response system for discussion of a single experiment.

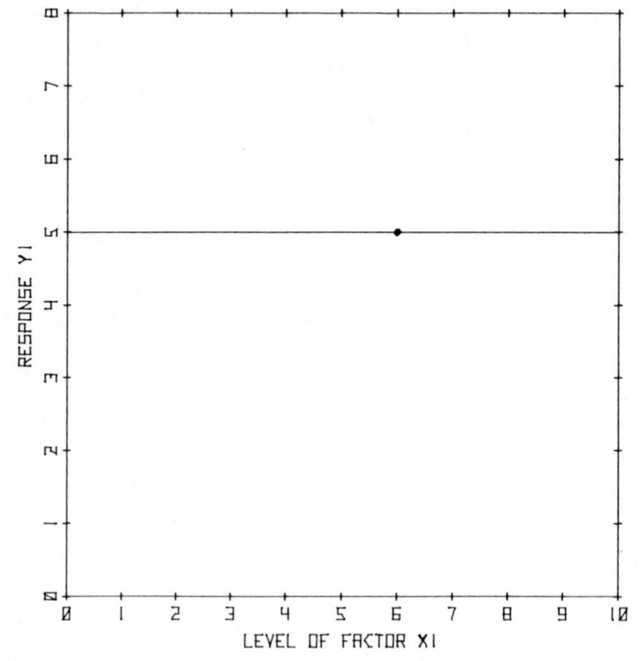

Figure 4.2. Graph of the deterministic model $y_{11} = \beta_0$.

of this form might be appropriate for the description of many fundamental properties, such as the half-life of a radionuclide or the speed of light in vacuum. The single observation of response from the system can be used to calculate the value of the parameter β_0: that is, $\beta_0 = y_{11}$. For example, if $y_{11} = 5$, then $\beta_0 = y_{11} = 5$. This model is *deterministic* in the sense that it does not take into account the possibility of uncertainty in the observed response.

4.2. A probabilistic model

A *probabilistic* or *statistical* model that does provide for uncertainty associated with the system is illustrated in Figure 4.3. For this example, it is assumed that the underlying response is zero and that any value of response other than zero is caused by some random process. This model might appropriately describe the horizontal velocity (speed *and* direction) of a single gas molecule in a closed system, or white noise in an electronic amplifier – in each case, the average value is expected to be zero, and deviations are assumed to be random. The model is

$$y_{11} = r_{11} \tag{4.2}$$

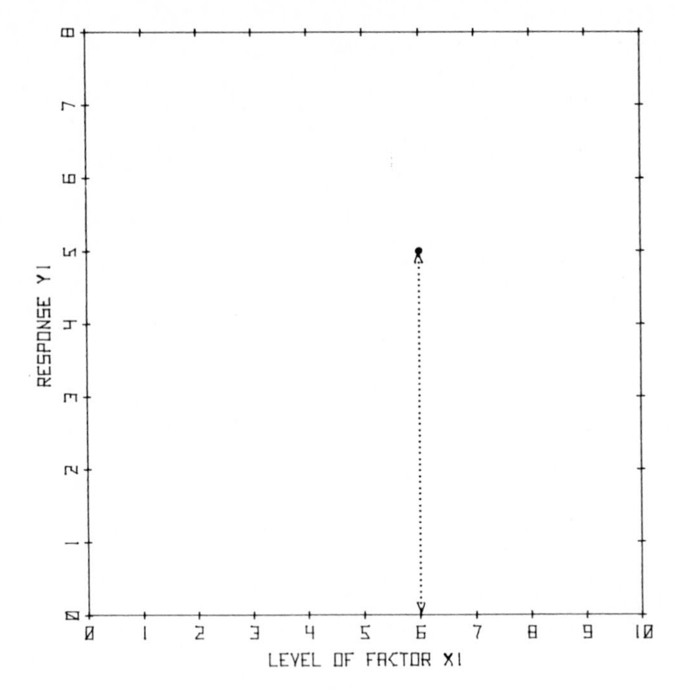

Figure 4.3. Graph of the probabilistic model $y_{11} = 0 + r_{11}$.

but to explicitly show the absence of a β_0 term, we will write the model as

$$y_{11} = 0 + r_{11} \tag{4.3}$$

where r_{11} is the residual or deviation between the response actually observed (y_{11}) and the response predicted by the model (the predicted response is given the symbol $\hat{y}_{11}$ and is equal to zero in the present model). Note that this residual is not the difference between a measured response and the average response (see Section 3.3). If $y_{11} = 5$, then $r_{11} = y_{11} - \hat{y}_{11} = 5 - 0 = 5$.

The term r_{11} is *not* a parameter of the model but is a single value sampled from the population of possible deviations. The magnitude of r_{11} might be used to provide an estimate of a parameter associated with that population of residuals, the *population variance of residuals*, σ_r^2. The *population standard deviation of residuals* is σ_r. The *estimates* of these two parameters are designated s_r^2 and s_r, respectively. If DF_r is the number of degrees of freedom associated with the residuals, then

$$s_r^2 = \sum_{i=1}^{n} r_{1i}^2 / DF_r \tag{4.4}$$

For this model and for only one experiment, $n = 1$ and $\mathrm{DF}_r = 1$ (we have not calculated a mean and thus have not taken away any degrees of freedom) so that $s_r^2 = r_{11}^2$ and $s_r = r_{11}$. *In this example*, $s_r = r_{11} = 5$.

4.3. A proportional model

A third possible model is a deterministic one that suggests a *proportional* relationship between the response and the factor.

$$y_{11} = \beta_1 x_{11} \tag{4.5}$$

where β_1 is a parameter that expresses the effect of the factor x_1 on the response y_1. This approximation to the response surface is shown graphically in Figure 4.4. In spectrophotometry, Beer's Law ($A = abc$) is a model of this form where β_1 represents the product of absorptivity a and path length b, and x_{11} corresponds to the concentration c (in grams per liter) of some chemical species that gives rise to the observed absorbance A. For a single observation of response at a given level of the factor x_1, the parameter β_1 can be calculated: $\beta_1 = y_{11}/x_{11} = 5/6$ for the example shown in Figure 4.4.

When using models of this type with real systems, caution should be exercised in assuming a *causal* relationship between the response y_1 and the factor x_1: *a single experiment is not sufficient to identify which of any number of controlled or uncontrolled factors might be responsible for the measured response* (see Section 1.2 on masquerading factors). In the case of Beer's Law, for example, the observed

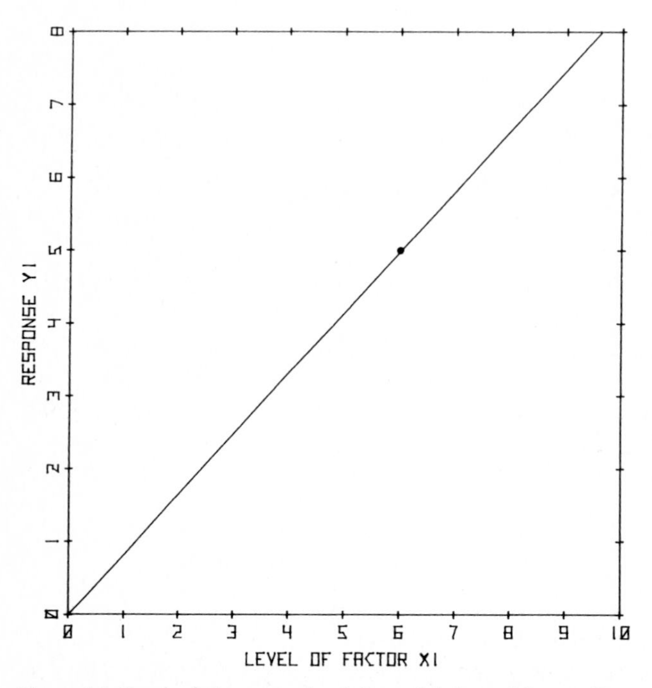

Figure 4.4. Graph of the proportional deterministic model $y_{11} = \beta_1 x_{11}$.

absorbance might be caused not by the supposed chemical species but rather by some unknown impurity in the solvent.

4.4. Multiparameter models

A more general statistical (probabilistic) model of the system takes into account both an offset (β_0) and uncertainty (r_{11}).

$$y_{11} = \beta_0 + r_{11} \tag{4.6}$$

This model offers greater flexibility for obtaining an approximation to the true response surface. However, with only a single observation of response, it is not possible to assign unique values to both β_0 and r_{11}. Figure 4.5 illustrates this difficulty: it is possible to partition the response y_{11} into an infinite number of combination of β_0 and r_{11} such that each combination of β_0 and r_{11} adds up to the observed value of response. Examples in this case are $\beta_0 = 3.2$ and $r_{11} = 1.8$, and $\beta_0 = 1.9$ and $r_{11} = 3.1$.

The reason there are an infinite number of combinations of β_0 and r_{11} that satisfy this model is that there are more degrees of freedom in the set of items to be

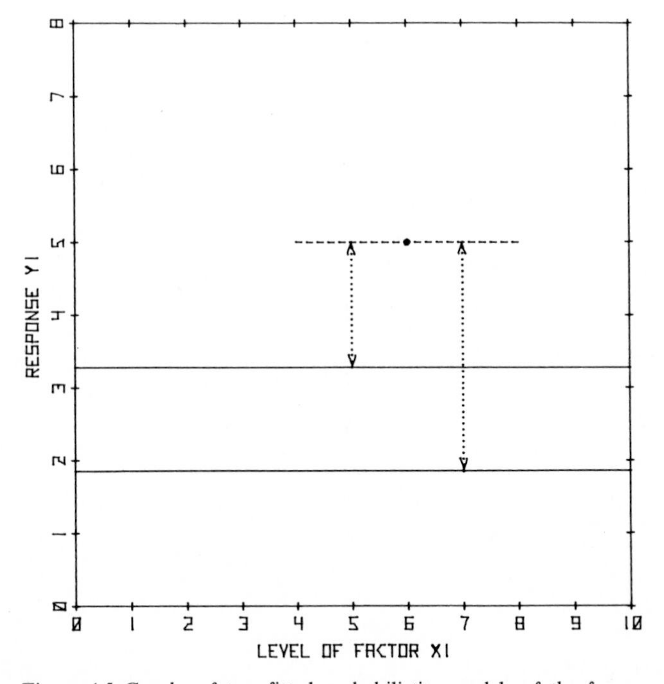

Figure 4.5 Graphs of two fitted probabilistic models of the form $y_{11} = \beta_0 + r_{11}$: $y_{11} = 3.2 + 1.8$, and $y_{11} = 1.9 + 3.1$.

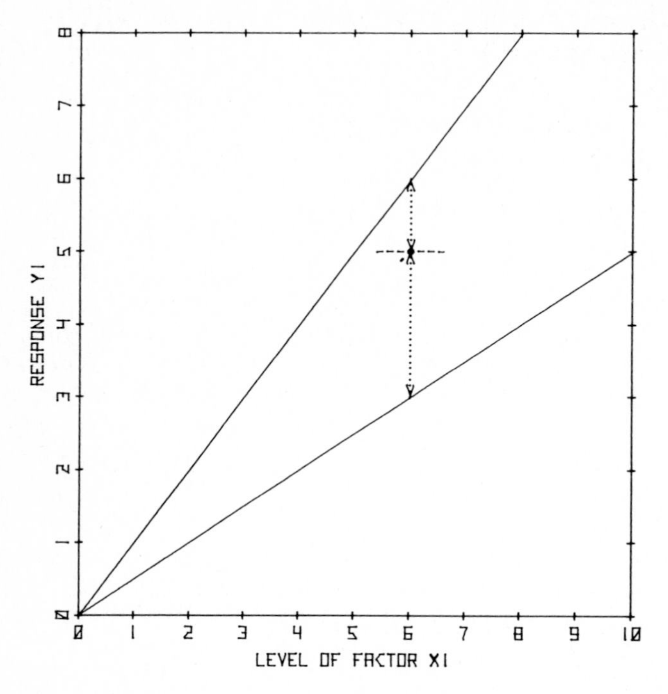

Figure 4.6. Graphs of two fitted probabilistic models of the form $y_{11} = \beta_1 x_{11} + r_{11}$.

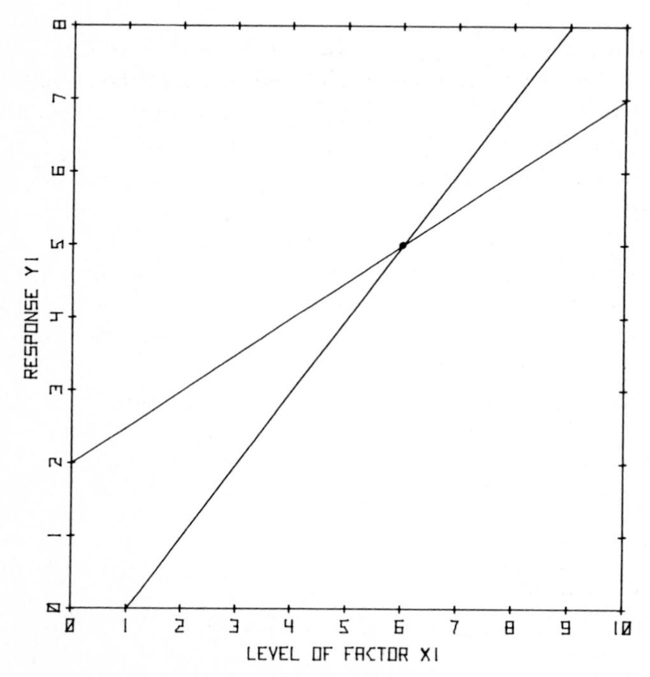

Figure 4.7. Graphs of two fitted deterministic models of the form $y_{11} = \beta_0 + \beta_1 x_{11}$.

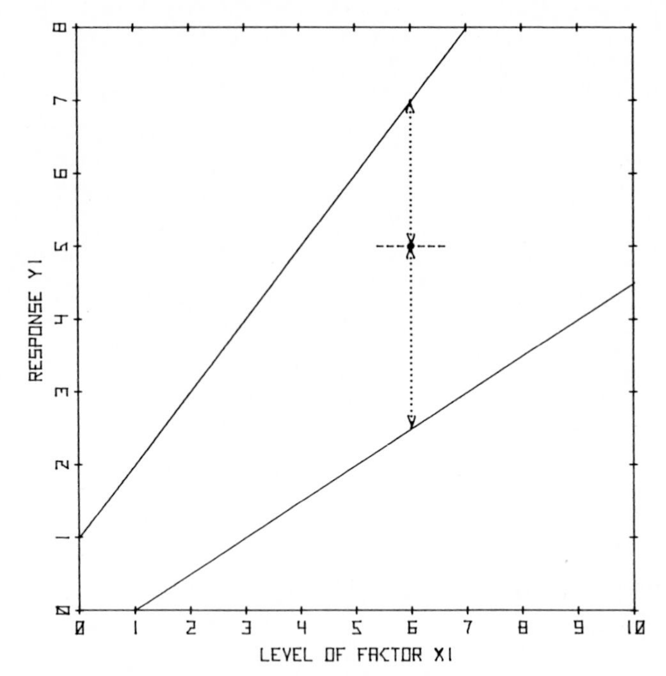

Figure 4.8. Graphs of two fitted probabilistic models of the form $y_{11} = \beta_0 + \beta_1 x_{11} + r_{11}$.

estimated (two – β_0 and r_{11}) than there are degrees of freedom in the data set (one). It is as if someone asked you for two integer values and told you that their sum must equal a certain value (see Equation 4.6 and Section 3.2): there still remains one degree of freedom that could be utilized to choose an infinite number of pairs of integers that would satisfy the single constraint. If we could reduce by one the number of degrees of freedom in the set of items to be estimated (the degrees of freedom would then be equal to one), then the single value of response would be sufficient to estimate β_0 and r_{11}. To accomplish this, let us impose the constraint that the absolute value of the estimated residual be as small as possible. For this particular example, the residual is smallest when it is equal to zero, and the estimated value of β_0 would then be 5.

As shown in Figures 4.6–4.8, this difficulty also exists for other multiparameter models containing more degrees of freedom in the set of items to be estimated than there are degrees of freedom in the data set. The other models are

$$y_{11} = \beta_1 x_{11} + r_{11} \tag{4.7}$$

$$y_{11} = \beta_0 + \beta_1 x_{11} \tag{4.8}$$

and

$$y_{11} = \beta_0 + \beta_1 x_{11} + r_{11} \tag{4.9}$$

respectively. The probabilistic model represented by Equation 4.7 can be fit to the single experimental result by imposing the previous constraint that the residual be as small as possible; the models represented by Equations 4.8 and 4.9 cannot.

Each of the models discussed in this chapter can be made to fit the data perfectly. It is thus evident that a single observation of response does not provide sufficient information to decide which model represents the best approximation to the true response surface. In the absence of additional knowledge about the system, it is not possible to know if the single observed response should be attributed to a constant effect ($y_{11} = \beta_0$), to uncertainty ($y_{11} = 0 + r_{11}$), to a proportional effect ($y_{11} = \beta_1 x_{11}$), to some combination of these, or perhaps even to some other factor.

Because it is not possible to choose among several "explanations" of the observed response, a single experiment, by itself, provides no information about the effect of the factor x_1 on the behavior of the system.

References

Campbell, S.K. (1974). *Flaws and Fallacies in Statistical Thinking.* Prentice-Hall, Englewood Cliffs, New Jersey.

Cochran, W.G., and Cox, G.M. (1957). *Experimental Designs*, 2nd ed. Wiley, New York.

Daniel, C. (1976). *Applications of Statistics to Industrial Experimentation.* Wiley, New York.

Daniel, C., and Wood, F.S. (1971). *Fitting Equations to Data.* Wiley, New York.

Fisher, R.A. (1966). *The Design of Experiments.* 8th ed. Hafner, New York.

Himmelblau, D.M. (1970). *Process Analysis by Statistical Methods.* Wiley, New York.

Mandel, J. (1964). *The Statistical Analysis of Experimental Data.* Wiley, New York.

Moore, D.S. (1979). *Statistics. Concepts and Controversies.* Freeman, San Francisco, California.

Natrella, M.G. (1963). *Experimental Statistics* (Nat. Bur. of Stand. Handbook 91). US Govt. Printing Office, Washington, DC.

Neter, J., and Wasserman, W. (1974). *Applied Linear Statistical Models. Regression, Analysis of Variance, and Experimental Designs.* Irwin, Homewood, Illinois.

Shewhart, W.A. (1939). *Statistical Method from the Viewpoint of Quality Control.* Lancaster Press, Pennsylvania.

Wilson, E.B., Jr. (1952). *An Introduction to Scientific Research.* McGraw-Hill, New York.

Youden, W.J. (1951). *Statistical Methods for Chemists.* Wiley, New York.

Exercises

4.1. Inevitable events

Suppose a researcher develops a drug that is intended to cure the common cold. He gives the drug to a volunteer who has just contracted a viral cold. One week later the volunteer no longer has a cold, and the researcher announces to the press that his drug is a success. Comment. Suggest a better experimental design.

4.2. Coincidental events

It has been reported that one night a woman in New Jersey plugged her iron into an electrical outlet and then looked out the window to see all of the lights in New

York City go out. She concluded it was her action that caused the blackout, and she called the power company to apologize for overloading the circuits. Comment.

4.3. Confounding

A researcher drank one liter of a beverage containing about one-third gin, two-thirds tonic water, and a few milliliters of lime juice. When asked if he would like a refill, he replied yes, but he requested that his host leave out the lime juice because it was making his speech slur and his eyes lose focus. Comment. Suggest an experimental design that might provide the researcher with better information.

4.4. Mill's methods

In 1843, John Stuart Mill first published his *System of Logic*, a book that went through eight editions, each revised by Mill to "attempt to improve the work by additions and corrections, suggested by criticism or by thought..." Although much of Mill's logic and philosophy has decreased in prestige over the years, his influence on methods in science is today still very much felt, primarily through the canons of experimental methods he set down in his *Logic*. The canons themselves were not initially formulated by Mill, but were given his name because of the popularization he gave to them.

Mill's First Canon states: "If two or more instances of the phenomenon under investigation have only one circumstance in common, the circumstance in which alone all the instances agree is the cause (or effect) of the given phenomenon".

Why is it necessary that there be "two or more" instances? Is this canon infallible? Why or why not?

4.5. Antecedent and consequent

"Smoking causes cancer". Some might argue that "cancer causes smoking". Comment. Is it always clear which is cause and which is effect?

4.6. Terminology

Define or describe the following: parameter, deterministic, probabilistic, statistical, residual, proportional, cause, effect.

4.7. Deterministic models

Write two models that are deterministic (e.g., $F = ma$). Are there any limiting assumptions behind these deterministic models? Are there any situations in which measured values would not be predicted exactly by the deterministic models?

4.8. Sources of variation

Suppose you have chosen the model $y_{11} = \beta_0 + r_{11}$ to describe measured values from a system. What is the source of variation in the data (r_{11})? Is it the system? Is it the measurement process? Is it both? Draw a general system theory diagram showing the relationship between a system of interest and an auxiliary system used to measure the system of interest (see Figures 2.14 and 3.1).

4.9. Single-point calibrations

Many measurement techniques are based upon physical or chemical relationships of the form $c = km$, where m is a measured value, c is the property of the material being evaluated, and k is a proportionality constant. As an example, the weight of an object is often measured by suspending it from a spring; the length of extension of the spring, m, is proportional, $1/k$, to the mass of the object, c. Such measurement techniques are often calibrated by the single-point method: i.e., measuring m for a known value of c and calculating $k = c/m$. Comment on the assumptions underlying this method of calibration. Are the assumptions always valid? Suggest a method of calibration that would allow the testing of these assumptions.

4.10. Confounding

A student wishes to measure the absorptivity (a) of compound X in water at a concentration (c) of 2 milligrams per liter in a one centimeter pathlength (b) glass cuvet at a wavelength of 150 nanometers. The measured absorbance A is found to be 4.0. What value would the student obtain for the absorptivity a? The true value is 0. Why might the student have obtained the wrong value? How could he have designed a set of experiments to avoid this mistake?

4.11. Measurement systems

An individual submitted a sample of material to a laboratory for the quantitative determination of a constituent of the material. The laboratory reported the value, 10.238%. When asked if this was a single datum or the mean of several data, the laboratory representative replied, "Oh, that's just a single measurement. We have a lot of variability with this method, so we never repeat the measurements". Comment.

4.12. Mean of zero

Name four systems for which the model $y_{11} = 0 + r_{11}$ is appropriate (see Section 4.2).

4.13. Choice of model

Equation 4.4 is a general equation for calculating the variance of residuals. Equation 3.4 is a specific equation for calculating the variance of residuals. What is the *model* that gives rise to Equation 3.4?

4.14. Dimensions, constraints, and degrees of freedom

How many straight lines can pass through a single point? Through two non-coincident points? Through three non-colinear points? How many flat planes can pass through a single point? Through two non-coincident points? Through three non-colinear points? Through four non-coplanar points?

4.15. Models

Graph the following models over the factor domain $-10 \leqslant x_1 \leqslant 10$:

a) $y_{1i} = 5.0$

b) $y_{1i} = 0.5x_{1i}$

c) $y_{1i} = 5.0 + 0.5x_{1i}$
d) $y_{1i} = 0.05x_{1i}^2$
e) $y_{1i} = 5.0 - 0.05x_{1i}^2$
f) $y_{1i} = 0.5x_{1i} - 0.05x_{1i}^2$
g) $y_{1i} = 0.5x_{1i} + 0.05x_{1i}^2$
h) $y_{1i} = 5.0 + 0.5x_{1i} - 0.05x_{1i}^2$

4.16. Calibration

Many electrochemical systems exhibit behavior that can be described by a relationship of the form $E = E^0 - k\log[X]$, where E is an observed voltage (measured with respect to a known reference electrode), E^0 is the voltage that would be observed under certain standard conditions, k is a proportionality constant, and $[X]$ represents the concentration of an electroactive species. Rewrite this model in terms of y, β's, and $x = \log[X]$.

The "glass electrode" used with so-called "pH meters" to determine pH ($= -\log[H^+]$) responds to $\log[H^+]$ according to the above relationship. Comment on the practice of calibrating pH meters at only one $[H^+]$. Why is "two-point" calibration preferred?

4.17. Hypotheses

A hungry student who had never cooked before followed the directions in a cookbook and baked a cake that was excellent. Based on this one experiment, he inwardly concluded that he was a good cook. Two days later he followed the same recipe, but the cake was a disaster. What might have gone wrong? How confident should he have been (after baking only one cake) that he was a good cook? How many cakes did it take to disprove this hypothesis? What are some other hypotheses he could have considered after his initial success?

4.18. Disproving hypotheses

Sometimes a single experiment can be designed to prove that a factor *does not* have an effect on a response. Suppose a stranger comes up to you and states that he keeps the tigers away by snapping his fingers. When you tell him it is a ridiculous hypothesis, he becomes defensive and tells you the reason there aren't any tigers around is because he has been snapping his fingers. Design a single experiment (involving the stranger, not you) to test the hypothesis that snapping one's fingers will keep tigers away.

4.19. Consequent and antecedent

"Thiotimoline" is a substance that has been (fictitiously) reported to dissolve a short time *before* being added to water. What inevitable effects could be caused by such a substance? [See Asimov, I. (1948). "The Endochronic Properties of Resublimated Thiotimoline", *Astounding Science Fiction*.]

CHAPTER 5

Two Experiments

We consider now the possible interpretations of the results of *two* experiments on a system for which a single factor, x_1, is to be investigated. What can be learned about the effect of the factor x_1 on the behavior of the system from these two pieces of information?

For the moment, we will investigate the *experimental design* in which each experiment is carried out at a different level of the single factor. Later, in Section 5.6, we will consider the case in which both experiments are performed at the same level.

Before we begin, it is important to point out a common misconception that involves the definition of a *linear model*. Many individuals understand the term "linear model" to mean (and to be limited to) straight line relationships of the form

$$y_{1i} = \beta_0 + \beta_1 x_{1i} \tag{5.1}$$

where y_1 is generally considered to be the dependent variable, x_1 is the independent variable, and β_0 and β_1 are the parameters of the model (intercept and slope, respectively). Although it is true that Equation 5.1 is a linear model, the reason is *not* that its graph is a straight line. Instead, it is a linear model because it is constructed of additive terms, each of which contains one and only one multiplicative parameter. That is, the model is first-order or linear *in the parameters*. This definition of "linear model" includes models that are not first-order in the independent variables. The model

$$y_{1i} = \beta_0 + \beta_1 x_{1i} + \beta_2 10^{x_{1i}} + \beta_3 \log(x_{1i}) + \beta_4 x_{1i}^2 \tag{5.2}$$

is a linear model by the above definition. The model

$$y_{1i} = \beta_1 [\exp(-\beta_2 x_{1i})] \tag{5.3}$$

however, is a nonlinear model because it contains more than one parameter in a single term. For some nonlinear models it is possible to make transformations on the dependent and independent variables to "linearize" the model. Taking the natural logarithm of both sides of Equation 5.3, for example, gives a model that is linear in the parameters β_1' and β_2: $\log_e(y_{1i}) = \beta_1' - \beta_2 x_{1i}$, where $\beta_1' = \log_e(\beta_1)$.

This book is limited to models that are linear in the parameters.

5.1. Matrix solution for simultaneous linear equations

In the previous chapter, it was seen that a single observation of response from a system does not provide sufficient information to fit a multiparameter model. A

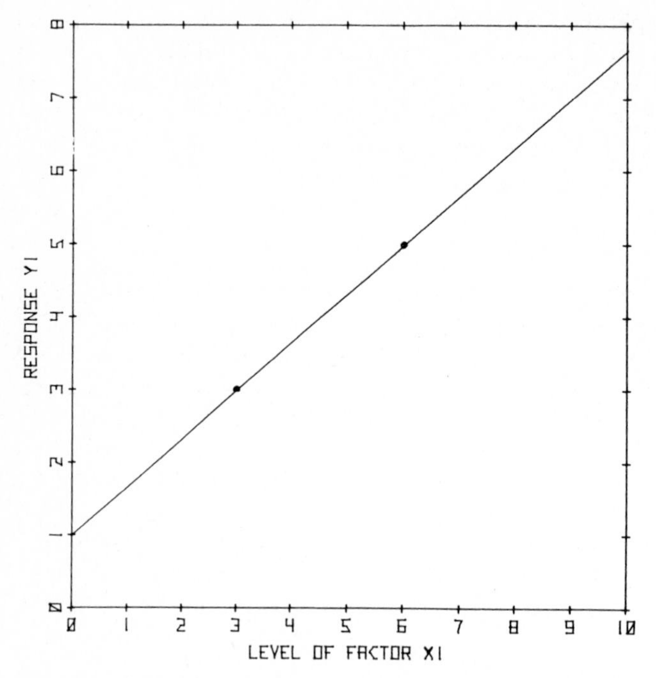

Figure 5.1. Graph of the deterministic model $y_{1i} = \beta_0 + \beta_1 x_{1i}$ fitted to the results of two experiments at different levels of the factor x_1.

geometric interpretation of this limitation for the linear model $y_{1i} = \beta_0 + \beta_1 x_{1i}$ (Equation 5.1) is that a straight line drawn through a single point is not fixed, but is free to rotate about that point (see Figure 4.7). The limitation is overcome when *two* experiments are carried out at *different* levels (x_{11} and x_{12}) of the factor x_1 as shown in Figure 5.1. The best approximation to the true response surface is then uniquely defined by the solutions (β_0 and β_1) to the set of *simultaneous linear equations* that can be obtained by writing the linear model (Equation 5.1) for each of the two experiments.

$$y_{11} = 1 \times \beta_0 + x_{11}\beta_1, \qquad y_{12} = 1 \times \beta_0 + x_{12}\beta_1 \tag{5.4}$$

where y_{11} is the response obtained when factor x_1 is set at the level x_{11}, and y_{12} is the response at x_{12}. Again, the parameter β_0 is interpreted as the *intercept* on the response axis (at $x_1 = 0$); the parameter β_1 is interpreted as the *slope* of the response surface with respect to the factor x_1 – i.e., it expresses the first-order (straight line) effect of the factor x_1 on the response y_1.

Equation 5.4 may be rewritten as

$$1 \times \beta_0 + x_{11}\beta_1 = y_{11}, \qquad 1 \times \beta_0 + x_{12}\beta_1 = y_{12} \tag{5.5}$$

Expressed in matrix form (see Appendix A), this becomes

$$\begin{bmatrix} 1 \times \beta_0 + x_{11}\beta_1 \\ 1 \times \beta_0 + x_{12}\beta_1 \end{bmatrix} = \begin{bmatrix} y_{11} \\ y_{12} \end{bmatrix} \tag{5.6}$$

Recall that in matrix multiplication, an element in the ith *row* (a row is a horizontal arrangement of numbers) and jth *column* (a column is a vertical arrangement of numbers) of the product matrix is obtained by multiplying the ith row of one matrix by the jth column of a second matrix, element by corresponding element and summing the products. Thus, Equation 5.6 may be rewritten as

$$\begin{bmatrix} 1 & x_{11} \\ 1 & x_{12} \end{bmatrix} \begin{bmatrix} \beta_0 \\ \beta_1 \end{bmatrix} = \begin{bmatrix} y_{11} \\ y_{12} \end{bmatrix} \tag{5.7}$$

Note that the leftmost matrix contains the coefficients of the parameters as they appear in the model from left to right; *each row contains information about a different experiment*, and *each column contains information about a different parameter*. The next matrix contains the parameters, from top to bottom as they appear in the model from left to right. The matrix on the right of Equation 5.7 contains the corresponding responses; each row represents a different experiment. Let the leftmost 2-row by 2-column (2×2) matrix be designated the *matrix of parameter coefficients*, $\boldsymbol{X}$; let the 2-row by 1-column (2×1) matrix of β's be designated the *matrix of parameters*, $\boldsymbol{B}$; and let the rightmost 2×1 matrix be designated the *matrix of measured responses*, $\boldsymbol{Y}$. Equation 5.5 may now be expressed in concise matrix notation as

$$\boldsymbol{XB} = \boldsymbol{Y} \tag{5.8}$$

The matrix solution for the parameters of the simultaneous linear equations is stated here without proof:

$$\boldsymbol{B} = \boldsymbol{X}^{-1}\boldsymbol{Y} \tag{5.9}$$

where $\boldsymbol{X}^{-1}$ is the *inverse* (see Appendix A) of the matrix of parameter coefficients. If the elements of the 2×2 $\boldsymbol{X}$ matrix are designated a, b, c, and d, then

$$\boldsymbol{X} = \begin{bmatrix} a & b \\ c & d \end{bmatrix} \tag{5.10}$$

and

$$\boldsymbol{X}^{-1} = \begin{bmatrix} d/D & -b/D \\ -c/D & a/D \end{bmatrix} \tag{5.11}$$

where

$$D = a \times d - c \times b \tag{5.12}$$

is the *determinant* of the 2×2 $\boldsymbol{X}$ matrix. Thus, the notation of Equation 5.9 is equivalent to

$$\begin{bmatrix} \beta_0 \\ \beta_1 \end{bmatrix} = \begin{bmatrix} d/D & -b/D \\ -c/D & a/D \end{bmatrix} \begin{bmatrix} y_{11} \\ y_{12} \end{bmatrix} \tag{5.13}$$

and

$$\beta_0 = d(y_{11}/D) - b(y_{12}/D) \tag{5.14}$$

$$\beta_1 = -c(y_{11}/D) + a(y_{12}/D) \tag{5.15}$$

The matrix approach to the solution of a set of simultaneous linear equations is entirely general. Requirements for a solution are that there be a number of equations exactly equal to the number of parameters to be calculated and that the determinant D of the $\boldsymbol{X}$ matrix be nonzero. This latter requirement can be seen from Equations 5.14 and 5.15. Elements a and c of the $\boldsymbol{X}$ matrix associated with the present model are both equal to unity (see Equations 5.10 and 5.7); thus, with this model, the condition for a nonzero determinant (see Equation 5.12) is that element b (x_{11}) not equal element d (x_{12}). When the experimental design consists of two experiments carried out at different levels of the factor x_1 ($x_{11} \neq x_{12}$; see Figure 5.1), the condition is satisfied.

To illustrate the matrix approach to the solution of a set of simultaneous linear equations, let us use the data points in Figure 5.1: $x_{11} = 3$, $y_{11} = 3$, $x_{12} = 6$, and $y_{12} = 5$. Equation 5.5 becomes

$$1 \times \beta_0 + 3 \times \beta_1 = 3, \qquad 1 \times \beta_0 + 6 \times \beta_1 = 5 \tag{5.16}$$

In matrix form, the equation $\boldsymbol{XB} = \boldsymbol{Y}$ is

$$\begin{bmatrix} 1 & 3 \\ 1 & 6 \end{bmatrix} \begin{bmatrix} \beta_0 \\ \beta_1 \end{bmatrix} = \begin{bmatrix} 3 \\ 5 \end{bmatrix} \tag{5.17}$$

The determinant of the matrix of parameter coefficients is $D = (1 \times 6 - 1 \times 3) = 3$. Inverting the $\boldsymbol{X}$ array gives

$$\boldsymbol{X}^{-1} = \begin{bmatrix} 6/3 & -3/3 \\ -1/3 & 1/3 \end{bmatrix} = \begin{bmatrix} 2 & -1 \\ -1/3 & 1/3 \end{bmatrix} \tag{5.18}$$

and the solution for $\boldsymbol{B} = \boldsymbol{X}^{-1}\boldsymbol{Y}$ is

$$\boldsymbol{B} = \begin{bmatrix} \beta_0 \\ \beta_1 \end{bmatrix} = \begin{bmatrix} 2 & -1 \\ -1/3 & 1/3 \end{bmatrix} \begin{bmatrix} 3 \\ 5 \end{bmatrix} = \begin{bmatrix} 2 \times 3 - 1 \times 5 \\ -3/3 + 5/3 \end{bmatrix} = \begin{bmatrix} 1 \\ 2/3 \end{bmatrix} \tag{5.19}$$

Thus, the intercept (β_0) is 1 and the slope with respect to the factor x_1 (β_1) is 2/3. Substitution of these values into Equation 5.16 serves as a check.

$$1 \times 1 + 3 \times (2/3) = 3, \qquad 1 \times 1 + 6 \times (2/3) = 5 \tag{5.20}$$

This particular experimental design involving two experiments at two different levels of a single factor has allowed the exact fitting of the model $y_{1i} = \beta_0 + \beta_1 x_{1i}$. Note that both of the β's are parameters of the model and use up the total degrees of freedom ($DF = 2$). It is not possible to estimate any uncertainty due to random processes that might be taking place; there are no degrees of freedom available for calculating s_r^2 (see Equation 4.4).

5.2. Matrix least squares

Consider now the probabilistic model illustrated in Figure 5.2 and expressed as

$$y_{1i} = \beta_0 + r_{1i} \tag{5.21}$$

Note that the response is not a function of any factors. For this model, an *estimate* of β_0 (the estimate is given the symbol b_0) is the mean of the two responses, y_{11} and y_{12}.

$$b_0 = \bar{y}_1 = (y_{11} + y_{22})/2 = (3 + 5)/2 = 4 \tag{5.22}$$

Of the total two degrees of freedom, one degree of freedom has been used to estimate the parameter β_0, leaving one degree of freedom for the estimation of the variance of the residuals, σ_r^2.

$$s_r^2 = \left[\sum_{i=1}^{n} (y_{1i} - b_0)^2\right]/(2-1) = (1^2 + 1^2)/1 = 2 \tag{5.23}$$

and

$$s_r = 1.41 \tag{5.24}$$

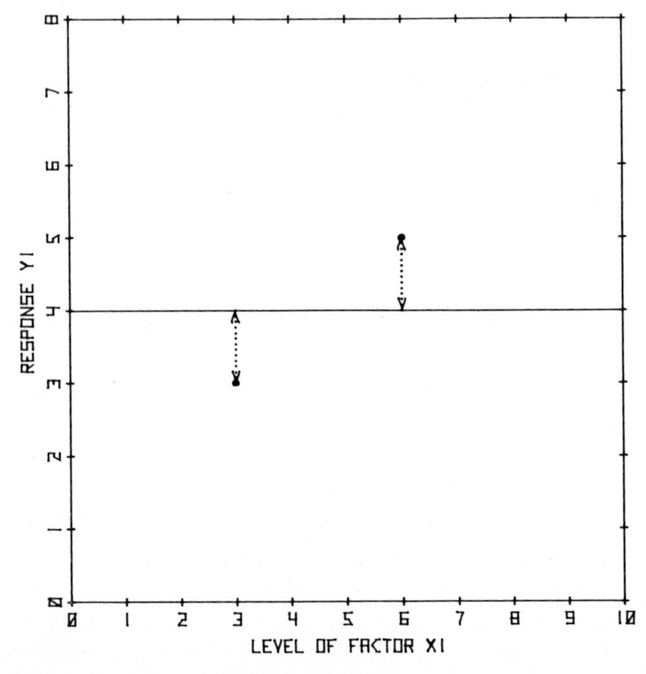

Figure 5.2. Graph of the fitted probabilistic model $y_{1i} = \beta_0 + r_{1i}$.

Suppose these solutions had been attempted using simultaneous linear equations. One reasonable set of linear equations might appear to be

$$1 \times \beta_0 + 1 \times r_{11} = y_{11}, \qquad 1 \times \beta_0 + 1 \times r_{12} = y_{12} \tag{5.25}$$

There is a problem with this approach, however – a problem with the residuals. The residuals are neither parameters of the model nor parameters associated with the uncertainty. They are *quantities* related to a parameter that expresses the variance of the residuals, σ_r^2. The problem, then, is that the simultaneous equations approach in Equation 5.25 would attempt to uniquely calculate *three* items (β_0, r_{11}, and r_{12}) using only *two* experiments, clearly an impossible task. What is needed is an additional constraint to reduce the number of items that need to be estimated. A unique solution will then exist.

We will use the constraint that the sum of squares of the residuals be minimal. The following is a brief development of the matrix approach to the *least squares* fitting of linear models to data. The approach is entirely general for all linear models.

Again, let $\boldsymbol{X}$ be the matrix of parameter coefficients defined by the model to be fit and the coordinates of the experiments in factor space. Let $\boldsymbol{Y}$ be the response matrix associated with those experiments. Let $\boldsymbol{B}$ be the matrix of parameters, and let a new matrix $\boldsymbol{R}$ be the *matrix of residuals*. Equation 5.25 may now be rewritten in matrix notation as

$$\boldsymbol{XB} + \boldsymbol{R} = \boldsymbol{Y} \tag{5.26}$$

$$\begin{bmatrix} 1 \\ 1 \end{bmatrix} [B_0] + \begin{bmatrix} r_{11} \\ r_{12} \end{bmatrix} = \begin{bmatrix} y_{11} \\ y_{12} \end{bmatrix} \tag{5.27}$$

(Recall that the coefficient of β_0 is equal to one in each case; therefore, only 1's appear in the $\boldsymbol{X}$ matrix.)

Equation 5.26 can be rearranged to give

$$\boldsymbol{R} = \boldsymbol{Y} - \boldsymbol{XB} \tag{5.28}$$

It is now useful to note that the $\boldsymbol{R}$ matrix multiplied by its *transpose*

$$\boldsymbol{R}' = [r_{11} \quad r_{12}] \tag{5.29}$$

gives the *sum of squares of residuals*, abbreviated SS_r.

$$SS_r = \boldsymbol{R'R} = [r_{11} \quad r_{12}] \begin{bmatrix} r_{11} \\ r_{12} \end{bmatrix} = [r_{11}^2 + r_{12}^2] \tag{5.30}$$

Although a complete proof of the following is beyond the scope of this presentation, it can be shown that partial differentiation of the sum of squares of residuals with respect to the $\boldsymbol{B}$ matrix gives, in a simple matrix expression, all of the partial derivatives of the sum of squares of residuals with respect to each of the β's.

$$\boldsymbol{R'R} = (\boldsymbol{Y} - \boldsymbol{XB})'(\boldsymbol{Y} - \boldsymbol{XB}) \tag{5.31}$$

$$\partial(\boldsymbol{R'R})/\partial\boldsymbol{B} = \partial[(\boldsymbol{Y} - \boldsymbol{XB})'(\boldsymbol{Y} - \boldsymbol{XB})]/\partial\boldsymbol{B} = \boldsymbol{X}'(\boldsymbol{Y} - \boldsymbol{XB}) = \boldsymbol{X'Y} - \boldsymbol{X'XB} \tag{5.32}$$

If this matrix of partial derivatives is set equal to zero (at which point the sum of squares of residuals with respect to each β will be minimal), the matrix equation

$$0 = \boldsymbol{X}'\boldsymbol{Y} - \boldsymbol{X}'\boldsymbol{X}\hat{\boldsymbol{B}} \tag{5.33}$$

is obtained where $\hat{\boldsymbol{B}}$ is the *matrix of parameter estimates* giving this minimum sum of squares. Rearranging gives

$$\boldsymbol{X}'\boldsymbol{X}\hat{\boldsymbol{B}} = \boldsymbol{X}'\boldsymbol{Y} \tag{5.34}$$

The $(\boldsymbol{X}'\boldsymbol{X})$ array can be eliminated from the left side of the matrix equation if both sides are multiplied by its inverse, $(\boldsymbol{X}'\boldsymbol{X})^{-1}$.

$$(\boldsymbol{X}'\boldsymbol{X})^{-1}(\boldsymbol{X}'\boldsymbol{X})\hat{\boldsymbol{B}} = (\boldsymbol{X}'\boldsymbol{X})^{-1}(\boldsymbol{X}'\boldsymbol{Y}) \tag{5.35}$$

$$\hat{\boldsymbol{B}} = (\boldsymbol{X}'\boldsymbol{X})^{-1}(\boldsymbol{X}'\boldsymbol{Y}) \tag{5.36}$$

This is the general matrix solution for the set of b's that gives the minimum sum of squares of residuals. Again, the solution is valid for all models that are linear in the parameters.

Let us use the matrix least squares method to obtain an algebraic expression for the estimate of β_0 in the model $y_{1i} = \beta_0 + r_{1i}$ (see Figure 5.2) with two experiments at two different levels of the factor x_1. The initial $\boldsymbol{X}$, $\boldsymbol{B}$, $\boldsymbol{R}$, and $\boldsymbol{Y}$ arrays are given in Equation 5.27. Other matrices are

$$\boldsymbol{X}' = [1 \quad 1] \tag{5.37}$$

$$(\boldsymbol{X}'\boldsymbol{X}) = [1 \quad 1]\begin{bmatrix} 1 \\ 1 \end{bmatrix} = [1 \times 1 + 1 \times 1] = [2] \tag{5.38}$$

The inverse of a 1×1 matrix is the reciprocal of the single value. Thus,

$$(\boldsymbol{X}'\boldsymbol{X})^{-1} = [1/2] \tag{5.39}$$

$$(\boldsymbol{X}'\boldsymbol{Y}) = [1 \quad 1]\begin{bmatrix} y_{11} \\ y_{12} \end{bmatrix} = [1 \times y_{11} + 1 \times y_{12}] = [y_{11} + y_{12}] \tag{5.40}$$

The least squares estimate of the single parameter of the model is

$$\hat{\boldsymbol{B}} = [b_0] = (\boldsymbol{X}'\boldsymbol{X})^{-1}(\boldsymbol{X}'\boldsymbol{Y}) = [1/2][y_{11} + y_{12}] = (y_{11} + y_{12})/2 \tag{5.41}$$

and shows that for this model the mean response is the *best least squares estimate* of β_0, the estimate for which the sum of squares of residuals is minimal. The data points in the present example are $x_{11} = 3$, $y_{11} = 3$, $x_{12} = 6$, and $y_{12} = 5$, and the least squares estimate of β_0 is

$$b_0 = (3 + 5)/2 = 4 \tag{5.42}$$

Figure 5.3 plots the squares of the individual residuals and the sum of squares of residuals, SS_r, for this data set as a function of different values of b_0, demonstrating that $b_0 = 4$ is the estimate of β_0 that does provide the best fit in the least squares sense.

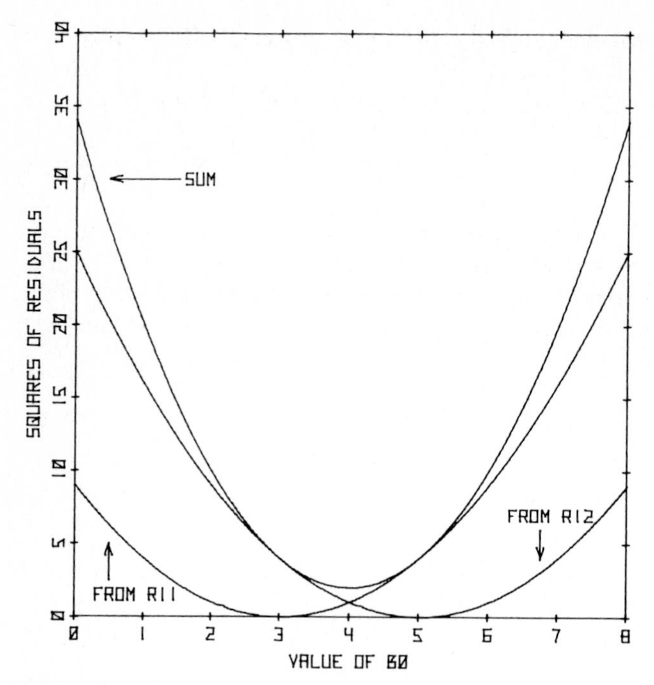

Figure 5.3. Squares of the individual residuals and the sum of squares of residuals, as functions of different values of b_0.

Equations 5.28 and 5.30 provide a general matrix approach to the calculation of the sum of squares of residuals. This sum of squares, SS_r, divided by its associated number of degrees of freedom, DF_r, is the sample estimate, s_r^2, of the population variance of residuals, σ_r^2.

$$s_r^2 = SS_r/DF_r = (\boldsymbol{R}'\boldsymbol{R})/DF_r = [(\boldsymbol{Y} - \boldsymbol{X}'\hat{\boldsymbol{B}})'(\boldsymbol{Y} - \boldsymbol{X}\hat{\boldsymbol{B}})]/DF_r \tag{5.43}$$

For two experiments and the model $y_{1i} = \beta_0 + r_{1i}$, the degrees of freedom of residuals is one, and the estimate of the variance of residuals is calculated as follows:

$$\boldsymbol{R} = \boldsymbol{Y} - \boldsymbol{X}\hat{\boldsymbol{B}} = \begin{bmatrix} 3 \\ 5 \end{bmatrix} \begin{bmatrix} -1 \\ 1 \end{bmatrix} [4] = \begin{bmatrix} 3 \\ 5 \end{bmatrix} \begin{bmatrix} -4 \\ 4 \end{bmatrix} = \begin{bmatrix} -1 \\ 1 \end{bmatrix} \tag{5.44}$$

$$\boldsymbol{R}'\boldsymbol{R} = [-1 \quad 1] \begin{bmatrix} -1 \\ 1 \end{bmatrix} = [-1 \times (-1) + 1 \times 1] = 2 \tag{5.45}$$

$$s_r^2 = SS_r/DF_r = 2/1 = 2 \tag{5.46}$$

$$s_r = 1.41 \tag{5.47}$$

5.3. The straight line model constrained to pass through the origin

Another statistical model that might be fit to two experiments at two different levels is

$$y_{1i} = \beta_1 x_{1i} + r_{1i} \tag{5.48}$$

and is illustrated in Figure 5.4. The model is a probabilistic proportional model that is constrained to pass through the origin; there is no β_0 term, so only a zero intercept is allowed. In this model, the response is assumed to be directly proportional to the level of the factor x_1; any deviations are assumed to be random. The matrix least squares solution is obtained as follows.

$$\boldsymbol{X} = \begin{bmatrix} x_{11} \\ x_{12} \end{bmatrix} \tag{5.49}$$

$$\boldsymbol{B} = [\beta_1] \tag{5.50}$$

$$\boldsymbol{Y} = \begin{bmatrix} y_{11} \\ y_{12} \end{bmatrix} \tag{5.51}$$

$$\boldsymbol{R} = \begin{bmatrix} r_{11} \\ r_{12} \end{bmatrix} \tag{5.52}$$

$$\boldsymbol{X}' = [x_{11} \quad x_{12}] \tag{5.53}$$

$$(\boldsymbol{X}'\boldsymbol{X}) = [x_{11} \quad x_{12}] \begin{bmatrix} x_{11} \\ x_{12} \end{bmatrix} = [x_{11}^2 + x_{12}^2] \tag{5.54}$$

$$(\boldsymbol{X}'\boldsymbol{X})^{-1} = [1/(x_{11}^2 + x_{12}^2)] \tag{5.55}$$

$$(\boldsymbol{X}'\boldsymbol{Y}) = [x_{11} \quad x_{12}] \begin{bmatrix} y_{11} \\ y_{12} \end{bmatrix} = [x_{11}y_{11} + x_{12}y_{12}] \tag{5.56}$$

$$\hat{\boldsymbol{B}} = [b_1] = (\boldsymbol{X}'\boldsymbol{X})^{-1}(\boldsymbol{X}'\boldsymbol{Y}) = [1/(x_{11}^2 + x_{12}^2)][x_{11}y_{11} + x_{12}y_{12}]$$
$$= (x_{11}y_{11} + x_{12}y_{12})/(x_{11}^2 + x_{12}^2) \tag{5.57}$$

For the data points in Figure 5.4 ($x_{11} = 3$, $y_{11} = 3$, $x_{12} = 6$, $y_{12} = 5$), the estimates of β_1 and s_r^2 are

$$b_1 = (3 \times 3 + 6 \times 5)/(3^2 + 6^2) = 39/45 = 13/15 \tag{5.58}$$

$$s_r^2 = (\boldsymbol{R}'\boldsymbol{R})/1 = [6/15 \quad -3/15] \begin{bmatrix} 6/15 \\ -3/15 \end{bmatrix} = 36/225 + 9/225 = 1/5 \tag{5.59}$$

A plot of r_{11}^2, r_{12}^2, and the sum of squares of residuals vs. b_1 is shown in Figure 5.5; SS_r is clearly minimal when $b_1 = 13/15$.

Note that the residuals do not add up to zero when the sum of squares of residuals is minimal for this example. (The residuals do add up to zero in Equation 5.44 for the model involving the β_0 parameter.) To understand why they are not

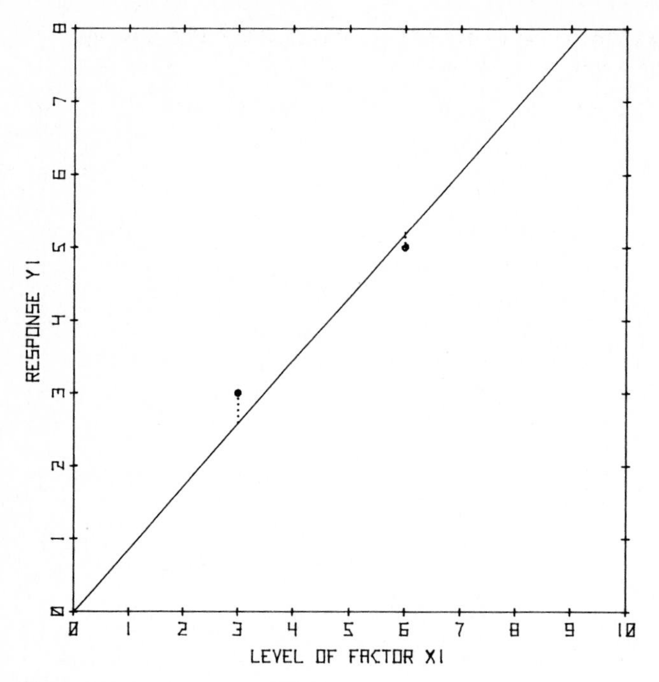

Figure 5.4. Graph of the fitted probabilistic model $y_{1i} = \beta_1 x_{1i} + r_{1i}$.

equal and opposite in this example, look at Figure 5.4 and mentally increase the slope until the two residuals are equal in magnitude (this situation corresponds to the point in Figure 5.5 where the r_{11}^2 and r_{12}^2 curves cross between $b_1 = 13/15$ and $b_1 = 14/15$). If the slope of the straight line in Figure 5.4 is now decreased, the magnitude of r_{11}^2 will increase, but the magnitude of r_{12}^2 will decrease much faster (for the data shown) and this will tend to decrease the sum of squares (see Figure 5.5). However, as the slope is further decreased and r_{12}^2 gets smaller, the relative decrease in the SS_r caused by r_{12}^2 also gets smaller (see Figure 5.5); the increase in the SS_r caused by an increasing r_{11}^2 finally dominates, and at $b_1 = 13/15$ the sum of squares of residuals starts to increase again.

In general, the *sum of residuals* (not to be confused with the sum of *squares* of residuals) will equal zero for models containing a β_0 term; for models not containing a β_0 term, the sum of residuals usually will not equal zero.

5.4. Matrix least squares for the case of an exact fit

Figure 5.1 shows the unconstrained straight line model $y_{1i} = \beta_0 + \beta_1 x_{1i}$ passing exactly through the two experimental points. The matrix least squares approach can be used in situations such as this where there is an exact fit. Using the data $x_{11}=3$, $y_{11}=3$, $x_{12}=6$, $y_{12}=5$, the appropriate arrays are

$$\boldsymbol{X}=\begin{bmatrix}1 & 3\\ 1 & 6\end{bmatrix} \tag{5.60}$$

$$\boldsymbol{B}=\begin{bmatrix}\beta_0\\ \beta_1\end{bmatrix} \tag{5.61}$$

$$\boldsymbol{Y}=\begin{bmatrix}3\\ 5\end{bmatrix} \tag{5.62}$$

$$\boldsymbol{X}'=\begin{bmatrix}1 & 1\\ 3 & 6\end{bmatrix} \tag{5.63}$$

$$(\boldsymbol{X}'\boldsymbol{X})=\begin{bmatrix}1 & 1\\ 3 & 6\end{bmatrix}\begin{bmatrix}1 & 3\\ 1 & 6\end{bmatrix}=\begin{bmatrix}1\times 1+1\times 1 & 1\times 3+1\times 6\\ 3\times 1+6\times 1 & 3\times 3+6\times 6\end{bmatrix}=\begin{bmatrix}2 & 9\\ 9 & 45\end{bmatrix} \tag{5.64}$$

$$D=(2\times 45-9\times 9)=9 \tag{5.65}$$

$$(\boldsymbol{X}'\boldsymbol{X})^{-1}=\begin{bmatrix}45/9 & -9/9\\ -9/9 & 2/9\end{bmatrix}=\begin{bmatrix}5 & -1\\ -1 & 2/9\end{bmatrix} \tag{5.66}$$

$$(\boldsymbol{X}'\boldsymbol{Y})=\begin{bmatrix}1 & 1\\ 3 & 6\end{bmatrix}\begin{bmatrix}3\\ 5\end{bmatrix}=\begin{bmatrix}1\times 3+1\times 5\\ 3\times 3+6\times 5\end{bmatrix}=\begin{bmatrix}8\\ 39\end{bmatrix} \tag{5.67}$$

$$\hat{\boldsymbol{B}}=\begin{bmatrix}b_0\\ b_1\end{bmatrix}=(\boldsymbol{X}'\boldsymbol{X})^{-1}(\boldsymbol{X}'\boldsymbol{Y})=\begin{bmatrix}5 & -1\\ -1 & 2/9\end{bmatrix}\begin{bmatrix}8\\ 39\end{bmatrix}=\begin{bmatrix}40-39\\ -8+78/9\end{bmatrix}=\begin{bmatrix}1\\ 2/3\end{bmatrix} \tag{5.68}$$

Thus, the intercept $b_0=1$ and the slope $b_1=2/3$ are identical to those values obtained using the simultaneous linear equations approach (see Equation 5.19 and Figure 5.1). Because there is an exact fit, the residuals are equal to zero.

$$\boldsymbol{R}=\boldsymbol{Y}-\boldsymbol{X}\hat{\boldsymbol{B}}=\begin{bmatrix}3\\ 5\end{bmatrix}-\begin{bmatrix}1 & 3\\ 1 & 6\end{bmatrix}\begin{bmatrix}1\\ 2/3\end{bmatrix}$$

$$=\begin{bmatrix}3\\ 5\end{bmatrix}-\begin{bmatrix}1\times 1+3\times(2/3)\\ 1\times 1+6\times(2/3)\end{bmatrix}=\begin{bmatrix}3-3\\ 5-5\end{bmatrix}=\begin{bmatrix}0\\ 0\end{bmatrix} \tag{5.69}$$

The sum of squares of residuals must also be equal to zero.

$$SS_r=\boldsymbol{R}'\boldsymbol{R}=[0 \quad 0]\begin{bmatrix}0\\ 0\end{bmatrix}=0 \tag{5.70}$$

The variance of residuals $s_r^2=SS_r/DF_r$ is undefined: the data set contains two values ($n=2$) but one degree of freedom has been lost for *each* of the two

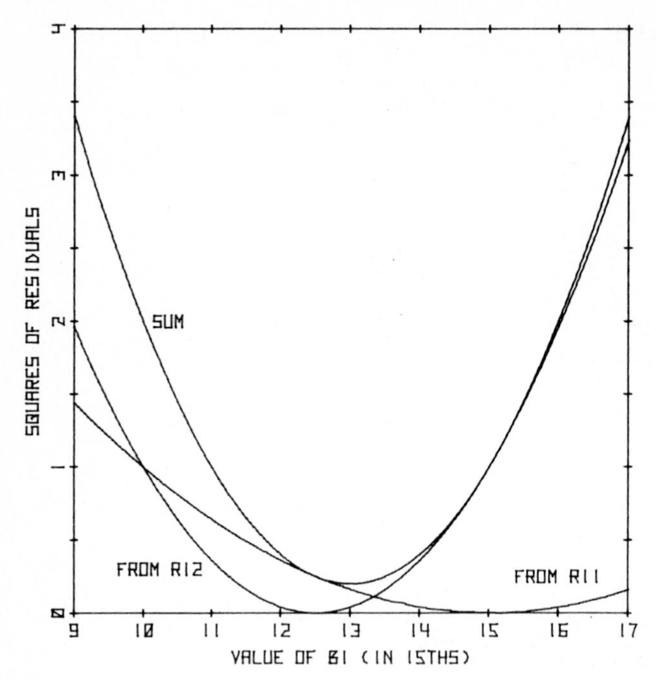

Figure 5.5. Squares of the individual residuals and the sum of squares of residuals, as functions of different values of b_1.

parameters β_0 and β_1. Thus, $DF_r = 2 - 2 = 0$ and there are no degrees of freedom available for calculating the variance of residuals.

5.5. Judging the adequacy of models

Let us return now to a question asked at the beginning of this chapter, "What can be learned about the behavior of the system from two experiments, each carried out at a different level of a single factor"?

For the data used in this section, the unconstrained straight line model $y_{1i} = \beta_0 + \beta_1 x_{1i} + r_{1i}$ fits exactly. Although the constrained model $y_{1i} = \beta_1 x_{1i} + r_{1i}$ does not fit exactly, it does "explain" the observed data better than the simple model $y_{1i} = \beta_0 + r_{1i}$ (s_r^2 values of 1/5 and 2, respectively). Thus, the response of the system would seem to increase as the level of the factor x_1 increases.

Is it justifiable to conclude that the factor x_1 has an influence on the output y_1? The answer requires a knowledge of the *purely experimental uncertainty* of the response, the variability that arises from causes *other than intentional changes in the factor levels*. This *variance due to purely experimental uncertainty* is given the symbol σ_{pe}^2, and its estimate is denoted s_{pe}^2.

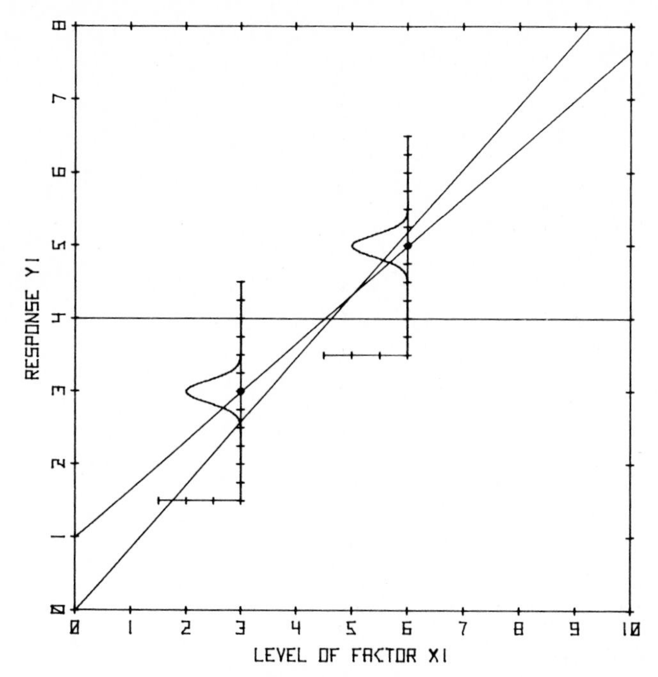

Figure 5.6. Graphs of three models ($y_{1i} = \beta_0 + \beta_1 x_{1i}$, $y_{1i} = \beta_0 + r_{1i}$, and $y_{1i} = \beta_1 x_{1i} + r_{1i}$) fit to two data points having small purely experimental uncertainty.

If the purely experimental uncertainty is small compared with the difference in response for the two experiments (see Figure 5.6), then the observed responses would be relatively precise indicators of the true response surface and the simple model $y_{1i} = \beta_0 + r_{1i}$ would probably be an inadequate representation of the system's behavior; the constrained model $y_{1i} = \beta_1 x_{1i} + r_{1i}$ or the unconstrained model $y_{1i} = \beta_0 + \beta_1 x_{1i} + r_{1i}$ might be better estimates of the true response surface. However, if the purely experimental uncertainty is large compared with the differences in response for the two observations (see Figure 5.7), then the observed responses might be relatively poor indicators of the underlying response surface and there would be less reason to question the adequacy of the model $y_{1i} = \beta_0 + r_{1i}$; the other two models would still fit the data better, but there would be less reason to believe they offered *significantly* better descriptions of the true response surface.

Unfortunately, two experiments at two different levels of a single factor cannot provide an estimate of the purely experimental uncertainty. The difference in the two observed responses might be due to experimental uncertainty, or it might be caused by a sloping response surface, or it might be caused by both. For this particular experimental design, the effects are confused (or *confounded*) and there is no way to separate the relative importance of these two sources of variation.

If the purely experimental uncertainty were known, it would then be possible to judge the adequacy of the model $y_{1i} = \beta_0 + r_{1i}$: if s_r^2 were very much greater than

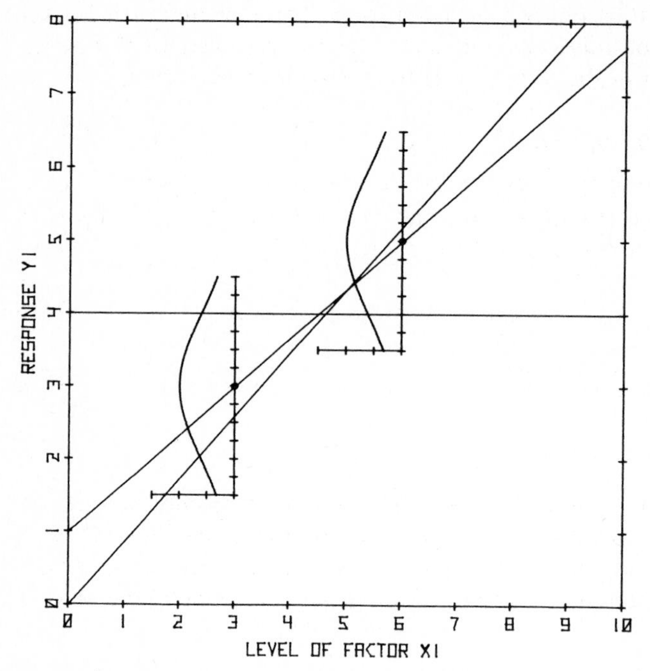

Figure 5.7. Graphs of three models ($y_{1i} = \beta_0 + \beta_1 x_{1i}$, $y_{1i} = \beta_0 + r_{1i}$, and $y_{1i} = \beta_1 x_{1i} + r_{1i}$) fit to two data points having large purely experimental uncertainty.

s^2_{pe} (see Figure 5.6), then it would be unlikely the large residuals for that model occurred by chance, and we would conclude that the model *does not* adequately describe the true behavior of the system. However, if s^2_r were approximately the same as s^2_{pe} (see Figure 5.7), then we would conclude that the model was adequate. (The actual decision compares s^2_{pe} to a variance slightly different from s^2_r, but the reasoning is similar.)

The estimation of purely experimental uncertainty is essential for testing the adequacy of a model. The material in Chapter 3 and especially in Figure 3.1 suggests one of the important principles of experimental design: *the purely experimental uncertainty can be obtained only by setting all of the controlled factors at fixed levels and replicating the experiment.*

5.6. Replication

Replication is the independent performance of two or more experiments at the same set of levels of all controlled factors. Replication allows both the calculation of a mean response, $\bar{y}_1$, and the estimation of the purely experimental uncertainty, s^2_{pe}, at that set of factor levels.

We now consider the fitting of several models to two experiments carried out at the same levels of all controlled factors with the purpose, again, of learning something about the effect of the factor x_1 on the behavior of the system.

The model $y_{1i} = \beta_0 + \beta_1 x_{1i} + r_{1i}$

It is instructive to try to fit the unconstrained model $y_{1i} = \beta_0 + \beta_1 x_{1i} + r_{1i}$ to the results of two experiments carried out at a single level of the factor x_1. Let the data be $x_{11} = 6$, $y_{11} = 3$, $x_{12} = 6$, $y_{12} = 5$ (see Figure 5.8). Then,

$$\boldsymbol{X} = \begin{bmatrix} 1 & x_{11} \\ 1 & x_{12} \end{bmatrix} = \begin{bmatrix} 1 & 6 \\ 1 & 6 \end{bmatrix} \tag{5.71}$$

$$(\boldsymbol{X}'\boldsymbol{X}) = \begin{bmatrix} 1 & 1 \\ 6 & 6 \end{bmatrix} \begin{bmatrix} 1 & 6 \\ 1 & 6 \end{bmatrix} = \begin{bmatrix} 2 & 12 \\ 12 & 72 \end{bmatrix} \tag{5.72}$$

But,

$$D = a \times d - c \times b = 2 \times 72 - 12 \times 12 = 144 - 144 = 0 \tag{5.73}$$

Because the determinant is equal to zero, the $(\boldsymbol{X}'\boldsymbol{X})$ matrix cannot be inverted, and a unique solution does not exist. An "interpretation" of the zero determinant is that the slope β_1 and the response intercept β_0 are both undefined (see Equations 5.14 and 5.15). This interpretation is consistent with the experimental design used

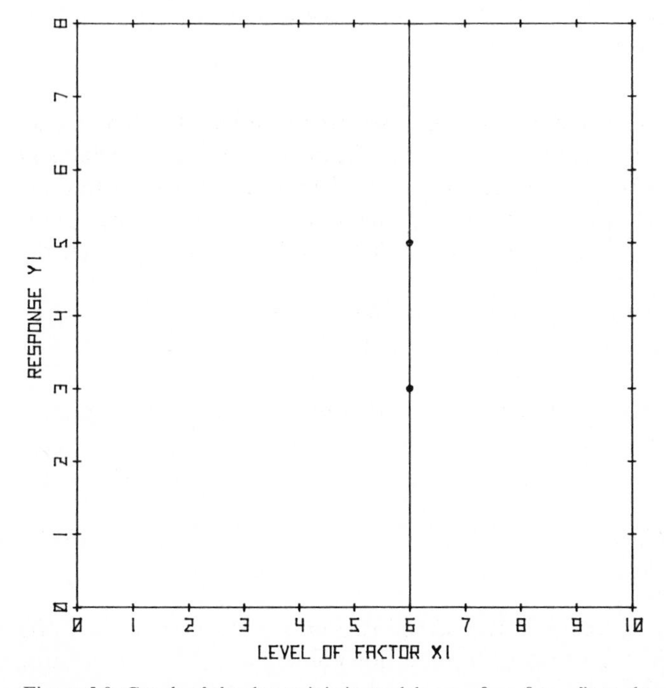

Figure 5.8. Graph of the deterministic model $y_{1i} = \beta_0 + \beta_1 x_{1i}$ fit to the results of two experiments at the same level of the factor x_1.

and the model attempted: the best straight line through the two points would have infinite slope (a vertical line) and the response intercept would not exist (see Figure 5.8).

The failure of this model when used with these replicate data points may also be considered in terms of degrees of freedom. Here, the total number of degrees of freedom is two, and we are trying to estimate two parameters, β_0 and β_1; this is not inconsistent with the arguments put forth in Sections 4.4, 5.1, and 5.4. However, replication *requires* that some degrees of freedom be used for the estimation of the purely experimental uncertainty. In general, if m replicates are carried out at a given set of factor levels, then $m-1$ degrees of freedom must go into the estimation of σ^2_{pe}. In the present example, one degree of freedom must be given to the estimation of purely experimental uncertainty; thus, only one other parameter can be estimated, and it is not possible to obtain a unique solution for the two parameters (β_0 and β_1) of the model $y_{1i} = \beta_0 + \beta_1 x_{1i} + r_{1i}$.

When replication is carried out at one set of experimental conditions (i.e., at a point in factor space), then

$$s^2_{pe} = \left(\left(\sum_{i=1}^{m} y_{1i} - \bar{y}_{1i}\right)^2\right)/(m-1) \tag{5.74}$$

where $\bar{y}_{1i}$ and y_{1i} refer only to the replicate responses at the single point in factor space (responses at other points in factor space are not used in this calculation). For this data set ($x_{11} = x_{12} = 6$), the purely experimental variance is

$$s^2_{pe} = \left[(3-4)^2 + (5-4)^2\right]/(2-1) = (1+1)/1 = 2 \tag{5.75}$$

The model $y_{1i} = \beta_1 x_{1i} + r_{1i}$

Figure 5. contains the same two replicate experiments shown in Figure 5.8, but here the response surface for the model $y_{1i} = \beta_1 x_{1i} + r_{1i}$ is shown. The least squares solution is obtained as follows.

$$\boldsymbol{X} = \begin{bmatrix} 6 \\ 6 \end{bmatrix} \tag{5.76}$$

$$(\boldsymbol{X}'\boldsymbol{X}) = [6 \quad 6]\begin{bmatrix} 6 \\ 6 \end{bmatrix} = [72] \tag{5.77}$$

$$(\boldsymbol{X}'\boldsymbol{X})^{-1} = [1/72] \tag{5.78}$$

$$(\boldsymbol{X}'\boldsymbol{Y}) = [6 \quad 6]\begin{bmatrix} 3 \\ 5 \end{bmatrix} = [48] \tag{5.79}$$

$$\hat{\boldsymbol{B}} = [b_1] = [48][1/72] = 2/3 \tag{5.80}$$

$$\boldsymbol{R} = \boldsymbol{Y} - \boldsymbol{X}\hat{\boldsymbol{B}} = \begin{bmatrix} 3 \\ 5 \end{bmatrix} - \begin{bmatrix} 6 \\ 6 \end{bmatrix}[2/3] = \begin{bmatrix} 3 \\ 5 \end{bmatrix}\begin{bmatrix} -4 \\ 4 \end{bmatrix} = \begin{bmatrix} -1 \\ 1 \end{bmatrix} \tag{5.81}$$

$$s^2_r = (\boldsymbol{R}'\boldsymbol{R})/1 = [-1 \quad 1]\begin{bmatrix} -1 \\ 1 \end{bmatrix} = 2 \tag{5.82}$$

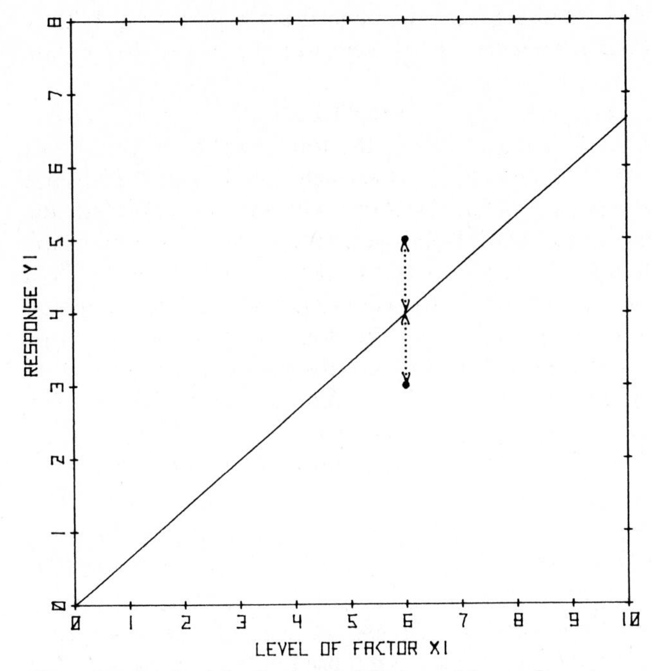

Figure 5.9. Graph of the fitted probabilistic model $y_{1i} = \beta_1 x_{1i} + r_{1i}$.

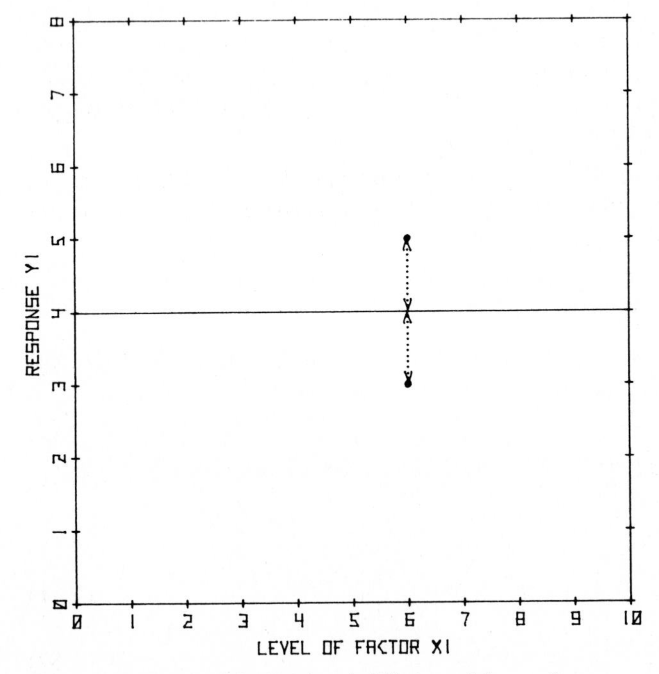

Figure 5.10. Graph of the fitted probabilistic model $y_{1i} = \beta_0 + r_{1i}$.

Note that b_1 is used with the fixed level of $x_1 = 6$ to estimate the mean response for the replicate observations.

The model $y_{1i} = \beta_0 + r_{1i}$

Let us fit the probabilistic model, $y_{1i} = \beta_0 + r_{1i}$, to the same data (see Figure 5.10). If the least squares approach to the fitting of this model is employed, the appropriate matrices and results are exactly those given in Section 5.2 where the same model was fit to the different factor levels: $x_{11} = 3$, $y_{11} = 3$, $x_{12} = 6$, $y_{12} = 5$. This identical mathematics should not be surprising: the model does not include a term for the factor x_1 and thus the matrix of parameter coefficients, $\boldsymbol{X}$, should be the same for both sets of data. The parameter β_0 is again estimated to be 4, and σ_r^2 is estimated to be 2.

The model $y_{1i} = 0 + r_{1i}$

Before leaving this chapter we will consider one final model, the purely probabilistic model $y_{1i} = 0 + r_{1i}$ (see Section 4.2 and Equation 4.3). Whether obtained at different levels of the factor x_1 or at the same level (replicates), there is a possibility that the two observed responses, $y_{11} = 3$ and $y_{12} = 5$, belong to a population for which the mean (μ) is zero. The fact that the two numbers we have obtained give a sample mean ($\bar{y}_1$) of 4 might have occurred simply by chance. Thus, when judging the adequacy of various models, we should not overlook the possibility that the purely probabilistic model is adequate.

It is not possible to fit this model using matrix least squares techniques: the matrix of parameter coefficients, $\boldsymbol{X}$, does not exist – it is a 0×0 matrix and has no elements because there are no parameters in the model. However, the matrix of residuals, $\boldsymbol{R}$, is defined. It should not be surprising that for this model, $\boldsymbol{R} = \boldsymbol{Y}$; that is, the matrix of residuals is identical to the matrix of responses.

References

Box, G.E.P., Hunter, W.G., and Hunter, J.S. (1978). *Statistics for Experimenters. An Introduction to Design, Data Analysis, and Model Building*. Wiley, New York.

Cochran, W.G., and Cox, G.M. (1957). *Experimental Designs*, 2nd. ed. Wiley, New York.

Daniel, C., and Wood, F.S. (1971). *Fitting Equations to Data*. Wiley, New York.

Draper, N.R., and Smith, H. (1966). *Applied Regression Analysis*. Wiley, New York.

Himmelblau, D.M. (1970). *Process Analysis by Statistical Methods*. Wiley, New York.

Mandel, J. (1964). *The Statistical Analysis of Experimental Data*. Wiley, New York.

Meyer, S.L. (1975). *Data Analysis for Scientists and Engineers*. Wiley, New York.

Natrella, M.G. (1963). *Experimental Statistics* (Nat. Bur. of Stand. Handbook 91). US Govt. Printing Office, Washington, DC.

Neter, J., and Wasserman, W. (1974). *Applied Linear Statistical Models. Regression, Analysis of Variance, and Experimental Designs*. Irwin, Homewood, Illinois.

Youden, W.J., (1951). *Statistical Methods for Chemists*. Wiley, New York.

Exercises

5.1. Linear models

Which of the following models are linear? Which are nonlinear? Can any of the non-linear models be transformed into linear models? How?

a) $y_{1i} = 0 + r_{1i}$
b) $y_{1i} = \beta_0 + \beta_1 x_{1i} + \beta_{11} x_{1i}^2 + \beta_{111} x_{1i}^3 + r_{1i}$
c) $y_{1i} = \beta_0 + \beta_1 \log x_{1i} + \beta_{11} \log x_{1i}^2 + r_{1i}$
d) $y_{1i} = \beta_1 \beta_2 x_{1i} x_{2i} + r_{1i}$
e) $y_{1i} = \beta_1 (1 - \exp[-\beta_2 x_{1i}]) + r_{1i}$
f) $y_{1i} = \beta_0 + r_{1i}$
g) $y_{1i} = \beta_0 + \beta_1 x_{1i} + \beta_2 x_{2i} + \beta_{11} x_{1i}^2 + \beta_{22} x_{2i}^2 + \beta_{12} x_{1i} x_{2i} + r_{1i}$

5.2. Simultaneous equations

Use simultaneous equations (Section 5.1) to solve for β_0 and β_1 in the following set of equations:

$9 = \beta_0 + 3\beta_1$
$13 = \beta_0 + 5\beta_1$

5.3. Simultaneous equations

Use simultaneous equations (Section 5.1) to solve for β_0, β_1, and β_2 in the following set of equations:

$16 = \beta_0 + 5\beta_1 + 2\beta_2$
$11 = \beta_0 + 3\beta_1 + 1\beta_2$
$32 = \beta_0 + 6\beta_1 + 7\beta_2$

(See Appendix A for a method of finding the inverse of a 3×3 matrix.)

5.4. Matrix least squares

Use matrix least squares (Section 5.2) to estimate b_0 and b_1 in Exercise 5.2.

5.5. Matrix least squares

Use matrix least squares (Section 5.2) to estimate b_0, b_1, and b_2 in Exercise 5.3.

5.6. Sum of squares of residuals

Construct a graph similar to figure 5.3 showing the individual squares of residuals and the sum of squares of residuals as a function of b_0 ($b_1 = 2$) for Exercise 5.4. Construct a second graph as a function of b_1 ($b_0 = 3$).

5.7. Covariance

Construct five graphs showing the sum of squares of residuals as a function of b_0 for Exercise 5.4. Graph 1: $b_1 = 1$; Graph 2: $b_1 = 2$ (see Exercise 5.6); Graph 3: $b_1 = 3$; Graph 4: $b_1 = 4$; Graph 5: $b_1 = 5$. Why doesn't the minimum occur at the same value of b_0 in all graphs? Which graph gives the overall minimum sum of squares of residuals?

5.8. Matrix least squares

Use matrix least squares to fit the model $y_{1i} = \beta_1 x_{1i} + r_{1i}$ to the data in Exercise 5.2 ($x_{11} = 3$, $y_{11} = 9$, $x_{12} = 5$, $y_{12} = 13$). Does the sum of residuals (not the sum of squares of residuals) equal zero? Graph the individual squares of residuals and the sum of squares of residuals as a function of b_1.

5.9. Matrix least squares

Use matrix least squares to fit the data in Section 3.1 to the model $y_{1i} = \beta_0 + r_{1i}$. Compare $\boldsymbol{R'R}/DF_r$ with the variance in Equation 3.5.

5.10. Residuals

Plot the following data pairs (x_{1i}, y_{1i}) on a piece of graph paper: (0, 3.0), (1, 5.0), (2, 6.8), (3, 8.4), (4, 9.8), (5, 11.0), (6, 12.0), (7, 12.8), (8, 13.4), (9, 13.8), (10, 14.0). Use a ruler or other straightedge to draw a "good" straight line ($y_{1i} = \beta_0 + \beta_1 x_{1i} + r_{1i}$) through the data. Measure the residual for each point (record both sign and magnitude) and on a second piece of graph paper plot the *residuals* as a function of x_1. Is there a pattern to the residuals? [See, for example, Draper, N.R., and Smith, H. (1966). *Applied Regression Analysis*, Chapter 3, Wiley, New York.] Suggest a better empirical linear model that might fit the data more closely. How well does the model $y_{1i} = 3 + 2.1x_{1i} - 0.1x_{1i}^2$ fit?

5.11. Purely experimental uncertainty

Most textbooks refer to σ_{pe}^2 as the "variance due to pure error", or the "pure error variance". In this textbook, σ_{pe}^2 is called the "variance due to purely experimental uncertainty", or the "purely experimental uncertainty variance". What assumptions might underlie each of these systems of naming? [See Problem 6.14; see also Mandel, J. (1964). *The Statistical Analysis of Experimental Data*, pp. 123-127. Wiley, New York.]

5.12. Replication

A researcher carries out the following set of experiments. Which are replicates?

i	x_{1i}	x_{2i}	y_{1i}
1	3	7	12.6
2	7	1	4.5
3	2	3	2.9
4	7	5	11.5
5	2	4	11.4
6	6	3	3.2
7	3	7	12.9
8	2	3	3.2
9	5	1	4.3
10	2	3	2.8

If the factor x_1 is known to have no effect on y_1, and if x_1 is ignored, then which of the experiments can be considered to be replicates?

5.13. Purely experimental uncertainty

Refer to Figure 3.1. How can purely experimental uncertainty be decreased? What are the advantages of making the purely experimental uncertainty small? What are the disadvantages of making the purely experimental uncertainty small?

5.14. Purely experimental uncertainty

Assume that the nine values of measured response in Section 3.1 are replicates. What is the mean value of these replicates? How many degrees of freedom are removed by calculation of the mean? How many degrees of freedom remain for the estimation of σ_{pe}^2? What is the value of σ_{pe}^2?

5.15. Replication

Consider the following four experiments: $x_{11}=3$, $y_{11}=2$, $x_{12}=3$, $y_{12}=4$, $x_{13}=6$, $y_{13}=6$, $x_{14}=6$, $y_{14}=4$. At how many levels of x_1 have experiments been carried out? What is the mean value of y_1 at each level of x_1? How many degrees of freedom are removed by calculation of these two means? How many degrees of freedom remain for the calculation of σ_{pe}^2? If n is the number of experiments in this set and f is the number of levels of x_1, then what is the relationship between n and f that expresses the number of degrees of freedom available for calculating the purely experimental uncertainty? Why is $s_{pe}^2=2$ for this set of data?

5.16. Matrix algebra

Given $\boldsymbol{XB}=\boldsymbol{Y}$ for a set of simultaneous equations, show that $\boldsymbol{B}=\boldsymbol{X}^{-1}\boldsymbol{Y}$ (see Appendix A).

5.17. Matrix algebra

Given $\boldsymbol{XB}=\boldsymbol{Y}$ for an overdetermined set of equations, rigorously show that $\hat{\boldsymbol{B}}=(\boldsymbol{X}'\boldsymbol{X})^{-1}(\boldsymbol{X}'\boldsymbol{Y})$ gives the minimum sum of squares. [See, for example, Kempthorne, O. (1952). *The Design and Analysis of Experiments*. Wiley, New York.]

5.18. $(\boldsymbol{X}'\boldsymbol{X})^{-1}$ matrix

Calculate the $(\boldsymbol{X}'\boldsymbol{X})$ matrix, its determinant, and the $(\boldsymbol{X}'\boldsymbol{X})^{-1}$ matrix for the model $y_{1i}=\beta_0+\beta_1 x_{1i}+r_{1i}$ fit to each of the following three sets of data: a) $x_1=1.9, 2.1$; b) $x_1=0, 4$; c) $x_1=0, 0, 4, 4$. From this information, can you offer any insight into conditions that make the elements of $(\boldsymbol{X}'\boldsymbol{X})^{-1}$ small? Design a set of experiments to prove or disprove these insights.

5.19. Importance of replication

Suppose a researcher who has never gambled before goes to Las Vegas, bets \$10 on red in a roulette game, and wins \$20. Based on the results of that one experiment, he makes a conclusion and bets the \$20 and all of the rest of his money on red again. He loses. Comment about the usefulness of restrained replication.

Suppose a researcher who has been sheltered in his laboratory for ten years goes outside and says hello to the first person he meets. The stranger stops, takes the researcher's money, and runs away. The researcher makes a conclusion, becomes a recluse, and never speaks to another human again. Comment about the usefulness of bold replication.

5.20. Matrix least squares

Use matrix least squares to fit the model $y_{1i} = \beta_0 + \beta_1 x_{1i} + r_{1i}$ to the following eleven data points:

i	x_{1i}	y_{1i}
1	0	0.1
2	6	11.7
3	9	19.0
4	4	8.6
5	5	10.0
6	6	15.9
7	1	1.8
8	10	19.9
9	2	4.1
10	3	5.9
11	7	14.1

5.21. Replication

Measurement laboratories are often sent blind duplicates of material for analysis (i.e., the measurement laboratory is not told they are duplicates). Why?

CHAPTER 6

Hypothesis Testing

In Section 5.5 a question was raised concerning the adequacy of models when fit to experimental data (see also Section 2.4). It was suggested that any test of the adequacy of a given model must involve an estimate of the purely experimental uncertainty. In Section 5.6 it was indicated that replication provides the information necessary for calculating s_{pe}^2, the estimate of σ_{pe}^2. We now consider in more detail how this information can be used to test the adequacy of linear models.

6.1. The null hypothesis

Figure 6.1 shows the replicate data points $x_{11} = 6$, $y_{11} = 3$, $x_{12} = 6$, $y_{12} = 5$, and their relationships to the two models, $y_{1i} = 0 + r_{1i}$ and $y_{1i} = \beta_0 + r_{1i}$. We might ask, "Does the purely probabilistic model $y_{1i} = 0 + r_{1i}$ adequately fit the data, or does the model $y_{1i} = \beta_0 + r_{1i}$ offer a *significantly* better fit"?

Another way of phrasing this question is to ask if the parameter β_0 is significantly different from zero. Two possibilities exist:

(1) if β_0 is significantly different from zero, then the model $y_{1i} = \beta_0 + r_{1i}$ offers a significantly better fit.

(2) if β_0 is not significantly different from zero, then the model $y_{1i} = 0 + r_{1i}$ cannot be rejected; it provides an adequate fit to the data.

In statistical terms, we seek to disprove the *null* hypothesis that the difference between β_0 and zero is "null". The null hypothesis is written

$$H_0: \quad \beta_0 = 0 \tag{6.1}$$

The *alternative hypothesis* is written

$$H_a: \quad \beta_0 \neq 0 \tag{6.2}$$

If we can demonstrate to our satisfaction that the null hypothesis is false, then we can reject that hypothesis and accept the alternative hypothesis that $\beta_0 \neq 0$.

However, *failure to disprove the null hypothesis does not mean we can reject the alternative hypothesis and accept the null hypothesis.* This is a subtle but extremely important point in hypothesis testing, especially when hypothesis testing is used to identify factors in research and development projects (see Section 1.2 and Table 1.1).

In everyday language, the words "accept" and "reject" are usually used as exact complements (see Figure 6.2): if something is not accepted, it is rejected; if

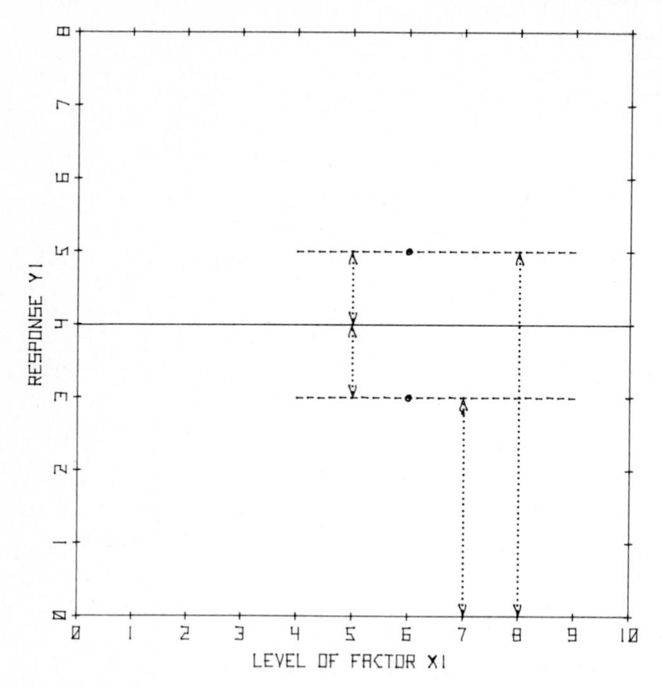

Figure 6.1. Relationships of two replicate data points ($x_{11}=6$, $y_{11}=3$, $x_{12}=6$, $y_{12}=5$) to two probabilistic models, $y_{1i}=\beta_0+r_{1i}$ and $y_{1i}=0+r_{1i}$.

something is not rejected, it is accepted. As an example, if someone orders a new car and it arrives with a gross defect, the purchaser will immediately reject it. If the new car arrives in good condition and seems to perform well in a test drive, the purchaser will probably accept the new car.

However, the word "accept" might not be entirely correct in this automotive example – it could imply a level of commitment that does not exist. Perhaps a better choice of words would be to say that the owner "puts up with" or "tolerates" the new car. As long as the new car performs reasonably well and does not exhibit a serious malfunction, the owner will not reject it. Truly "accepting" the car, however, might occur only after many years of trouble-free operation.

In this example, as in statistical hypothesis testing, there is a more or less broad region of "tolerance" or "indecision" between the commitments of "acceptance" and "rejection" (see Figure 6.3). In seeking to disprove the null hypothesis, we are exploring the region between "undecided" and "rejection". The exploration of the region between "undecided" and "acceptance" involves a very different type of

Figure 6.2. Common usage of the terms "accept" and "reject".

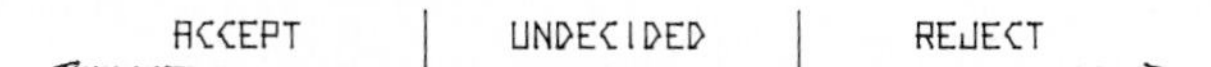

Figure 6.3. Statistical usage of the terms "accept" and "reject", showing the intermediate region of "undecided".

experimentation and requires a different set of criteria for hypothesis testing. In this book we will limit the testing of hypotheses to the "undecided-rejection" region, realizing, again, that failure to reject the null hypothesis means only to tolerate it as adequate, not necessarily to fully accept it.

6.2. Confidence intervals

Figure 6.4 shows the relationship between b_0 (the least squares estimate of β_0) and zero for the data in Figure 6.1. In this example, the parameter β_0 has been estimated on the basis of only two experimental results; if another independent set of two experiments were carried out on the same system, we would very likely obtain different values for y_{13} and y_{14} and thus obtain a different estimate for β_0 from the second set of experiments. It is for this reason that the estimation of β_0 is usually subject to uncertainty.

It can be shown that if the uncertainties associated with the measurements of the response are approximately normally distributed (see Equation 3.8), then parameter estimates obtained from these measurements are also normally distributed. The standard deviation of the estimate of a parameter will be called the *standard uncertainty*, s_b, of the parameter estimate (it is usually called the "standard error") and can be calculated from the $(\boldsymbol{X}'\boldsymbol{X})^{-1}$ matrix if an estimate of σ_{pe}^2 is available.

For the data in Figure 6.1, the standard uncertainty in b_0 (s_{b_0}) is estimated as

$$s_{b_0} = \left[s_{pe}^2(\boldsymbol{X}'\boldsymbol{X})^{-1}\right]^{1/2} \tag{6.3}$$

The matrix least squares solution for the model $y_{1i} = \beta_0 + r_{1i}$ and the data in Figure 6.1 was obtained in Section 5.2 (see Equations 5.26, 5.27, and 5.37–5.47) and gave the results $b_0 = 4$, $s_r^2 = 2$. Because the two experiments in this example are replicates, $s_{pe}^2 = s_r^2$ and Equation 6.3 becomes

$$s_{b_0} = [2 \times (1/2)]^{1/2} = (1)^{1/2} = 1 \tag{6.4}$$

Both the parameter estimate b_0 and the standard uncertainty of its estimate s_{b_0}

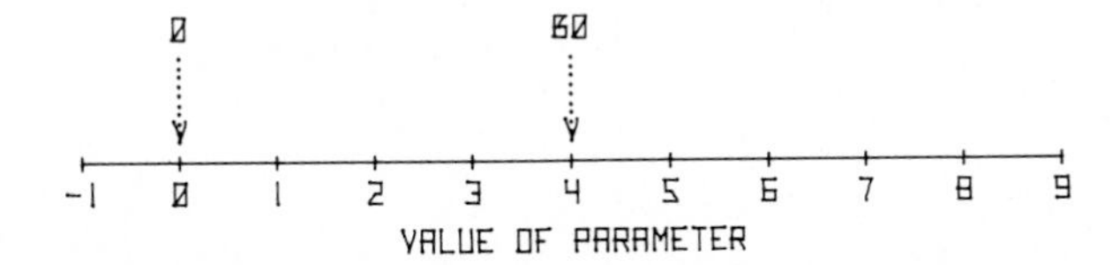

Figure 6.4. Relationship between b_0 and 0 for the model $y_{1i} = \beta_0 + r_{1i}$ fit to the data of Figure 6.1.

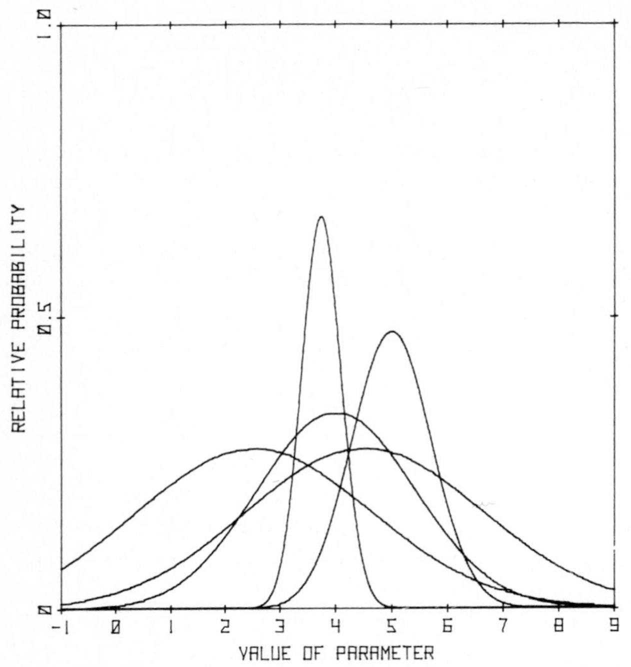

Figure 6.5. Five gaussian curves, each estimated from an independent set of two experiments on the same system, showing uncertainty of estimating b_0 and s_{b_0}.

depend upon the $\boldsymbol{Y}$ matrix of experimental responses (see Equations 5.36, 5.28, and 5.43): if one set of experiments yields responses that agree closely, the standard uncertainty of b_0 will be small; if a different set of experiments happens to produce responses that are dissimilar, the standard uncertainty of b_0 will be large. Thus, not only is the estimation of the parameter β_0 itself subject to uncertainty, but the estimation of its standard uncertainty is *also* subject to uncertainty.

The estimation of β_0 is therefore doubly uncertain, first because the *value* of β_0 is not known with certainty, and second because the exact *distribution* of estimates for β_0 is unknown. Figure 6.5 illustrates the problem: each curve represents a pair of estimates of b_0 and s_{b_0} obtained from an independent set of two experiments on the same system.

Suppose we could use b_0 and s_{b_0} from only one set of experiments to construct a *confidence interval* about b_0 such that there is a given probability that the interval contains the population value of β_0 (see Section 3.4). The interpretation of such a confidence interval is this: if we find that the interval includes the value zero, then (with our data) we cannot disprove the null hypothesis that $\beta_0 = 0$; that is, on the basis of the estimates b_0 and s_{b_0}, it is not improbable that the true value of β_0 could be zero. Suppose, however, we find that the confidence interval does not contain the value zero: because we know that if β_0 were really equal to zero this lack of overlap could happen by chance only very rarely, we must conclude that it is highly unlikely the true value of β_0 is zero; the null hypothesis is rejected (with an admitted finite risk of being wrong) and we accept the alternative hypothesis that β_0 is significantly different from zero.

The other piece of information (in addition to b_0 and s_{b_0}) required to establish a confidence interval for a parameter estimate was not available until 1908 when W.S. Gosset, an English chemist who used the pseudonym "Student", provided a solution to the statistical problem. The resulting values are known as "critical values of Student's t" and may be obtained from so-called "t-tables" (see Appendix B for values at the 95% level of confidence).

Using an appropriate t-value, we can now estimate a confidence interval (CI) for b_0 obtained from the data in Figure 6.1:

$$CI = b_0 \pm t \times s_{b_0} \tag{6.5}$$

$$CI = b_0 \pm t \times s_{b_0} = 4 \pm 12.71 \times 1 = 4 \pm 12.71 \tag{6.6}$$

where 12.71 is the tabulated value of t for one degree of freedom at the 95% level of confidence. Thus, we expect with 95% confidence that the population value of β_0 lies within the interval

$$-8.71 \leqslant \beta_0 \leqslant +16.71 \tag{6.7}$$

(see Figure 6.6). Because this confidence interval contains the value zero, we cannot disprove the null hypothesis that $\beta_0 = 0$. There is no reason to believe (at the 95% level of confidence) that β_0 is significantly different from zero, and we must therefore retain as adequate the model $y_{1i} = 0 + r_{1i}$.

6.3. The *t*-test

A somewhat different computational procedure is often used in practice to carry out the test described in the previous section. The procedure involves two questions: "What is the minimum calculated interval about b_0 that will include the value zero"? and, "Is this minimum calculated interval greater than the confidence interval estimated using the tabular critical value of t"? If the calculated interval is

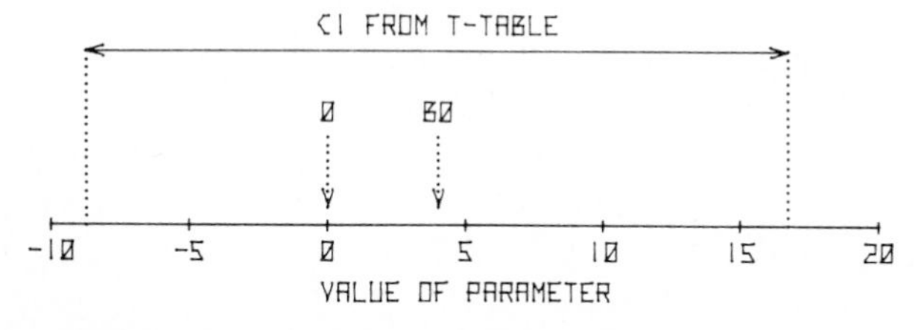

Figure 6.6. Confidence interval (95% level) for b_0.

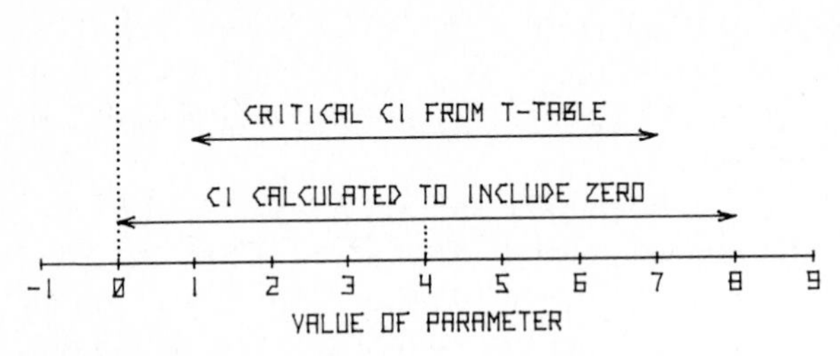

Figure 6.7. Confidence intervals for t-test in which the null hypothesis H_0: $\beta = 0$ would be disproved at the specified level of confidence.

larger than the critical confidence interval (see Figure 6.7), a significant difference between β_0 and zero probably exists and the null hypothesis is disproved. If the calculated interval is *smaller* than the critical confidence interval (see Figure 6.8), there is insufficient reason to believe that a significant difference exists and the null hypothesis cannot be rejected.

Each of these confidence intervals (the calculated interval and the critical interval) can be expressed in terms of b_0, s_{b_0}, and some value of t (see Equation 6.5). Because the same values of b_0 and s_{b_0} are used for the construction of these intervals, the information about the relative widths of the intervals is contained in the two values of t. One of these, t_{crit}, is simply obtained from the table of critical value of t – it is the value used to obtain the critical confidence interval shown in Figure 6.7 or 6.8. The other, t_{calc}, is calculated from the minimum confidence interval about b_0 that will include the value zero and is obtained from a rearrangement of Equation 6.5.

$$t_{calc} = |b_0 - 0|/s_{b_0} \tag{6.8}$$

The "t-test" involves the comparison of these two values:

(1) If $t_{calc} > t_{crit}$ (see Figure 6.7), the minimum confidence interval is greater than the critical confidence interval and there is strong reason to believe that β_0 is significantly different from zero.

(2) If $t_{calc} < t_{crit}$ (see Figure 6.8), the minimum confidence interval is less than the

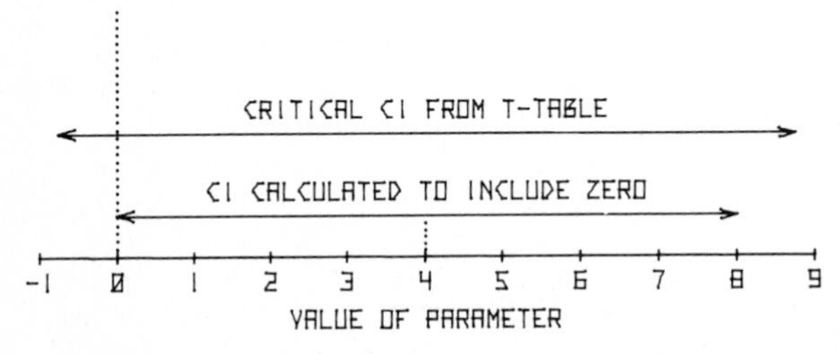

Figure 6.8. Confidence intervals for t-test in which the null hypothesis H_0: $\beta = 0$ could *not* be disproved at the specified level of confidence.

critical confidence interval and there is no reason to believe β_0 is significantly different from zero.

In short, if $t_{calc} > t_{crit}$, then H_0 is rejected at the given level of confidence and the alternative hypothesis, H_a, is accepted.

As an example, consider again the data in Figure 6.1. From the t-table in Appendix B, $t_{crit} = 12.71$ (95% level of confidence). Based upon the estimates b_0 and s_{b_0},

$$t_{calc} = |4 - 0|/1 = 4 \tag{6.9}$$

Thus, $t_{calc} < t_{crit}$ ($4 < 12.71$) and there is no reason to believe, at the 95% level of confidence and for the data in Figure 6.1, that β_0 is significantly different from zero. The model $y_{1i} = 0 + r_{1i}$ cannot be rejected; it provides an adequate fit to the data.

6.4. Sums of squares

Let us now consider a slightly different question. Rather than inquiring about the significance of the specific parameter β_0, we might ask instead, "Does the model $y_{1i} = 0 + r_{1i}$ provide an adequate fit to the experimental data, or does this model show a significant lack of fit"?

We begin by examining more closely the sum of squares of residuals between the measured response, y_{1i}, and the predicted response, $\hat{y}_{1i}$ ($\hat{y}_{1i} = 0$ for all i of this model), which is given by

$$SS_r = \boldsymbol{R'R} = \sum_{i=1}^{n} r_{1i}^2 = \sum_{i=1}^{n} (y_{1i} - \hat{y}_{1i})^2 = \sum_{i=1}^{n} (y_{1i} - 0)^2 \tag{6.10}$$

If the model $y_{1i} = 0 + r_{1i}$ does describe the true behavior of the system, we would expect replicate experiments to have a mean value of zero ($\bar{y}_{1i} = 0$); the sum of squares due to purely experimental uncertainty would be expected to be

$$SS_{pe} = \sum_{i=1}^{n} (y_{1i} - \bar{y}_{1i})^2 = \sum_{i=1}^{n} (y_{1i} - 0)^2 \tag{6.11}$$

if all n of the experiments are replicates. Thus, for the experimental design of Figure 6.1, if $y_{1i} = 0 + r_{1i}$ is an adequate model, we would expect the sum of squares of residuals to be approximately equal to the sum of squares due to purely experimental uncertainty (Equations 6.10 and 6.11).

$$SS_r \cong SS_{pe} \tag{6.12}$$

Suppose now that the model $y_{1i} = 0 + r_{1i}$ *does not* adequately describe the true behavior of the system. Then we would not expect replicate experiments to have a mean value of zero ($\bar{y}_{1i} \neq 0$; see Equation 6.11) and the sum of squares due to purely experimental uncertainty would *not* be expected to be approximately equal

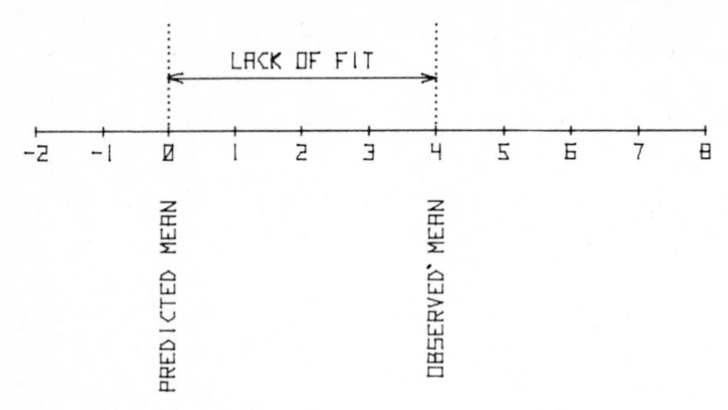

Figure 6.9. Illustration of the lack of fit as the difference between an observed mean and the corresponding predicted mean.

to the sum of squares of residuals; instead, SS_r would be larger than SS_{pe}. This imperfect agreement between what the model predicts and the mean of replicate experiments is called *lack of fit of the model to the data*; an interpretation is suggested in Figure 6.9. When replicate experiments have been carried out at the same set of levels of all controlled factors, the best estimate of the response at that set of levels is the mean of the observed response, $\bar{y}_{1i}$. The difference between this mean value of response and the corresponding value predicted by the model is a measure of the lack of fit of the model to the data.

If (*and only if*) replicate experiments have been carried out on a system, it is possible to partition the sum of squares of residuals, SS_r, into two components (see Figure 6.10): one component is the already familiar sum of squares due to purely experimental uncertainty, SS_{pe}; the other component is associated with variation attributed to the lack of fit of the model to the data and is called the *sum of squares due to lack of fit*, SS_{lof}.

$$SS_{lof} = \sum_{i=1}^{n} (\bar{y}_{1i} - \hat{y}_{1i})^2 \tag{6.13}$$

The proof of the partitioning of SS_r into SS_{lof} and SS_{pe} is based upon the following identity for a single measured response.

$$r_{1i} = (y_{1i} - \hat{y}_{1i}) = (y_{1i} - \bar{y}_{1i}) + (\bar{y}_{1i} - \hat{y}_{1i}) \tag{6.14}$$

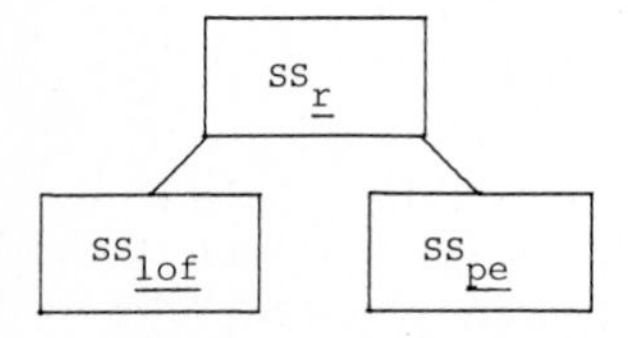

Figure 6.10. Sums of squares tree illustrating the relationship $SS_r = SS_{lof} + SS_{pe}$.

Given a set of m replicate measurements,

$$SS_r = \sum_{i=1}^{m} r_{1i}^2 = \sum_{i=1}^{m} [(y_{1i} - \bar{y}_{1i}) + (\bar{y}_{1i} - \hat{y}_{1i})]^2 \quad (6.15)$$

$$SS_r = \sum_{i=1}^{m} \left[(y_{1i} - \bar{y}_{1i})^2 + 2(y_{1i} - \bar{y}_{1i})(\bar{y}_{1i} - \hat{y}_{1i}) + (\bar{y}_{1i} - \hat{y}_{1i})^2\right] \quad (6.16)$$

$$SS_r = \sum_{i=1}^{m} (y_{1i} - \hat{y}_{1i})^2 + \sum_{i=1}^{m} \left[2(y_{1i} - \bar{y}_{1i})(\bar{y}_{1i} - \hat{y}_{1i}) + (y_{1i} - \bar{y}_{1i})^2\right] \quad (6.17)$$

By Equation 6.13,

$$SS_r = SS_{lof} + \sum_{i=1}^{m} \left[2(\bar{y}_{1i} - \hat{y}_{1i})y_{1i} - 2(\bar{y}_{1i} - \hat{y}_{1i})\bar{y}_{1i} + y_{1i}^2 - 2y_{1i}\bar{y}_{1i} + \bar{y}_{1i}^2\right] \quad (6.18)$$

$$SS_r = SS_{lof} + \sum_{i=1}^{m} \left\{y_{1i}^2 + \left[2(\bar{y}_{1i} - \hat{y}_{1i})y_{1i} - 2(\bar{y}_{1i} - \hat{y}_{1i})\bar{y}_{1i}\right] - 2y_{1i}\bar{y}_{1i} + \bar{y}_{1i}^2\right\} \quad (6.19)$$

Because $\hat{y}_{1i}$ is the same for each of these replicate measurements,

$$SS_r = SS_{lof} + \sum_{i=1}^{m} (y_{1i}^2 - 2y_{1i}\bar{y}_{1i} + \bar{y}_{1i}^2) + 2(\bar{y}_{1i} - \hat{y}_{1i}) \sum_{i=1}^{m} y_{1i} - 2(\bar{y}_{1i} - \hat{y}_{1i}) \sum_{i=1}^{m} \bar{y}_{1i} \quad (6.20)$$

However,

$$\sum_{i=1}^{m} y_{1i} = \sum_{i=1}^{m} \bar{y}_{1i} \quad (6.21)$$

and thus

$$SS_r = SS_{lof} + \sum_{i=1}^{m} (y_{1i}^2 - 2y_{1i}\bar{y}_{1i} + \bar{y}_{1i}^2) \quad (6.22)$$

$$SS_r = SS_{lof} + \sum_{i=1}^{m} (y_{1i} - \bar{y}_{1i})^2 \quad (6.23)$$

which by Equation 6.11 is equivalent to

$$SS_r = SS_{lof} + SS_{pe} \quad (6.24)$$

Although it is beyond the scope of this presentation, it can be shown that if the model $y_{1i} = 0 + r_{1i}$ is a true representation of the behavior of the system, then the three sums of squares SS_r, SS_{lof}, and SS_{pe} divided by the associated degrees of freedom (2, 1, and 1 respectively for this example) will all provide unbiased estimates of σ_{pe}^2 and there will not be significant differences among these estimates. If $y_{1i} = 0 + r_{1i}$ is *not* the true model, the parameter estimate s_{pe}^2 will still be a good estimate of the purely experimental uncertainty, σ_{pe}^2 (the estimate of purely experimental uncertainty is independent of any model – see Sections 5.5 and 5.6). The parameter estimate s_{lof}^2, however, will be inflated because it now includes a non-random contribution from a nonzero difference between the mean of the observed replicate responses, $\bar{y}_{1i}$, and the responses predicted by the model, $\hat{y}_{1i}$ (see

Equation 6.13). The less likely it is that $y_{1i} = 0 + r_{1i}$ is the true model, the more biased and therefore larger should be the term s^2_{lof} compared to s^2_{pe}.

6.5. The *F*-test

The previous section suggests a method for testing the adequacy of the model $y_{1i} = 0 + r_{1i}$:

(1) If $y_{1i} = 0 + r_{1i}$ is not the true model, then s^2_{lof} will not be an unbiased estimate of σ^2_{pe} and there should be a difference between s^2_{lof} and s^2_{pe}: we would expect s^2_{lof} to be greater than s^2_{pe}; that is, $s^2_{\text{lof}} - s^2_{\text{pe}} > 0$.

(2) If $y_{1i} = 0 + r_{1i}$ is the true model, then s^2_{lof} and s^2_{pe} should both be good estimates of σ^2_{pe} and there should not be a significant difference between them: we would expect s^2_{lof} to be equal to s^2_{pe}; that is, $s^2_{\text{lof}} - s^2_{\text{pe}} = 0$.

Thus, we can test the null hypothesis

$$\text{H}_0: \quad s^2_{\text{lof}} - s^2_{\text{pe}} = 0 \tag{6.25}$$

with the alternative hypothesis being

$$\text{H}_a: \quad s^2_{\text{lof}} - s^2_{\text{pe}} > 0 \tag{6.26}$$

The test of this hypothesis makes use of the calculated *Fisher variance ratio*, *F*.

$$F_{DF_n, DF_d} = s^2_{\text{lof}}/s^2_{\text{pe}} \tag{6.27}$$

where F_{DF_n, DF_d} represents the calculated value of the variance ratio with DF_n degrees of freedom associated with the numerator and DF_d degrees of freedom associated with the denominator. The variance s^2_{lof} always appears in the numerator; the test does not merely ask if the two variances are different, but instead seeks to answer the question, "Is the variance due to lack of fit significantly greater than the variance due to purely experimental uncertainty"? The critical value of F at a given level of probability may be obtained from an appropriate F-table (see Appendix C for values at the 95% level of confidence). The calculated and critical F values are compared:

(1) If $F_{\text{calc}} > F_{\text{crit}}$, s^2_{lof} is significantly greater than s^2_{pe} and the null hypothesis can be rejected at the specified level of confidence. The model $y_{1i} = 0 + r_{1i}$ would thus exhibit a significant lack of fit.

(2) If $F_{\text{calc}} < F_{\text{crit}}$, there is no reason to believe s^2_{lof} is significantly greater than s^2_{pe}, and the null hypothesis cannot be rejected at the specified level of confidence. We would therefore retain as adequate the model $y_{1i} = 0 + r_{1i}$.

For the numerical data in Figure 6.1 and for the model $y_{1i} = 0 + r_{1i}$, the variance of residuals is obtained as follows.

$$SS_{\text{r}} = \boldsymbol{R}'\boldsymbol{R} = [3 \quad 5]\begin{bmatrix} 3 \\ 5 \end{bmatrix} = 34 \tag{6.28}$$

$$DF_{\text{r}} = n = 2 \tag{6.29}$$

$$s^2_{\text{r}} = SS_{\text{r}}/DF_{\text{r}} = 34/2 = 17 \tag{6.30}$$

The purely experimental uncertainty variance is estimated as

$$SS_{pe} = \sum_{i=1}^{m} (y_{1i} - \bar{y}_{1i})^2 = (3-4)^2 + (5-4)^2 = 2 \quad (6.31)$$

$$DF_{pe} = m - 1 = 2 - 1 = 1 \quad (6.32)$$

$$s_{pe}^2 = SS_{pe}/DF_{pe} = 2/1 = 2 \quad (6.33)$$

Finally, the variance due to lack of fit may be obtained from

$$SS_{lof} = \sum_{i=1}^{n} (\bar{y}_{1i} - \hat{y}_{1i})^2 = (4-0)^2 + (4-0)^2 = 32 \quad (6.34)$$

An alternative method of obtaining SS_{lof} provides the same result and demonstrates the additivity of sums of squares and degrees of freedom.

$$SS_{lof} = SS_r - SS_{pe} = 34 - 2 = 32 \quad (6.35)$$

$$DF_{lof} = DF_r - DF_{pe} = 2 - 1 = 1 \quad (6.36)$$

$$s_{lof}^2 = SS_{lof}/DF_{lof} = 32/1 = 32 \quad (6.37)$$

Finally,

$$F_{calc} = F_{DF_{lof}, DF_{pe}} = F_{1,1} = s_{lof}^2/s_{pe}^2 = 32/2 = 16 \quad (6.38)$$

At first glance, this ratio might appear to be highly significant. However, the critical value of F at the 95% level of confidence for one degree of freedom in the numerator and one degree of freedom in the denominator is 161 (see Appendix C). Thus, the critical value is not exceeded and the null hypothesis is not rejected. We must retain as adequate the model $y_{1i} = 0 + r_{1i}$.

It is interesting to note that for the example we have been using (data from Figure 6.1 and the models $y_{1i} = 0 + r_{1i}$ and $y_{1i} = \beta_0 + r_{1i}$), the critical value of F is equal to the square of the critical value of t, and the calculated value of F is equal to the square of the calculated value of t. For a given level of confidence, Student's t values are, in fact, identical to the square root of the corresponding F values with one degree of freedom in the numerator. *For these simple models and this particular experimental design*, the F-test for the adequacy of the model $y_{1i} = 0 + r_{1i}$ is equivalent to the t-test for the significance of the parameter b_0 in the model $y_{1i} = \beta_0 + r_{1i}$.

However, for more complex models and different experimental designs, *the two tests are not always equivalent*. The t-test can be used to test the significance of a single parameter. The F-test can also be used to test the significance of a single parameter, but as we shall see, it is more generally useful as a means of testing the significance of a set of parameters, or testing the lack of fit of a multiparameter model.

6.6. Level of confidence

In many situations it is appropriate to decide before a test is made the risk one is willing to take that the null hypothesis will be disproved when it is actually true. If an experimenter wishes to be wrong no more than one time in twenty, the risk α is set at 0.05 and the test has "95% confidence". The calculated value of t or F is compared to the critical 95% threshold value found in tables: if the calculated value is equal to or greater than the tabular value, the null hypothesis can be rejected with a confidence *equal to or greater than* 95%.

If the null hypothesis can be rejected on the basis of a 95% confidence test, then the risk of falsely rejecting the null hypothesis is *at most* 0.05, but might be much less. We don't know how much less it is unless we look up the critical value for, say, 99% confidence and find that the null hypothesis cannot be rejected at that high a level of confidence; the risk would then be somewhere between 0.01 and 0.05, but further narrowing with additional tables between 95% and 99% would be necessary to more precisely define the exact risk.

Similarly, if the null hypothesis cannot be rejected at the 95% level of confidence it does not mean that the quantity being tested is insignificant. Perhaps the null hypothesis could have been rejected at the 90% level of confidence. The quantity would still be rather significant, with a risk somewhere between 0.05 and 0.10 of having falsely rejected the null hypothesis.

In other situations it is not necessary to decide before a test is made the risk one is willing to take. Such a situation is indicated by the subtly different question, "What are my chances of being right if I reject this null hypothesis"? In this case, it is desirable to assign an exact level of confidence to the quantity being tested. Such a level of confidence would then designate the estimated level of risk associated with rejecting the null hypothesis.

There are today computer programs that will accept a calculated value of t or F and the associated degrees of freedom and return the corresponding level of confidence or, equally, the risk.

As an example, a calculated t value of 4 with one degree of freedom (see Equation 6.9) is significant at the 84.4% level of confidence ($\alpha = 0.156$). Similarly, a calculated F value of 16 with one degree of freedom in the numerator and one degree of freedom in the denominator (see Equation 6.38) is also significant at the 84.4% level of confidence.

It is common to see the 95% level of confidence arbitrarily used as a threshold for statistical decision making. However, the price to be paid for this very conservative level of certainty is that many null hypotheses will not be rejected when they are in fact false. A relevant example is the so-called "screening" of factors to see which ones are "significant" (see Section 1.2 and Table 1.1). In selecting for further investigation factors that are significant at a given level of probability, the investigator is assured that those factors will probably be useful in improving the response. But this "efficiency" is gained at the expense of omitting factors that might also be significant. Ideally, the question that should be asked is not, "What factors are significant at the P level of probability"? but rather, "What factors are *insignificant* at the P level of probability"? The number of factors retained when using the second criterion will in general be larger than the number retained when using the first; the investigator will, however, be assured that he is probably not omitting from investigation any factors that are important. Unfortunately, this type of statistical testing requires a very different type of experimentation and requires a different set of criteria for hypothesis testing (see Figure 6.3). An alternative is to relax the requirement for confidence in rejecting the null hypothesis.

The widespread use of the 95% level of confidence can be directly linked to Fisher's opinion that scientists should be allowed to be wrong no more than one time in 20. There are many areas of decision making, however, where the arbitrary threshold level of 95% confidence is possibly too low (e.g., a doctor's confidence that a patient will survive an unnecessary but life-threatening operation) or too high (e.g., one's confidence in winning even money on a bet). Proper assessment of the risk one is willing to take and an exact knowledge of the level of significance of calculated statistics can lead to better decision making.

References

Abramowitz, M., and Stegun, I.A. (1968). *Handbook of Mathematical Functions with Formulas, Graphs, and Mathematical Tables* (Nat. Bur. of Stand. Appl. Math. Ser., No. 55), 7th printing. US Govt. Printing Office, Washington, DC.

Arkin, H., and Colton, R.R. (1970). *Statistical Methods*, 5th ed. Barnes and Noble, New York.

Box, G.E.P., Hunter, W.G., and Hunter, J.S. (1978). *Statistics for Experimenters. An Introduction to Design, Data Analysis, and Model Building*. Wiley, New York.

Campbell, S.K. (1974). *Flaws and Fallacies in Statistical Thinking*. Prentice-Hall, Englewood Cliffs, New Jersey.

Daniel, C., and Wood, F.S. (1971). *Fitting Equations to Data*. Wiley, New York.

Davies, O.L., Ed. (1956). *The Design and Analysis of Industrial Experiments*. Hafner, New York.

Draper, N.R., and Smith, H. (1966). *Applied Regression Analysis*. Wiley, New York.

Fisher, R.A. (1966). *The Design of Experiments*, 8th ed. Hafner, New York.

Fisher, R.A. (1970). *Statistical Methods for Research Workers*, 14th ed. Hafner, New York.

Fisher, R.A., and Yates, F. (1963). *Statistical Tables for Biological, Agricultural and Medical Research*. Hafner, New York.

Hofstadter, D.R. (1979). *Gödel, Escher, Bach: an Eternal Golden Braid*. Basic Books, New York.

Huff, D. (1954). *How to Lie with Statistics*. Norton, New York.

Kempthorne, O. (1952). *The Design and Analysis of Experiments*. Wiley, New York.

Mandel, J. (1964). *The Statistical Analysis of Experimental Data*. Wiley, New York.

Mendenhall, W. (1968). *Introduction to Linear Models and the Design and Analysis of Experiments*. Duxbury Press, Belmont, California.

Meyer, S.L. (1975). *Data Analysis for Scientists and Engineers*. Wiley, New York.

Moore, D.S. (1979). *Statistics. Concepts and Controversies*. Freeman, San Francisco, California.

Natrella, M.G. (1963). *Experimental Statistics* (Nat. Bur. of Stand. Handbook 91). US. Govt. Printing Office, Washington, DC.

Neter, J., and Wasserman, W. (1974). *Applied Linear Statistical Models. Regression, Analysis of Variance, and Experimental Designs*. Irwin, Homewood, Illinois.

Shewhart, W.A. (1939). *Statistical Method from the Viewpoint of Quality Control*. Lancaster Press, Pennsylvania.
Wilson, E.B., Jr. (1952). *An Introduction to Scientific Research*. McGraw-Hill, New York.
Youden, W.J. (1951). *Statistical Methods for Chemists*. Wiley, New York.

Exercises

6.1. Hypothesis testing

Suppose a researcher believes a system behaves according to the model $y_{1i} = \beta_0 + \beta_1 x_{1i} + r_{1i}$ over the domain $0 \leqslant x_1 \leqslant 10$. Suggest a set of ten experiments that might be useful in either disproving the researcher's hypothesis or increasing the confidence that the researcher's hypothesis is correct. Give reasons for your chosen design.

6.2. Null hypothesis

Write a null hypothesis that might be useful for testing the hypothesis that $b_0 = 13.62$. What is the alternative hypothesis?

6.3. Rejection of null hypothesis

Suppose you are told that a box contains one marble, and that it is either red or blue or green. You are asked to guess the color of the marble, and you guess red. What is your null hypothesis? What is the alternative hypothesis? You are now told that the marble in the box is not red. What might be the color of the marble?

6.4. Quality of information

If the person giving you information in Exercise 6.3 is lying, or might be lying, what might be the color of the marble? What is the difference between direct evidence and hearsay evidence?

6.5. Confidence intervals

Calculate the 95% confidence interval about $b_0 = 213.92$ if $s_{b_0} = 5.12$ is based on five degrees of freedom.

6.6. Level of confidence

If the null hypothesis is "rejected at the 95% level of confidence", why can we be *at least* 95% confident about accepting the alternative hypothesis?

6.7. Risk and level of confidence

The relationship between the risk, α, of falsely rejecting the null hypothesis and the level of confidence, P, placed in the alternative hypothesis is $P = 100(1 - \alpha)\%$. If the null hypothesis is rejected at the 87% level of confidence, what is the risk that the null hypothesis was rejected falsely?

6.8. Statistical significance

Criticize the following statement: "The results of this investigation were shown to be statistically significant".

6.9. Threshold of significance

State five null hypotheses, that could be profitably tested at less than the 95% level of confidence. State five null hypotheses that *you* would prefer be tested at greater than the 95% level of confidence.

6.10. Confidence intervals

Draw a figure similar to Figure 6.6 showing the following confidence intervals for the data in Figure 6.1 (see Equation 6.6): 50%, 80%, 90%, 95%, 99%, 99.9%.

6.11. One-sided t-test

If a null hypothesis is stated H_0: $b_0 = 0$ with an alternative hypothesis H_a: $b_0 \neq 0$, then the t-test based on Equation 6.8 will provide an exact decision. If the null hypothesis is rejected, then $b_0 \neq 0$ is accepted as the alternative hypothesis at the specified level of confidence. Values of b_0 significantly greater *or* less than zero would be accepted.

Suppose, however, that the alternative hypothesis is H_a: $b_0 > 0$. Values of b_0 significantly less than zero would not satisfy the alternative hypothesis. If we did disprove the null hypothesis *and* b_0 were greater than zero, then we should be "twice" as confident about accepting the alternative hypotheses (or we should have only half the risk of being wrong). If t_{crit} is obtained from a regular "two-tailed" t-table specifying a risk α, then the level of confidence in the test is $100(1 - \alpha/2)\%$.

If the null and alternative hypotheses in Exercise 6.5 are H_0: $b_0 = 0$ and H_a: $b_0 > 0$, what should α be so the level of confidence associated with rejecting H_0 and accepting H_a will be at least 95%.

6.12. t-tables

Go to the library and find eight or ten books that contain t-tables (critical values of t at specified levels of confidence and different degrees of freedom). Look at the value for 2 degrees of freedom and a "percentage point" (or "percentile", or "probability point", or "percent probability level", or "confidence level" or "probability") of 0.95 (or 95%). How many tables give $t = 2.920$? How many give $t = 4.303$? Do those that give $t = 2.920$ at the 0.95 confidence level give $t = 4.303$ at the 0.975 confidence level? How many of the tables indicate that they are one-tailed tables? How many indicate that they are two-tailed tables? How many give no indication? Are t-values at the $100(1 - \alpha)\%$ level of confidence from two-tailed tables the same as t-values at the $100(1 - \alpha/2)\%$ level of confidence from one-tailed tables? In the absence of other information, how can you quickly tell if a t-table is for one- or two-tailed tests? [Hint: look at the level of confidence associated with the t-value of 1.000 at one degree of freedom.]

6.13. Sums of squares

Assume the model $y_{1i} = 0 + r_{1i}$ is used to describe the nine data points in Section 3.1. Calculate directly the sum of squares of residuals, the sum of squares due to purely experimental uncertainty, and the sum of squares due to lack of fit. How many degrees of freedom are associated with each sum of squares? Do SS_{lof} and SS_{pe} add up to give SS_r? Calculate s_{lof}^2 and s_{pe}^2. What is the value of the Fisher F-ratio for lack of fit (Equation 6.27)? Is the lack of fit significant at or above the 95% level of confidence?

6.14. True values

In some textbooks, a confidence interval is described as the interval within which there is a certain probability of finding the *true* value of the estimated quantity. Does the term "true" used in this sense indicate the *statistical population value* (e.g., μ if one is estimating a mean) or the *bias-free value* (e.g., 6.21% iron in a mineral)? Could these two interpretations of "true value" be a source of misunderstanding in conversations between a statistician and a geologist?

6.15. Sums of squares and degrees of freedom

Fit the model $y_{1i} = \beta_0 + \beta_1 x_{1i} + r_{1i}$ to the data $x_{11} = 3$, $y_{11} = 2$, $x_{12} = 3$, $y_{12} = 4$, $x_{13} = 6$, $y_{13} = 6$, $x_{14} = 6$, $y_{14} = 4$. Calculate the sum of squares of residuals, the sum of squares due to purely experimental uncertainty, and the sum of squares due to lack of fit. How many degrees of freedom are associated with each sum of squares? Can s_{lof}^2 be calculated? Why or why not?

6.16. Null hypotheses

State the following in terms of null and alternative hypotheses: a) the mean of a data set is equal to zero; b) the mean of a data set is equal to six; c) the mean of a data set is equal to or less than six; d) the variance of one data set is not different from the variance of a second data set; e) the variance of one data set is equal to or less than the variance of a second data set.

6.17. Distributions of means and standard deviations

Write the following numbers on 31 slips of paper, one number to each piece of paper, and place them in a container: 7, 8, 9, 9, 10, 10, 10, 11, 11, 11, 11, 11, 12, 12, 12, 12, 12, 12, 12, 13, 13, 13, 13, 13, 14, 14, 14, 15, 15, 16, 17. Carry out the following experiment 25 times: mix up the slips of paper, randomly draw out five pieces of paper, calculate the mean and variance of residuals of the numbers drawn, and put the pieces of paper back in the container. Round the values of the mean to the nearest 0.5 unit and the values of the variances to the nearest 0.5 unit and plot as frequency histograms. Is the histogram for the means roughly gaussian? Is the histogram for the variance of residuals roughly gaussian? What is the smallest value s_r^2 could assume? What is the largest value? What value appeared most frequently for s_r^2? [The values of s_r^2 are distributed according to the *chi-square distribution*, χ^2, which is skewed. See, for example, Mandel, J. (1964). *The Statistical Analysis of Experimental Data*, p. 234. Wiley, New York.]

6.18. Equivalence of t- and F-values

Compare the t-values in Appendix B with the square root of the F-values in Appendix C with one degree of freedom in the numerator.

6.19. Null hypothesis

Suppose you see a dog and hypothesize, "This animal is a stray and has no home". What might you do to try to prove this hypothesis? What might you do to try to disprove it? Which would be easier, proving the hypothesis or disproving it?

6.20. Lack of fit

What is the relationship between Figures 3.5 and 6.9 (see Exercise 3.21).

6.21. Confidence intervals

The confidence interval of the mean is sometimes written $\bar{y} \pm ts/(n^{1/2})$. How is this related to Equation 6.5?

CHAPTER 7

The Variance-Covariance Matrix

In Section 6.2, the standard uncertainty of the parameter estimate b_0 was obtained by taking the square root of the product of the purely experimental uncertainty variance estimate, s_{pe}^2, and the $(X'X)^{-1}$ matrix (see Equation 6.3). A single number was obtained because the single-parameter model being considered ($y_{1i} = \beta_0 + r_{1i}$) produced a 1×1 $(X'X)^{-1}$ matrix.

For the general, multiparameter case, the product of the purely experimental uncertainty estimate, s_{pe}^2, and the $(X'X)^{-1}$ matrix gives the *estimated variance-covariance matrix*, V.

$$V = s_{pe}^2 (X'X)^{-1} \tag{7.1}$$

Each of the upper left to lower right diagonal elements of V is an *estimated variance of a parameter estimate*, s_b^2; these elements correspond to the parameters as they appear in the model from left to right. Each of the off-diagonal elements is an *estimated covariance between two of the parameter estimates*.

Thus, for a single-parameter model such as $y_{1i} = \beta_0 + r_{1i}$, the estimated variance-covariance matrix contains no covariance elements; the square root of the single variance element corresponds to the standard uncertainty of the single parameter estimate.

In this chapter, we will examine the variance-covariance matrix to see how the location of experiments in factor space (i.e., the experimental design) affects the individual variances and covariances of the parameter estimates. Throughout this section we will be dealing with the specific two-parameter first-order model $y_{1i} = \beta_0 + \beta_1 x_{1i} + r_{1i}$ only; the resulting principles are entirely general, however, and can be applied to all other linear models.

We will also assume that we have a prior estimate of σ_{pe}^2 for the system under investigation and that the variance is homoscedastic (see Section 3.3). Our reason for assuming the availability of an estimate of σ_{pe}^2 is to obviate the need for replication in the experimental design so that the effect of the location of the experiments in factor space can be discussed by itself.

7.1. Influence of the experimental design

Each element of the $(X'X)$ matrix is a summation of products (see Appendix A).

A common algebraic representation of the $(X'X)$ matrix for the straight-line model $y_{1i} = \beta_0 + \beta_1 x_{1i} + r_{1i}$ is

$$(X'X) = \begin{bmatrix} \sum_{i=1}^{n} 1 & \sum_{i=1}^{n} x_{1i} \\ \sum_{i=1}^{n} x_{1i} & \sum_{i=1}^{n} x_{1i}x_{1i} \end{bmatrix} \tag{7.2}$$

Let $s_{b_0}^2$ be the estimated variance associated with the parameter estimate b_0; let $s_{b_1}^2$ be the estimated variance associated with b_1; and let $s_{b_0b_1}^2$ (or $s_{b_1b_0}^2$) represent the estimated covariance between b_0 and b_1. Then

$$V = \begin{bmatrix} s_{b_0}^2 & s_{b_0b_1}^2 \\ s_{b_1b_0}^2 & s_{b_1}^2 \end{bmatrix} = s_{\text{pe}}^2 \begin{bmatrix} \sum_{i=1}^{n} 1 & \sum_{i=1}^{n} x_{1i} \\ \sum_{i=1}^{n} x_{1i} & \sum_{i=1}^{n} x_{1i}x_{1i} \end{bmatrix} \tag{7.3}$$

From Equation 7.3 it is evident that the experimental design (i.e., the level of each x_{1i}) has a direct effect upon the variances and covariances of the parameter estimates.

The effect on the variance-covariance matrix of two experiments located at different positions in factor space can be investigated by locating one experiment at $x_{11} = 1$ and varying the location of the second experiment. The first row of the matrix of parameter coefficients for the model $y_{1i} = \beta_0 + \beta_1 x_{1i} + r_{1i}$ can be made to correspond to the fixed experiment at $x_{11} = 1$.

$$X = \begin{bmatrix} 1 & x_{11} \\ - & - \end{bmatrix} = \begin{bmatrix} 1 & 1 \\ - & - \end{bmatrix} \tag{7.4}$$

Let us now locate a series of "second experiments" from $x_{12} = -5$ to $x_{12} = +5$. The X matrix for this changing experimental design may be represented as

$$X = \begin{bmatrix} 1 & 1 \\ 1 & (-5 \leqslant x_{12} \leqslant +5) \end{bmatrix} \tag{7.5}$$

Three numerical examples of the calculation of the $(X'X)^{-1}$ matrix are given here.

Example 7.1: $x_{12} = -4$

$$(X'X) = \begin{bmatrix} 1 & 1 \\ 1 & -4 \end{bmatrix} \begin{bmatrix} 1 & 1 \\ 1 & -4 \end{bmatrix} = \begin{bmatrix} 2 & -3 \\ -3 & 17 \end{bmatrix} \tag{7.6}$$

$$(X'X)^{-1} = \begin{bmatrix} 17/25 & 3/25 \\ 3/25 & 2/25 \end{bmatrix} \tag{7.7}$$

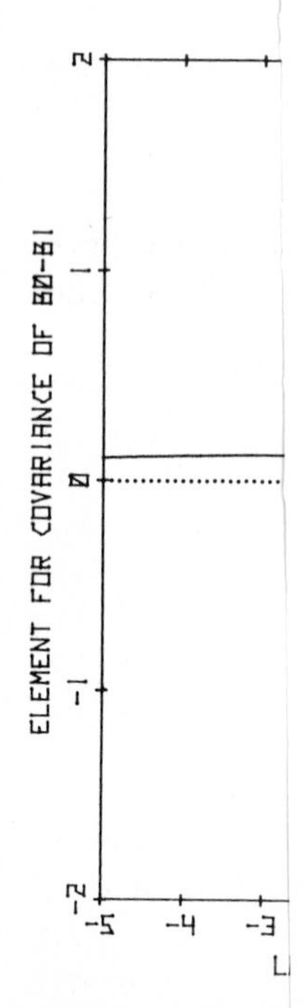

Example 7.2: $x_{12} = +0.5$

$$(\boldsymbol{X}'\boldsymbol{X}) = \begin{bmatrix} 1 & 1 \\ 1 & 0.5 \end{bmatrix} \begin{bmatrix} 1 & 1 \\ 1 & 0.5 \end{bmatrix} = \begin{bmatrix} 2 & 1.5 \\ 1.5 & 1.25 \end{bmatrix} \tag{7.8}$$

$$(\boldsymbol{X}'\boldsymbol{X})^{-1} = \begin{bmatrix} 5 & -6 \\ -6 & 8 \end{bmatrix} \tag{7.9}$$

Example 7.3: $x_{12} = +4$

$$(\boldsymbol{X}'\boldsymbol{X}) = \begin{bmatrix} 1 & 1 \\ 1 & 4 \end{bmatrix} \begin{bmatrix} 1 & 1 \\ 1 & 4 \end{bmatrix} = \begin{bmatrix} 2 & 5 \\ 5 & 17 \end{bmatrix} \tag{7.10}$$

$$(\boldsymbol{X}'\boldsymbol{X})^{-1} = \begin{bmatrix} 17/9 & -5/9 \\ -5/9 & 2/9 \end{bmatrix} \tag{7.11}$$

7.2. Effect on the variance of b_1

Let us first consider the effect of these experimental designs upon the uncertainty of estimating the parameter β_1. This parameter represents the slope of a straight line relationship between y_1 and x_1.

There is an intuitive feeling that for a given uncertainty in response, the farther apart the two experiments are located, the more precise the estimate of β_1 should be. An analogy is often made with the task of holding a meter stick at a particular angle using both hands: if the meter stick is grasped at the 5- and 95-cm marks, a small amount of up-and-down "jitter" or uncertainty in the hands will cause the stick to move from the desired angle, but only slightly; if the meter stick is held at the 45- and 55-cm marks, the same vertical jitter or uncertainty in the hands will be "amplified" and could cause the stick to wiggle quite far from the desired angle. In the first case, there would be only a slight uncertainty in the slope of the meter stick; in the second case, there would be much greater uncertainty in the slope of the meter stick.

These intuitive feelings are confirmed in the calculations of Examples 7.1–7.3. When the two experimental points are relatively far apart, as they are in Example 7.1 ($x_{11} = +1$, $x_{12} = -4$, distance apart = 5), the value of the element for the estimated variance of b_1 is relatively small – only 2/25 (the lower right hand element of the $(\boldsymbol{X}'\boldsymbol{X})^{-1}$ matrix in Example 7.1). When the experiments are located closer in factor space, as they are in Example 7.2 ($x_{11} = +1$, $x_{12} = +0.5$, distance apart = 0.5), the value of the element for the estimated variance of b_1 is relatively large – 8 in Example 7.2. When the experiments are once again located far apart in factor space, as they are in Example 7.3 ($x_{11} = +1$, $x_{12} = +4$, distance apart = 3),

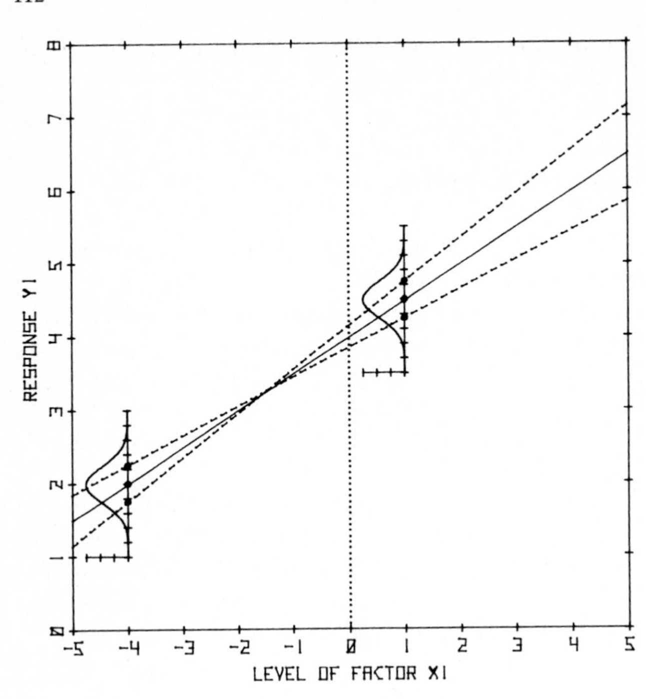

Figure 7.5. Illustration of positive covariance: if the slope (b_1) increases, the intercept (b_0) increases; if the slope decreases, the intercept decreases.

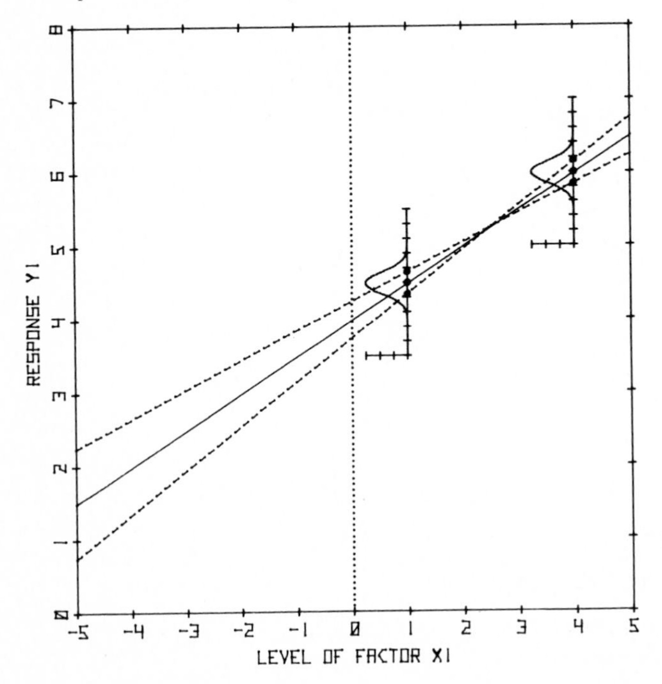

Figure 7.6. Illustration of negative covariance: if the slope (b_1) increases, the intercept (b_0) decreases; if the slope decreases, the intercept increases.

$y_{15} = 4.25$, and $y_{16} = 2.25$ are obtained by chance because of uncertainty in the response; the corresponding relationships are shown as dashed lines in Figure 7.5. For these sets of data, both the slope and the y_1-intercept increase and decrease together. The same tendency exists for most other sets of experimental data obtained at $x_1 = +1$ and $x_1 = -4$: as the slope through these points increases, the intercept at $x_1 = 0$ will probably increase; as the intercept increases, the slope will probably increase. Similarly, as one parameter decreases, the other will decrease. *As one parameter estimate varies, so varies the other: the covariance is positive.*

Assume now that we have carried out our two experiments at $x_{11} = +1$ and $x_{12} = +4$ (see Example 7.3). The off-diagonal element of the $(\boldsymbol{X}'\boldsymbol{X})^{-1}$ matrix is *negative* 5/9. Assume also that we have obtained responses of $y_{11} = 4.5$ and $y_{12} = 6.0$. The first-order relationship through these points is shown as a solid line in Figure 7.6. Again, for purposes of illustration, suppose the experiments are repeated twice, and values of $y_{13} = 4.35$, $y_{14} = 6.15$, $y_{15} = 4.65$, and $y_{16} = 5.85$ are obtained; the corresponding relationships are shown as dashed lines in Figure 7.6. For these sets of data, the slope and the y_1-intercept do not tend to increase and decrease together. Instead, *each parameter estimate varies oppositely as the other: the covariance is negative.*

Let us now assume that we have carried out our two experiments at $x_{11} = +1$ and $x_{12} = -1$ and that we have obtained responses of $y_{11} = 4.5$ and $y_{12} = 3.5$.

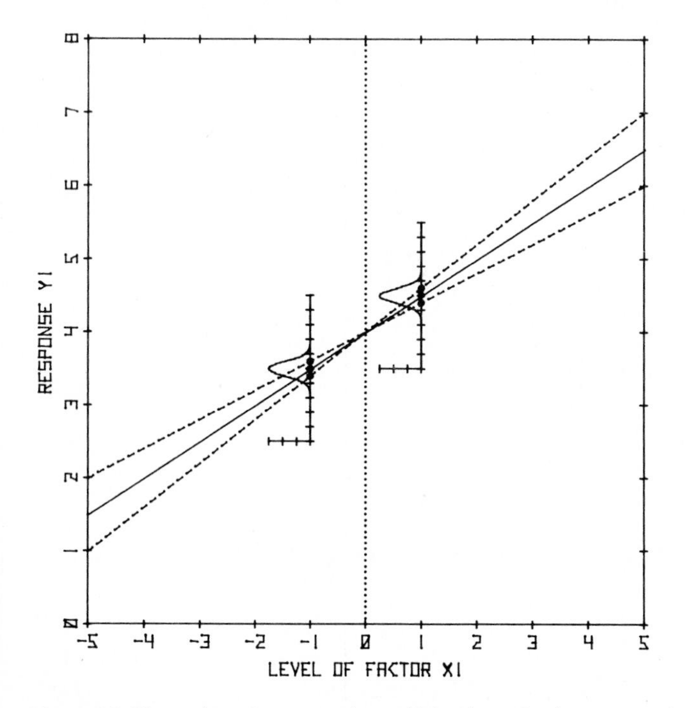

Figure 7.7. Illustration of zero covariance: if the slope (b_1) increases or decreases, the intercept (b_0) does not change.

When the experiments are repeated twice, we obtain $y_{13} = 4.4$, $y_{14} = 3.6$, $y_{15} = 4.6$, and $y_{16} = 3.4$; the corresponding relationships are shown in Figure 7.7. Note that in this example, *the intercept does not depend upon the slope, and the slope does not depend upon the intercept: the covariance is zero.*

As an illustration, if $x_{11} = +1$ and $x_{12} = -1$, then

Example 7.4: $x_{12} = -1$

$$(X'X) = \begin{bmatrix} 1 & 1 \\ 1 & -1 \end{bmatrix} \begin{bmatrix} 1 & 1 \\ 1 & -1 \end{bmatrix} = \begin{bmatrix} 2 & 0 \\ 0 & 2 \end{bmatrix} \tag{7.12}$$

$$(X'X)^{-1} = \begin{bmatrix} 1/2 & 0 \\ 0 & 1/2 \end{bmatrix} \tag{7.13}$$

This is also seen in Figure 7.4 in which the covariance goes to zero when the second experiment is located at $x_1 = -1$.

7.5. Optimal design

Consideration of the effect of experimental design on the elements of the variance-covariance matrix leads naturally to the area of optimal design. Let us suppose that our purpose in carrying out two experiments is to obtain good estimates of the intercept and slope for the model $y_{1i} = \beta_0 + \beta_1 x_{1i} + r_{1i}$. We might want to know what levels of the factor x_1 we should use to obtain the most precise estimates of β_0 and β_1.

The most precise estimate of β_0 is obtained when the two experiments are located such that the center of the design is at $x_1 = 0$ (see Figure 7.2). Any other arrangement of experiments would produce greater uncertainty in the estimate of β_0. With two experiments, this optimal value of $s_{b_0}^2$ would be $0.5 \times s_{pe}^2$ (see Equations 7.1 and 7.13).

The most precise estimate of β_1 is obtained when the two experiments are located as far apart as possible. If the two experiments could be located at $-\infty$ and $+\infty$, there would be no uncertainty in the slope.

Example 7.5: $x_{11} = -\infty$, $x_{12} = +\infty$

$$(X'X) = \begin{bmatrix} 1 & 1 \\ -\infty & +\infty \end{bmatrix} \begin{bmatrix} 1 & -\infty \\ 1 & +\infty \end{bmatrix} = \begin{bmatrix} 2 & 0 \\ 0 & \infty^2 \end{bmatrix} \tag{7.14}$$

$$(X'X)^{-1} = \begin{bmatrix} 1/2 & 0 \\ 0 & 0 \end{bmatrix} \tag{7.15}$$

In most real situations, boundaries exist (see Section 2.3) and it is not possible to obtain an experimental design giving zero uncertainty in the slope. However, the minimum uncertainty in the slope will be obtained when the two experiments are located as far from each other as practical.

References

Arkin, H., and Colton, R.R. (1970). *Statistical Methods*, 5th ed. Barnes and Noble, New York.

Box, G.E.P., Hunter, W.G., and Hunter, J.S. (1978). *Statistics for Experimenters. An Introduction to Design, Data Analysis, and Model Building*. Wiley, New York.

Daniel, C., and Wood, F.S. (1971). *Fitting Equations to Data*. Wiley, New York.

Draper, N.R., and Smith, H. (1966). *Applied Regression Analysis*. Wiley, New York.

Fisher, R.A. (1966). *The Design of Experiments*, 8th ed. Hafner, New York.

Fisher, R.A. (1970). *Statistical Methods for Research Workers*, 14th ed. Hafner, New York.

Himmelblau, D.M. (1970). *Process Analysis by Statistical Methods*. Wiley, New York.

Kempthorne, O. (1952). *The Design and Analysis of Experiments*. Wiley, New York.

Mendenhall, W. (1968). *Introduction to Linear Models and the Design and Analysis of Experiments*. Duxbury Press, Belmont, California.

Meyer, S.L. (1975). *Data Analysis for Scientists and Engineers*. Wiley, New York.

Natrella, M.G. (1963). *Experimental Statistics* (Nat. Bur. of Stand. Handbook 91). US Govt. Printing Office, Washington, DC.

Neter, J., and Wasserman, W. (1974). *Applied Linear Statistical Models. Regression, Analysis of Variance, and Experimental Design*. Irwin, Homewood, Illinois.

Youden, W.J. (1951). *Statistical Methods for Chemists*. Wiley, New York.

Exercises

7.1. Variance-covariance matrix

Calculate the variance-covariance matrix associated with the straight line relationship $y_{1i} = \beta_0 + \beta_1 x_{1i} + r_{1i}$ for the following data:

$$\boldsymbol{D} = \begin{bmatrix} -4 \\ -4 \\ 1 \\ 1 \\ -4 \\ 1 \end{bmatrix} \qquad \boldsymbol{Y} = \begin{bmatrix} 1.75 \\ 2.00 \\ 4.50 \\ 4.24 \\ 2.25 \\ 4.75 \end{bmatrix}$$

(Hint: there are no degrees of freedom for lack of fit; therefore, $s_{\text{pe}}^2 = s_{\text{r}}^2$ by Equation 6.24.)

7.2. Variance of b_1

What is the value of the lower right element of the $(\boldsymbol{X}'\boldsymbol{X})^{-1}$ matrix (the element for estimating the variance of b_1) when $x_{12} = 1$ in Figure 7.1. Why?

7.3. Variance-covariance relationships

Figure 7.2 implies that for the model $y_{1i} = \beta_0 + \beta_1 x_{1i} + r_{1i}$, if one experiment is placed at $x_1 = 0$, the position of the second experiment will have no influence on the estimated value of s_{b_0}. However, the position of the second experiment will affect the values of s_{b_1} and $s_{b_0b_1}$. Can you discover a relationship between s_{b_1} and $s_{b_0b_1}$ for this case?

7.4. Coding effects

Calculate the variance-covariance matrix associated with the straight line relationship $y_{1i} = \beta_0 + \beta_1 x_{1i} + r_{1i}$ for the following data set (assume $s_{pe}^2 = s_r^2$):

$$\boldsymbol{D} = \begin{bmatrix} 1 \\ 3 \\ 2 \\ 2 \end{bmatrix} \qquad \boldsymbol{Y} = \begin{bmatrix} 2 \\ 10 \\ 6 \\ 8 \end{bmatrix}$$

Subtract 2 from each value of $\boldsymbol{D}$ and recalculate. Plot the data points and straight line for each case. Comment on the similarities and differences in the two variance-covariance matrices.

7.5. Asymmetry

Why are the curves in Figures 7.3 and 7.4 not symmetrical?

7.6. Variance-covariance matrix

Equation 7.1 is one of the most important relationships in the area of experimental design. As we have seen in this chapter, the precision of estimated parameter values is contained in the variance-covariance matrix $\boldsymbol{V}$: the smaller the elements of $\boldsymbol{V}$, the more precise will be the parameter estimates. As we shall see in Chapter 10, the precision of estimating the response surface is also directly related to $\boldsymbol{V}$: the smaller the elements of $\boldsymbol{V}$, the less "fuzzy" will be our view of the estimated surface.

Equation 7.1 indicates that there are *two* requirements for a favorable variance-covariance matrix: small values of s_{pe}^2, and small elements of the $(\boldsymbol{X}'\boldsymbol{X})^{-1}$ matrix. How can s_{pe}^2 be made small? How can the elements of the $(\boldsymbol{X}'\boldsymbol{X})^{-1}$ matrix be made small?

7.7. Precision of estimation

If $s_{pe}^2 = 4$, how large should n be so that s_{b_0} will be less than or equal to 0.5?

7.8. Precision of estimation

If $s_{pe}^2 = 4$, how large should n be so that the 95% confidence interval associated with b_0 in Equation 6.5 will be no wider than $b_0 \pm 1.5$?

CHAPTER 8

Three Experiments

In a single-factor system, there are three possible designs for three experiments. We will discuss each of these possibilities in order, beginning with the situation in which all experiments are carried out at the same level of the single factor, and concluding with the case in which each experiment is carried out at a different level.

8.1. All experiments at one level

Let us suppose we have carried out the three experiments
$x_{11} = 6,\ y_{11} = 3.0,$
$x_{12} = 6,\ y_{12} = 5.0,$
$x_{13} = 6,\ y_{13} = 3.7.$

The model $y_{1i} = 0 + r_{1i}$

If we assume the model $y_{1i} = 0 + r_{1i}$, the data and uncertainties are as shown in Figure 8.1. We can test this model for lack of fit because there is replication in the experimental design which allows an estimate of the purely experimental uncertainty (with two degrees of freedom).

$$\bar{y}_{1i} = (3.0 + 5.0 + 3.7)/3 = 3.9 \tag{8.1}$$

$$SS_{pe} = \sum_{i=1}^{n} (y_{1i} - \bar{y}_{1i})^2 = 2.06 \tag{8.2}$$

$$s_{pe}^2 = SS_{pe}/(n-1) = 2.06/2 = 1.03 \tag{8.3}$$

The sum of squares of residuals, with three degrees of freedom, is given in this case by

$$SS_r = \sum_{i=1}^{n} r_{1i}^2 = \sum_{i=1}^{n} (y_{1i} - \hat{y}_{1i})^2 = 47.69 \tag{8.4}$$

The sum of squares due to lack of fit, with one degree of freedom, may be obtained by difference

$$SS_{lof} = SS_r - SS_{pe} = 47.69 - 2.06 = 45.63 \tag{8.5}$$

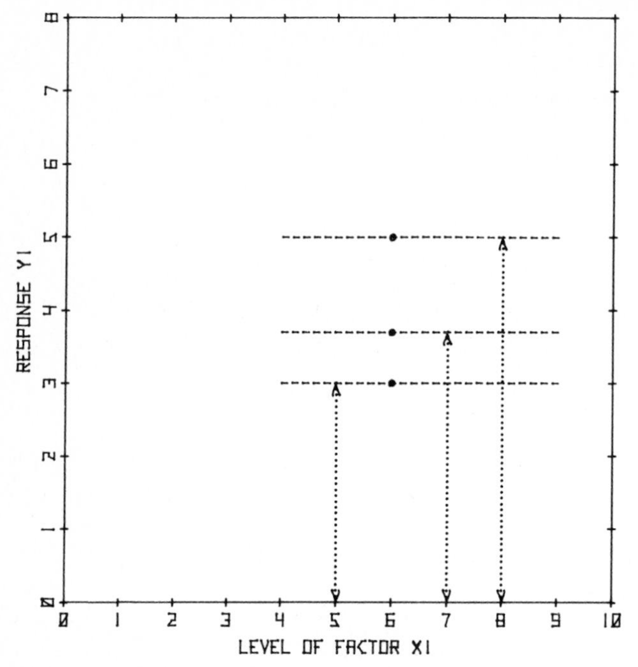

Figure 8.1. Relationship of three replicate data points to the model $y_{1i} = 0 + r_{1i}$.

or by direct calculation

$$SS_{\text{lof}} = \sum_{i=1}^{n} (\bar{y}_{1i} - \hat{y}_{1i})^2 = 45.63 \tag{8.6}$$

The estimate of the variance due to lack of fit is given by

$$s^2_{\text{lof}} = SS_{\text{lof}}/(3-2) = 45.63 \tag{8.7}$$

An appropriate test for lack of fit is the F-test described in Section 6.5.

$$F_{1,2} = s^2_{\text{lof}}/s^2_{\text{pe}} = 45.63/1.03 = 44.30 \tag{8.8}$$

The critical value of $F_{1,2}$ at the 95% level of confidence is 18.51 (see Appendix C). Thus, $F_{\text{calc}} > F_{\text{crit}}$ which allows us to state (with a risk of being wrong no more than one time in 20) that the model $y_{1i} = 0 + r_{1i}$ exhibits a significant lack of fit to the three data we have obtained. The level at which $F_{\text{calc}} = F_{\text{crit}}$ is 97.8% confidence, or a risk $\alpha = 0.022$. A better model is probably possible.

The model $y_{1i} = \beta_0 + r_{1i}$

If we fit the model $y_{1i} = \beta_0 + r_{1i}$ to this data, we will find that $b_0 = 3.9$. The sum of squares of residuals (with two degrees of freedom) is

$$SS_{\text{r}} = \boldsymbol{R'R} = [-0.9 \quad 1.1 \quad -0.2] \begin{bmatrix} -0.9 \\ 1.1 \\ -0.2 \end{bmatrix} = 0.81 + 1.21 + 0.04 = 2.06 \tag{8.9}$$

The variance of residuals is

$$s_r^2 = SS_r/(n-1) = 2.06/2 = 1.03 \tag{8.10}$$

The variance due to purely experimental uncertainty (two degrees of freedom) is

$$s_{pe}^2 = \sum_{i=1}^{n} (y_{1i} - \bar{y}_{1i})^2/(n-1) = 1.03 \tag{8.11}$$

Further,

$$s_{b_0} = \left[s_{pe}^2(\boldsymbol{X}'\boldsymbol{X})^{-1}\right]^{1/2} = [1.03 \times (1/3)]^{1/2} = [0.343]^{1/2} = 0.586 \tag{8.12}$$

We can test the significance of the parameter estimate b_0 by calculating the value of t required to construct a confidence interval extending from b_0 and including the value zero (see Section 6.3).

$$t_{calc} = |b_0 - 0|/s_{b_0} = |3.9 - 0|/0.586 = 6.66 \tag{8.13}$$

The critical value of t at the 95% level of confidence for two degrees of freedom is 4.30 (see Appendix B). Thus, $t_{calc} > t_{crit}$ which allows us to state (with a risk of being wrong no more than one time in 20) that the null hypothesis H_0: $\beta_0 = 0$ is disproved and that β_0 is different from zero. The level at which $t_{calc} = t_{crit}$ is 97.8% confidence, or a risk $\alpha = 0.022$.

It should be noted that in this case the sum of squares due to lack of fit must be zero because the sum of squares of residuals is equal to the sum of squares due to purely experimental uncertainty. This would seem to suggest that there is a perfect fit. In fact, however, s_{lof}^2 cannot be calculated and tested because there are no degrees of freedom for it: one degree of freedom has been used to estimate β_0 leaving only two degrees of freedom for the residuals; but the estimation of purely experimental uncertainty *requires* these two degrees of freedom, leaving no degrees of freedom for lack of fit (see Figure 6.10).

8.2. Experiments at two levels

Let us now assume that the first experiment was not carried out at $x_{11} = 6$, but rather was performed at $x_{11} = 3$. The three experiments are shown in Figure 8.2. One experiment is carried out at one level of the single factor x_1, and two experiments are carried out at a different level.

The model $y_{1i} = \beta_0 + r_{1i}$

If the model $y_{1i} = \beta_0 + r_{1i}$ is fit to this data (see Figure 8.2), we find, as before, that $b_0 = 3.9$ and $s_r^2 = 1.03$ (see Section 8.1). If we now calculate s_{pe}^2, we find that in this case we have only two replicates, whereas before we had three.

$$\bar{y}_{1i} = (5.0 + 3.7)/2 = 4.35 \tag{8.14}$$

$$SS_{pe} = \sum_{i=1}^{n} (y_{1i} - \bar{y}_{1i})^2 = 0.845 \tag{8.15}$$

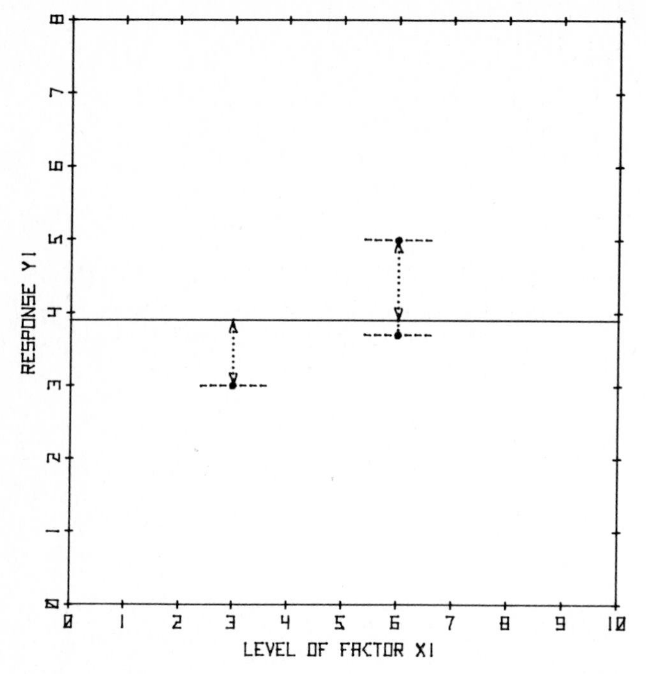

Figure 8.2. Relationship of three experiments at two levels to the fitted model $y_{1i} = \beta_0 + r_{1i}$.

with one degree of freedom.

$$s_{pe}^2 = SS_{pe}/(n-1) = 0.845/(2-1) = 0.845 \tag{8.16}$$

$$s_{b_0} = \left[s_{pe}^2(\boldsymbol{X'X})^{-1}\right]^{1/2} = [0.845 \times (1/3)]^{1/2} = [0.282]^{1/2} = 0.531 \tag{8.17}$$

$$t_{calc} = |b_0 - 0|/s_{b_0} = |3.9 - 0|/0.531 = 7.34 \tag{8.18}$$

The tabular value of t at the 95% level of confidence and one degree of freedom is 12.71 (see Appendix B). Thus, $t_{calc} < t_{crit}$ and we cannot conclude (at the 95% level of confidence) that β_0 is significantly different from zero for the data in this example. The level at which $t_{calc} = t_{crit}$ is 91.4% confidence, or a risk $\alpha = 0.086$.

The model $y_{1i} = \beta_0 + \beta_1 x_{1i} + r_{1i}$

We now fit the model $y_{1i} = \beta_0 + \beta_1 x_{1i} + r_{1i}$.

$$(\boldsymbol{X'X}) = \begin{bmatrix} 1 & 1 & 1 \\ 3 & 6 & 6 \end{bmatrix} \begin{bmatrix} 1 & 3 \\ 1 & 6 \\ 1 & 6 \end{bmatrix} = \begin{bmatrix} 3 & 15 \\ 15 & 81 \end{bmatrix} \tag{8.19}$$

$$D = 3 \times 81 - 15 \times 15 = 243 - 225 = 18 \tag{8.20}$$

$$(\boldsymbol{X'X})^{-1}=\begin{bmatrix} 81/18 & -15/18 \\ -15/18 & 3/18 \end{bmatrix}=\begin{bmatrix} 9/2 & -5/6 \\ -5/6 & 1/6 \end{bmatrix} \tag{8.21}$$

$$(\boldsymbol{X'Y})=\begin{bmatrix} 1 & 1 & 1 \\ 3 & 6 & 6 \end{bmatrix}\begin{bmatrix} 3.0 \\ 5.0 \\ 3.7 \end{bmatrix}=\begin{bmatrix} 11.7 \\ 61.2 \end{bmatrix} \tag{8.22}$$

$$\hat{\boldsymbol{B}}=\begin{bmatrix} b_0 \\ b_1 \end{bmatrix}=(\boldsymbol{X'X})^{-1}(\boldsymbol{X'Y})=\begin{bmatrix} 9/2 & -5/6 \\ -5/6 & 1/6 \end{bmatrix}\begin{bmatrix} 11.7 \\ 61.2 \end{bmatrix}=\begin{bmatrix} 1.65 \\ 0.45 \end{bmatrix} \tag{8.23}$$

The best least squares straight line is shown in Figure 8.3. Notice that the line goes through the point $x_{11}=3$, $y_{11}=3$ and through the average of the points $x_{12}=6$, $y_{12}=5.0$ and $x_{13}=6$, $y_{13}=3.7$. There is perhaps an intuitive feeling that this should be the case, and it is not surprising to see the residuals distributed as follows.

$$\boldsymbol{R}=\boldsymbol{Y}-\boldsymbol{X}\hat{\boldsymbol{B}}=\begin{bmatrix} 3.0 \\ 5.0 \\ 3.7 \end{bmatrix}-\begin{bmatrix} 1 & 3 \\ 1 & 6 \\ 1 & 6 \end{bmatrix}\begin{bmatrix} 1.65 \\ 0.45 \end{bmatrix}=\begin{bmatrix} 3.0 \\ 5.0 \\ 3.7 \end{bmatrix}-\begin{bmatrix} 3.0 \\ 4.35 \\ 4.35 \end{bmatrix}=\begin{bmatrix} 0 \\ 0.65 \\ -0.65 \end{bmatrix} \tag{8.24}$$

$$SS_r=\boldsymbol{R'R}=\begin{bmatrix} 0 & 0.65 & -0.65 \end{bmatrix}\begin{bmatrix} 0 \\ 0.65 \\ -0.65 \end{bmatrix}=0.845 \tag{8.25}$$

The sum of squares due to purely experimental uncertainty, already calculated for this data set (see Equation 8.15), is 0.845 with one degree of freedom. If we

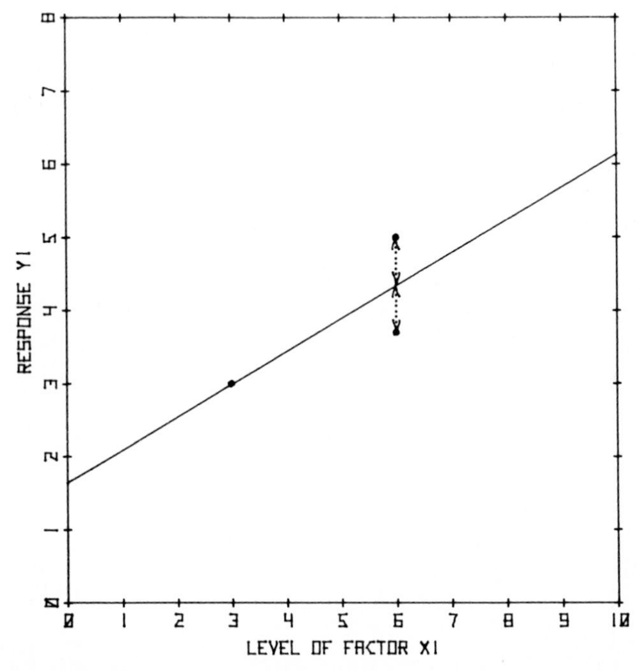

Figure 8.3. Relationship of three experiments at two levels to the fitted model $y_{1i}=\beta_0+\beta_1 x_{1i}+r_{1i}$.

calculate $SS_{lof} = SS_r - SS_{pe}$, then $SS_{lof} = 0$ and we might conclude that there is no lack of fit – that is, that the model fits the data perfectly. However, as seen before in Section 8.1, an analysis of the degrees of freedom shows that this is not a correct conclusion. The total number of degrees of freedom is three; of these, two have been used to calculate b_0 and b_1 and the third has been used to calculate s^2_{pe}. Thus, there are no degrees of freedom available for calculating s^2_{lof}. If a variance due to lack of fit cannot be calculated, there is no basis for judging the inadequacy of the model.

8.3. Experiments at three levels: first-order model

For this section, we will assume the model $y_{1i} = \beta_0 + \beta_1 x_{1i} + r_{1i}$. We will examine the $(\boldsymbol{X}'\boldsymbol{X})^{-1}$ matrix to determine the effect of carrying out a third experiment x_{13} between $x_1 = -5$ and $x_1 = +5$ given the very special case of two fixed experiments located at $x_{11} = +1$ and $x_{12} = -1$ (see Chapter 7 and Example 7.4). We give four numerical examples of the calculation of the $(\boldsymbol{X}'\boldsymbol{X})^{-1}$ matrix, one each for $x_{13} = -4$, 0, +0.5, and +4.

Example 8.1: $x_{13} = -4$

$$(\boldsymbol{X}'\boldsymbol{X}) = \begin{bmatrix} 1 & 1 & 1 \\ 1 & -1 & -4 \end{bmatrix} \begin{bmatrix} 1 & 1 \\ 1 & -1 \\ 1 & -4 \end{bmatrix} = \begin{bmatrix} 3 & -4 \\ -4 & 18 \end{bmatrix} \tag{8.26}$$

$$(\boldsymbol{X}'\boldsymbol{X})^{-1} = \begin{bmatrix} 9/19 & 2/19 \\ 2/19 & 3/38 \end{bmatrix} \tag{8.27}$$

Example 8.2: $x_{13} = 0$

$$(\boldsymbol{X}'\boldsymbol{X}) = \begin{bmatrix} 1 & 1 & 1 \\ 1 & -1 & 0 \end{bmatrix} \begin{bmatrix} 1 & 1 \\ 1 & -1 \\ 1 & 0 \end{bmatrix} = \begin{bmatrix} 3 & 0 \\ 0 & 2 \end{bmatrix} \tag{8.28}$$

$$(\boldsymbol{X}'\boldsymbol{X})^{-1} = \begin{bmatrix} 1/3 & 0 \\ 0 & 1/2 \end{bmatrix} \tag{8.29}$$

Example 8.3: $x_{13} = +0.5$

$$(\boldsymbol{X}'\boldsymbol{X}) = \begin{bmatrix} 1 & 1 & 1 \\ 1 & -1 & 0.5 \end{bmatrix} \begin{bmatrix} 1 & 1 \\ 1 & -1 \\ 1 & 0.5 \end{bmatrix} = \begin{bmatrix} 3 & 0.5 \\ 0.5 & 2.25 \end{bmatrix} \tag{8.30}$$

$$(\boldsymbol{X}'\boldsymbol{X})^{-1} = \begin{bmatrix} 9/26 & -1/13 \\ -1/13 & 6/13 \end{bmatrix} \tag{8.31}$$

Example 8.4: $x_{13} = +4$

$$(\boldsymbol{X}'\boldsymbol{X}) = \begin{bmatrix} 1 & 1 & 1 \\ 1 & -1 & 4 \end{bmatrix} \begin{bmatrix} 1 & 1 \\ 1 & -1 \\ 1 & 4 \end{bmatrix} = \begin{bmatrix} 3 & 4 \\ 4 & 18 \end{bmatrix} \tag{8.32}$$

$$(\boldsymbol{X}'\boldsymbol{X})^{-1} = \begin{bmatrix} 9/19 & -2/19 \\ -2/19 & 3/38 \end{bmatrix} \tag{8.33}$$

A more complete picture of the effect of the location of the third experiment on the variance of the slope estimate (b_1) is shown in Figure 8.4. From this figure, it is evident that the uncertainty associated with b_1 can be decreased by placing the third experiment far away from the other two. Placing the experiment at $x_{13} = 0$ seems to give the worst estimate of b_1. For the conditions giving rise to Figure 8.4, this is true, but it must not be concluded that the additional third experiment at $x_{13} = 0$ is detrimental: the variance of b_1 ($s^2_{b_1} = s^2_{pe}/2$) for the three experiments $x_{11} = -1$, $x_{12} = +1$, $x_{13} = 0$ is the same as the variance of b_1 for only two experiments at $x_{11} = -1$, $x_{12} = +1$ (see Equations 7.1 and 7.13).

Figure 8.5 shows that b_0 can be estimated most precisely when the third experiment is located at $x_{13} = 0$. This is reasonable, for the contribution of the third

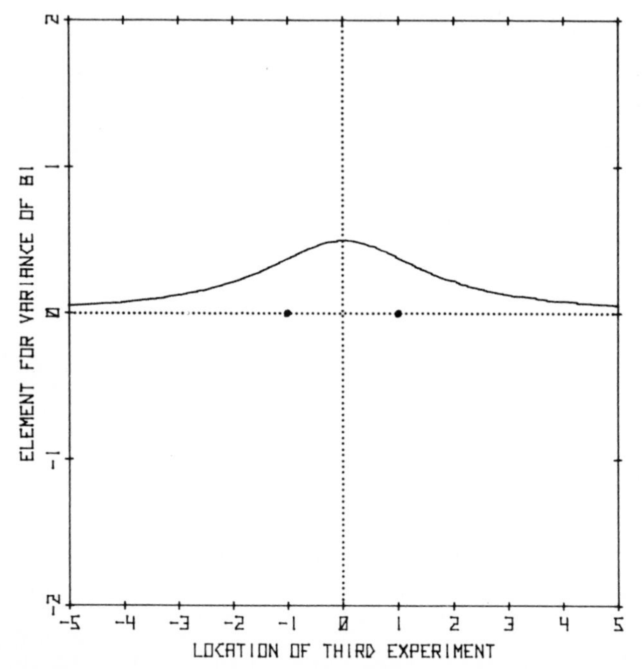

Figure 8.4. Value of the element of $(\boldsymbol{X}'\boldsymbol{X})^{-1}$ for the estimated variance of b_1 as a function of the location of a third experiment, two experiments fixed at $x_1 = -1$ and $x_1 = 1$.

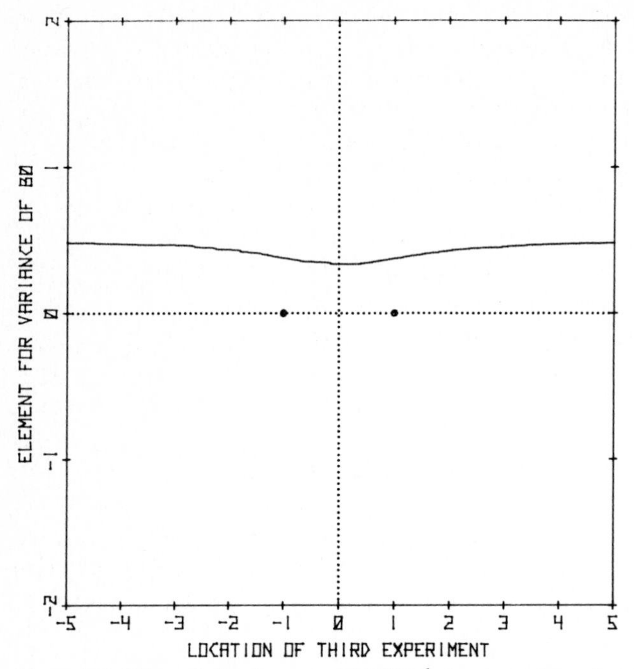

Figure 8.5. Value of the element of $(X'X)^{-1}$ for the estimated variance of b_0 as a function of the location of a third experiment, two experiments fixed at $x_1 = -1$ and $x_1 = 1$.

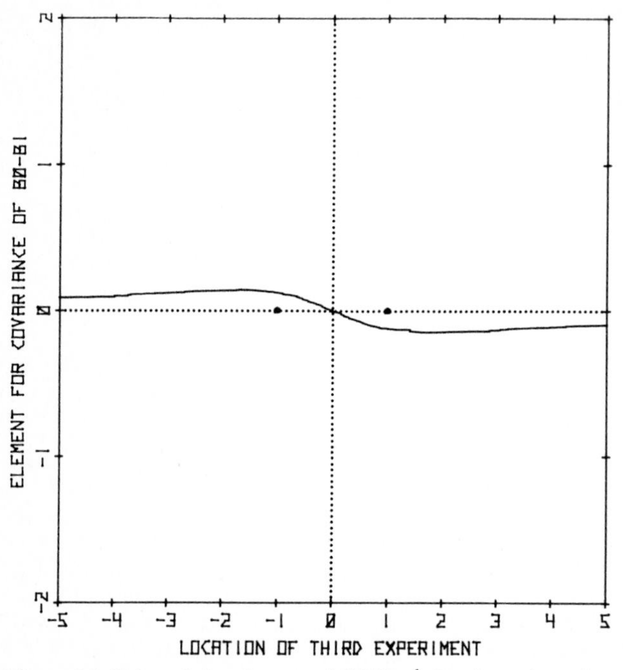

Figure 8.6. Value of the element of $(X'X)^{-1}$ for the estimated covariance between b_0 and b_1 as a function of the location of a third experiment, two experiments fixed at $x_1 = -1$ and $x_1 = 1$.

experiment at $x_1 = 0$ to the variance associated with b_0 involves no interpolation or extrapolation of a model: if the third experiment is carried out at $x_1 = 0$, then any discrepancy between y_{13} and the true intercept must be due to purely experimental uncertainty only. As the third experiment is moved away from $x_1 = 0$, $s_{b_0}^2$ does increase, but not drastically; the two stationary experiments remain positioned near $x_1 = 0$ and provide reasonably good estimates of b_0 by themselves.

Finally, the effect of the position of the third experiment upon the covariance associated with b_0 and b_1 is seen in Figure 8.6 to equal zero at $x_{13} = 0$. If the third experiment is located at $x_1 < 0$, then the estimates of the slope and intercept vary together in the same way (the covariance is positive; see Section 7.4). If the third experiment is located at $x_1 > 0$, the estimates of the slope and intercept vary together in opposite ways (the covariance is negative).

8.4. Experiments at three levels: second-order model

Let us now consider another case in which each experiment is carried out at a different level of the single factor x_1: $x_{11} = 3$, $x_{12} = 6$, and $x_{13} = 8$. The three points are distributed as shown in Figure 8.7.

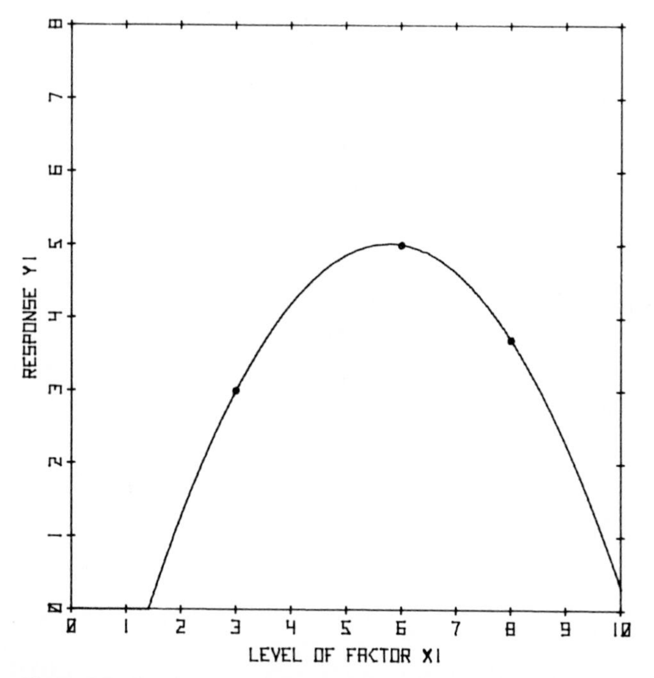

Figure 8.7. Graph of the probabilistic model $y_{1i} = \beta_0 + \beta_1 x_{1i} + \beta_{11} x_{1i}^2 + r_{1i}$ fit to the results of three experiments at different levels of the factor x_1.

It is possible to fit zero-, one-, two-, and three-parameter models to the data shown in Figure 8.7; for the moment, we will focus our attention on the three-parameter parabolic *second-order model*

$$y_{1i} = \beta_0 + \beta_1 x_{1i} + \beta_{11} x_{1i}^2 + r_{1i} \tag{8.34}$$

The subscript on the parameter β_{11} is used to indicate that it is associated with the coefficient corresponding to x_1 times x_1; that is, $x_1 x_1$, or x_1^2. This notation anticipates later usage where it will be seen that β_{12}, for example, is associated with the coefficient corresponding to x_1 times a second factor x_2, $x_1 x_2$.

Although the fit will be "perfect" and there will be no residuals, we will nevertheless use the matrix least squares method of fitting the model to the data. The equations for the experimental points in Figure 8.7 can be written as

$$\begin{aligned} 3 &= 1 \times \beta_0 + 3 \times \beta_1 + 3^2 \times \beta_{11} + r_{11} \\ 5 &= 1 \times \beta_0 + 6 \times \beta_1 + 6^2 \times \beta_{11} + r_{12} \\ 3.7 &= 1 \times \beta_0 + 8 \times \beta_1 + 8^2 \times \beta_{11} + r_{13} \end{aligned} \tag{8.35}$$

The matrix of parameter coefficients is thus of the form

$$\boldsymbol{X} = \begin{bmatrix} 1 & x_{11} & x_{11}^2 \\ 1 & x_{12} & x_{12}^2 \\ 1 & x_{13} & x_{13}^2 \end{bmatrix} = \begin{bmatrix} 1 & 3 & 9 \\ 1 & 6 & 36 \\ 1 & 8 & 64 \end{bmatrix} \tag{8.36}$$

Continuing with the data treatment procedures,

$$(\boldsymbol{X'X}) = \begin{bmatrix} 1 & 1 & 1 \\ 3 & 6 & 8 \\ 9 & 36 & 64 \end{bmatrix} \begin{bmatrix} 1 & 3 & 9 \\ 1 & 6 & 36 \\ 1 & 8 & 64 \end{bmatrix} = \begin{bmatrix} 3 & 17 & 109 \\ 17 & 109 & 755 \\ 109 & 755 & 5473 \end{bmatrix} \tag{8.37}$$

Procedures for the inversion of a 3×3 matrix are given in Appendix A.

$$(\boldsymbol{X'X})^{-1} = \begin{bmatrix} 13266/450 & -5373/450 & 477/450 \\ -5373/450 & 2269/450 & -206/450 \\ 477/450 & -206/450 & 19/450 \end{bmatrix} \tag{8.38}$$

$$(\boldsymbol{X'Y}) = \begin{bmatrix} 1 & 1 & 1 \\ 3 & 6 & 8 \\ 9 & 36 & 64 \end{bmatrix} \begin{bmatrix} 3 \\ 5 \\ 3.7 \end{bmatrix} = \begin{bmatrix} 11.7 \\ 68.6 \\ 443.8 \end{bmatrix} \tag{8.39}$$

$$\hat{\boldsymbol{B}} = \begin{bmatrix} b_0 \\ b_1 \\ b_{11} \end{bmatrix} = (\boldsymbol{X'X})^{-1}(\boldsymbol{X'Y}) = \begin{bmatrix} -3.740 \\ 3.037 \\ -0.2633 \end{bmatrix} \tag{8.40}$$

As a check,

$$\hat{\boldsymbol{Y}} = \boldsymbol{X}\hat{\boldsymbol{B}} = \begin{bmatrix} 1 & 3 & 9 \\ 1 & 6 & 36 \\ 1 & 8 & 64 \end{bmatrix} \begin{bmatrix} -3.740 \\ 3.037 \\ -0.2633 \end{bmatrix} = \begin{bmatrix} 3.001 \\ 5.003 \\ 3.705 \end{bmatrix} \tag{8.41}$$

which does reproduce the original data points. (The discrepancies are caused by rounding to obtain the four-significant-digit estimates of β_1 and β_{11} in Equation 8.40.)

The equation that best fits this data is therefore

$$y_{1i} = -3.740 + 3.037x_{1i} - 0.2633x_{1i}^2 + r_{1i} \tag{8.42}$$

Let us assume we have a prior estimate of σ_{pe}^2. We will now rewrite the $(\boldsymbol{X}'\boldsymbol{X})^{-1}$ matrix and consider its meaning in terms of the variances and covariances of the parameter estimates.

$$(\boldsymbol{X}'\boldsymbol{X})^{-1} = \begin{bmatrix} 29.48 & -11.94 & 1.06 \\ -11.94 & 5.04 & -0.46 \\ 1.06 & -0.46 & 0.04 \end{bmatrix} \tag{8.43}$$

The uncertainty in the estimate of β_0 (the value of y_1 at $x_1 = 0$) is relatively large ($s_{b_0}^2 = s_{pe}^2 \times 29.48$). Figure 8.7 suggests that this is reasonable – any uncertainty in the response will cause the parabolic curve to "wiggle", and this wiggle will have a rather severe effect on the intersection of the parabola with the response axis.

The uncertainty in the estimate of β_1 is smaller ($s_{b_1}^2 = s_{pe}^2 \times 5.04$), and the uncertainty in the estimate of β_{11} is smaller still ($s_{b_{11}}^2 = s_{pe}^2 \times 0.04$). The geometric interpretation of the parameters β_1 and β_{11} in this model is not straightforward, but β_1 essentially moves the apex of the parabola away from $x_1 = 0$, and β_{11} is a measure of the steepness of curvature. The geometric interpretation of the associated uncertainties in the parameter estimates is also not straightforward (for example, β_0, β_1, and β_{11} are expressed in different units). We will simply note that such uncertainties do exist, and note also that there is covariance between b_0 and b_1, between b_0 and b_{11}, and between b_1 and b_{11}.

8.5. Centered experimental designs and coding

There is a decided interpretive advantage to a different three-experiment design, a symmetrical design centered about $x_1 = 0$. Let us assume that two of the experimental points are located at $x_{11} = -1$ and $x_{12} = +1$. Figures 8.8–8.13 show the effects of moving the third experimental point from $x_{13} = -5$ to $x_{13} = +5$ on the elements of the $(\boldsymbol{X}'\boldsymbol{X})^{-1}$ matrix associated with the variances $s_{b_0}^2$, $s_{b_1}^2$, and $s_{b_{11}}^2$, and with the covariances $s_{b_0b_1}^2$, $s_{b_0b_{11}}^2$, and $s_{b_1b_{11}}^2$, respectively. Of particular interest to us is the case for which $x_{13} = 0$.

$$(\boldsymbol{X}'\boldsymbol{X}) = \begin{bmatrix} 1 & 1 & 1 \\ -1 & 1 & 0 \\ 1 & 1 & 0 \end{bmatrix} \begin{bmatrix} 1 & -1 & 1 \\ 1 & 1 & 1 \\ 1 & 0 & 0 \end{bmatrix} = \begin{bmatrix} 3 & 0 & 2 \\ 0 & 2 & 0 \\ 2 & 0 & 2 \end{bmatrix} \tag{8.44}$$

$$(\boldsymbol{X}'\boldsymbol{X})^{-1} = \begin{bmatrix} 1 & 0 & -1 \\ 0 & 1/2 & 0 \\ -1 & 0 & 3/2 \end{bmatrix} \tag{8.45}$$

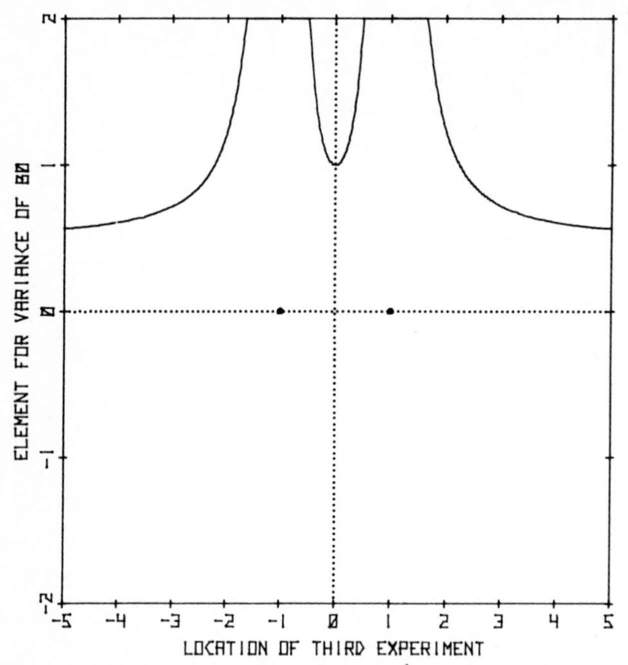

Figure 8.8. Value of the element of $(\boldsymbol{X}'\boldsymbol{X})^{-1}$ for the estimated variance of b_0 in the model $y_{1i} = \beta_0 + \beta_1 x_{1i} + \beta_{11} x_{1i}^2 + r_{1i}$ as a function of the location of a third experiment, two experiments fixed at $x_1 = -1$ and $x_1 = 1$.

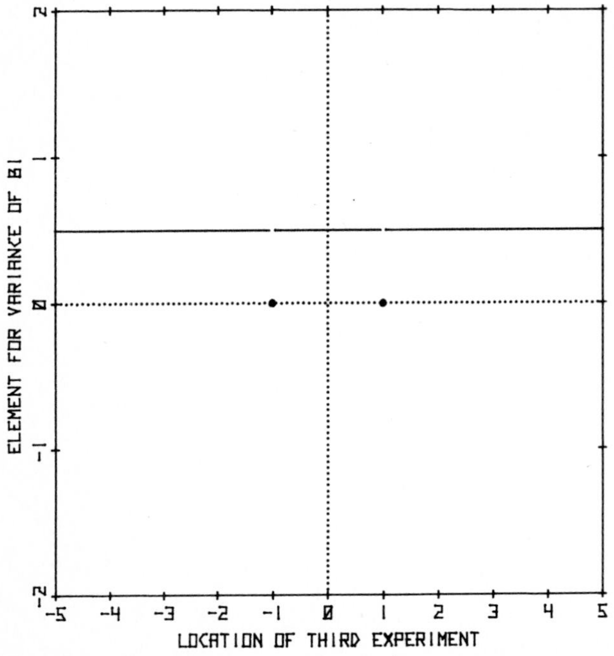

Figure 8.9. Value of the element of $(\boldsymbol{X}'\boldsymbol{X})^{-1}$ for the estimated variance of b_1.

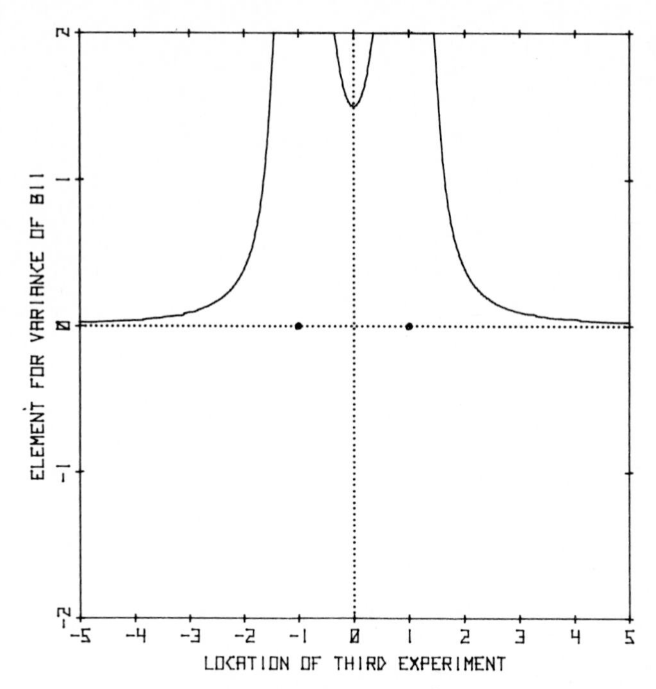

Figure 8.10. Value of the element of $(\boldsymbol{X'X})^{-1}$ for the estimated variance of b_{11}.

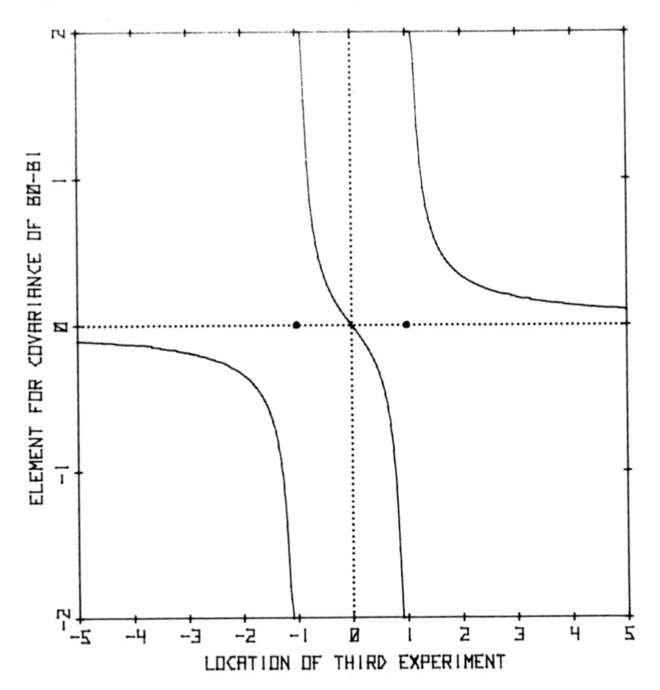

Figure 8.11. Value of the element of $(\boldsymbol{X'X})^{-1}$ for the estimated covariance between b_0 and b_1.

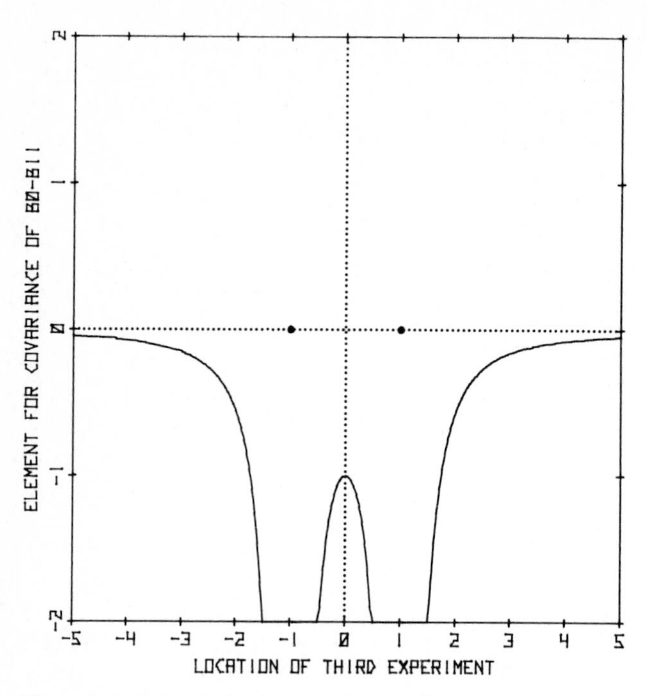

Figure 8.12. Value of the element of $(\boldsymbol{X}'\boldsymbol{X})^{-1}$ for the estimated covariance between b_0 and b_{11}.

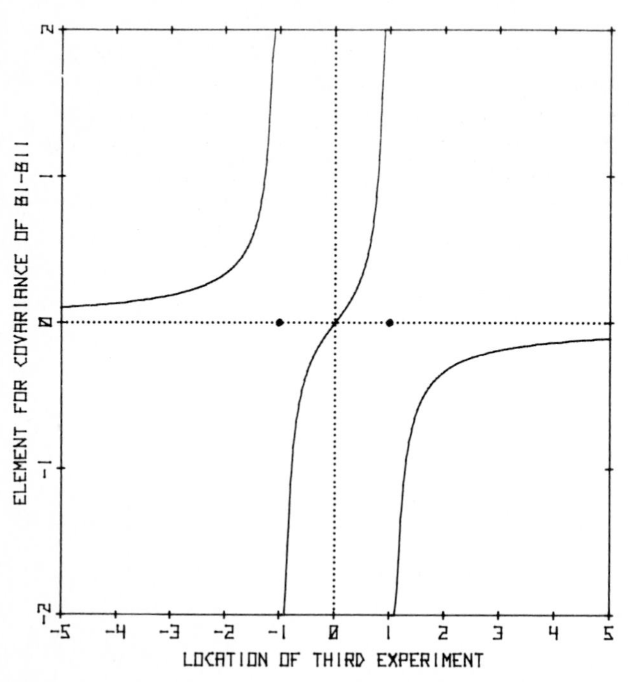

Figure 8.13. Value of the element of $(\boldsymbol{X}'\boldsymbol{X})^{-1}$ for the estimated covariance between b_1 and b_{11}.

Using this design, the covariances between the estimates of β_0 and β_1 and between the estimates of β_1 and β_{11} are zero. This is confirmed in Figures 8.11 and 8.13. Thus, the estimation of β_0 does not depend upon the estimated value of β_1 (and vice versa), and the estimated values of β_1 and β_{11} do not depend upon the estimated values of each other.

This advantage of a symmetrical experimental design centered at $x_1 = 0$ is usually realized in practice not by actually carrying out experiments about $x_1 = 0$ (in many practical cases, lower natural boundaries would prevent this; see Section 2.3), but instead by mathematically *translating* the origin of the factor space to the center of the design located in the desired region of factor space. Often, as a related procedure, the interval between experimental points is *normalized* to give a value of unity. When one or the other or both of these manipulations of the factor levels have been carried out, the experimental design is said to be *coded*. If c_{x_1} is the location of the center of the design along the factor x_1, and if d_{x_1} is the distance along x_1 between experimental points, then the coded factor levels (x^*_{1i}) are given by

$$x^*_{1i} = (x_{1i} - c_{x_1})/d_{x_1} \tag{8.46}$$

Another, often major, advantage of using coded factor levels is that the numerical values involved in matrix manipulations are smaller (especially the products and sums of products), and therefore are simpler to handle and do not suffer as much from round-off errors.

It is to be stressed, however, that the geometric interpretation of the parameter estimates obtained using coded factor levels is usually different from the interpretation of those parameter estimates obtained using uncoded factor levels. As an illustration, β^*_0 (the intercept in the coded system) represents the response *at the center of the experimental design*, whereas β_0 (the intercept in the uncoded system) represents the response *at the origin of the original coordinate system*; the two estimates (β^*_0 and β_0) are usually quite different numerically. This "difficulty" will not be important in the remainder of this book, and we will feel equally free to use either coded or uncoded factor levels as the examples require. Later, in Section 10.5, we will show how to translate coded parameter estimates back into uncoded parameter estimates.

8.6. Self interaction

Before leaving this chapter, we introduce the concept of *interaction* by pointing out an interpretation of the second-order character in the model we have been using. Equation 8.34 can be rewritten as

$$y_{1i} = \beta_0 + (\beta_1 + \beta_{11}x_{1i})x_{1i} + r_{1i} \tag{8.47}$$

Focusing our attention for a moment on the term $(\beta_1 + \beta_{11}x_{1i})x_{1i}$, it can be seen that over a small region of factor space, the first-order effect of the factor x_1 is given by the "slope", $\beta_1 + \beta_{11}x_{1i}$. But this "slope" depends upon the region of factor space one is observing! Another way of stating this is that *the effect of the factor depends upon the level of the factor*.

For the example of Figure 8.7, in the region where x_1 is small, the effect of x_1 is to *increase* the response y_1 as x_1 is increased; in the region where x_1 is large, the effect of x_1 is to *decrease* the response y_1 as x_1 is increased; in the region where x_1 is approximately equal to 5 or 6, there is very little effect of x_1 upon y_1. This dependence of a factor effect upon the level of the factor will be called "self interaction".

References

Cochran, W.G., and Cox, G.M. (1957). *Experimental Designs*, 2nd ed. Wiley, New York.

Daniel, C., and Wood, F.S. (1971). *Fitting Equations to Data*. Wiley, New York.

Draper, N.R., and Smith, H. (1966). *Applied Regression Analysis*. Wiley, New York.

Himmelblau, D.M. (1970). *Process Analysis by Statistical Methods*. Wiley, New York.

Mendenhall, W. (1968). *Introduction to Linear Models and the Design and Analysis of Experiments*. Duxbury Press, Belmont, California.

Natrella, M.G. (1963). *Experimental Statistics* (Nat. Bur. Stand. Handbook 91). US Govt. Printing Office, Washington, DC.

Neter, J., and Wasserman, W. (1974). *Applied Linear Statistical Models. Regression, Analysis of Variance, and Experimental Designs*. Irwin, Homewood, Illinois.

Exercises

8.1. Number of experimental designs

In a single-factor system, how many possible designs are there for four experiments (e.g., four experiments at a single factor level, three at one factor and one at another factor level, etc.)? For five experiments?

8.2. Calculation of SS_{lof}

Show by direct calculation that the numerical value in Equation 8.6 is correct.

8.3. Degrees of freedom

From discussions in this and previous chapters, formulate an algebraic equation that gives the number of degrees of freedom associated with lack of fit.

8.4. Effect of a fourth experiment

Recalculate Example 8.2 for four experiments, the fourth experiment located at $x_1 = 0$. Assuming s^2_{pe} is the same as for the three experiment case, what effect does this fourth experiment have on the precision of estimating b_0 and b_1?

8.5. Variance of b_1

The graph in Figure 8.9 is discontinuous at $x_1 = -1$ and $x_1 = 1$. Why? Are there corresponding discontinuities in Figures 8.8 and 8.10–8.13?

8.6. Matrix inversion

Verify that Equation 8.38 is the correct inverse of Equation 8.37 (see Appendix A).

8.7. Residuals

Why would Equation 8.41 be expected to reproduce the original data points of Equation 8.35?

8.8. Matrix inversion

Verify that Equation 8.45 is the correct inverse of Equation 8.44.

8.9. Matrix least squares

Complete the development of Equations 8.44 and 8.45 to solve for $\hat{B}$ assuming $y_{11} = 3$, $y_{12} = 5$, and $y_{13} = 4$.

8.10. Self interaction

Give three examples of self interaction. Does the sign of the slope invert in any of these examples?

8.11. Self interaction

Sketch a plot of "discomfort" vs. "extension" for Figure 1.11. Does the relationship you have plotted exhibit self interaction?

8.12. Self interaction

Indicate which of the following figures exhibit self interaction: Figures 1.2, 1.9, 1.15, 2.2, 2.5, 2.7, 2.8, 2.18, 4.2, 4.4, 5.3, 7.1, 7.4, 8.4, 8.9, 8.13.

8.13. Matrix least squares

Fit the model $y_{1i} = \beta_0 + \beta_1 x_{1i} + r_{1i}$ to the following data:

$$\boldsymbol{D} = \begin{bmatrix} -1 \\ 0 \\ 1 \end{bmatrix}, \qquad \boldsymbol{Y} = \begin{bmatrix} 2 \\ 6 \\ 8 \end{bmatrix}$$

Calculate s_r^2, s_{lof}^2, and s_{pe}^2.

8.14. Matrix least squares

A single matrix least squares calculation can be employed when the same linear model is used to fit each of several system responses. The $\boldsymbol{D}$, $\boldsymbol{X}$, $(\boldsymbol{X}'\boldsymbol{X})$, and $(\boldsymbol{X}'\boldsymbol{X})^{-1}$ matrices remain the same, but the $\boldsymbol{Y}$, $(\boldsymbol{X}'\boldsymbol{Y})$, $\hat{\boldsymbol{B}}$, and $\boldsymbol{R}$ matrices have additional columns, one column for each response. Fit the model $y_{ji} = \beta_0 + \beta_1 x_{1i} + r_{ji}$ to the following multiresponse data, $j = 1, 2, 3$:

$$\boldsymbol{D} = \begin{bmatrix} -1 \\ 0 \\ 1 \end{bmatrix}, \qquad \boldsymbol{Y} = \begin{bmatrix} 2 & 3 & 5 \\ 6 & 1 & 15 \\ 8 & -2 & 29 \end{bmatrix}$$

What are the b's for each response? Calculate $\boldsymbol{R}$. Calculate s_{rj}^2 for each of the three responses.

8.15. Coding

Code the data in Figure 8.7 so that the distance between the left point and the center point is 1.5 in the factor x_1^*, and the center point is located at $x_1^* = 3$. What are c_{x_1} and d_{x_1} of Equation 8.46 for this transformation?

8.16. Coding

"Autoscaling" is a technique for coding data so that the mean is zero and the standard deviation is unity. What should c_{y_1} and d_{y_1} be to autoscale the nine responses in Section 3.1? (Hint: see Figure 3.3.)

8.17. Covariance

Give interpretations of the covariance plots shown in Figures 8.11–8.13. Why is the covariance between β_0 and β_{11} not equal to zero when $x_{13} = 0$ (Figure 8.12)?

8.18. Covariance

Can you find an experimental design such that the off-diagonal elements of Equation 8.45 are all equal to zero? What would be the covariance between b_0 and b_1, between b_0 and b_{11}, and between b_1 and b_{11} for such a design? [See, for example, Box, G.E.P., and Hunter, J.S. (1957). "Multi-Factor Experimental Designs for Exploring Response Surfaces," *Ann. Math. Statist.*, 28, 195.]

CHAPTER 9

Analysis of Variance (ANOVA) for Linear Models

In Section 6.4, it was shown for replicate experiments at one factor level that the sum of squares of residuals, SS_r, can be partitioned into a sum of squares due to purely experimental uncertainty, SS_{pe}, and a sum of squares due to lack of fit, SS_{lof}. Each sum of squares divided by its associated degrees of freedom gives an estimated variance. Two of these variances, s_{lof}^2 and s_{pe}^2, were used to calculate a Fisher F-ratio from which the significance of the lack of fit could be estimated.

In this chapter we examine these and other sums of squares and resulting variances in greater detail. This general area of investigation is called the "*an*alysis *o*f *va*riance" (ANOVA) applied to linear models.

9.1 Sums of squares

There is an old "paradox" that suggests that it is not possible to walk from one side of the room to the other because you must first walk halfway across the room, then halfway across the remaining distance, then halfway across what still remains, and so on; because it takes an infinite number of these "steps", it is supposedly not possible to reach the other side. This seeming paradox is, of course, false, but the idea of breaking up a continuous journey into a (finite) number of discrete steps is useful for understanding the analysis of variance applied to linear models.

Each response y_{1i} may be viewed as a distance in response space (see Section 2.1). Statisticians find it useful to "travel" from the origin to each response in a number of discrete steps as illustrated in Figure 9.1. Each journey can be broken up as follows.

(1) From the origin (0) to the mean of the data set ($\bar{y}_1$). This serves to get us into the middle of all of the responses. It is allowed by the degree of freedom associated with β_0, the offset parameter.

(2) From the mean of the data set ($\bar{y}_1$) to the value predicted by the model ($\hat{y}_{1i}$). This distance is a measure of the effectiveness of the model in explaining the variation in the data set. It is allowed by the degrees of freedom associated with the coefficients of the factor effects (the other β's).

(3) From the value predicted by the model ($\hat{y}_{1i}$) to the mean of replicate responses (if any) at the same factor level ($\bar{y}_{1i}$). This distance is a measure of the lack of fit of the model to the data; if the model does a good job of predicting the

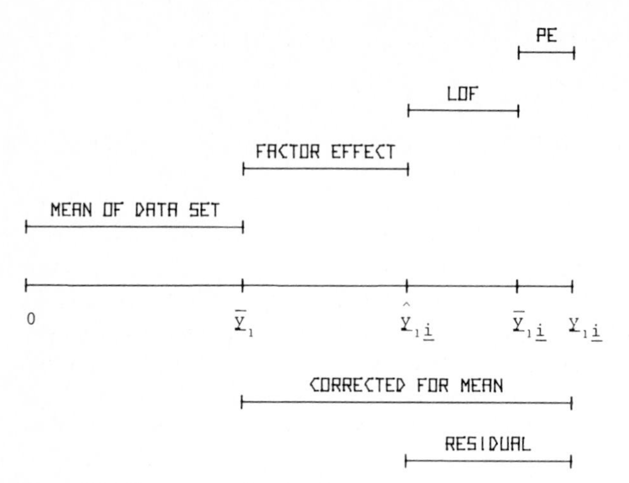

Figure 9.1. Discrete steps involved in "traveling" from the origin to a given response, y_{1i}.

response, this distance should be small.

(4) Finally, from the mean of replicate responses ($\bar{y}_{1i}$) to the response itself (y_{1i}). This distance is a measure of the purely experimental uncertainty. If the measurement of response is precise, this distance should be small.

Alternative itineraries can be planned with fewer stops along the way (see Figure 9.1). Two useful combinations are

(5) From the mean of the data set ($\bar{y}_1$) to the response itself (y_{1i}). This is a distance that has been "corrected" for the mean.

(6) From the value predicted by the model ($\hat{y}_{1i}$) to the response itself. This distance corresponds to the already familiar "residual".

In this section we will develop matrix representations of these distances, show simple matrix calculations for associated sums of squares, and demonstrate that certain of these sums of squares are additive.

Total sum of squares

The individual responses in a data set are conveniently collected in a *matrix of measured responses*, $\boldsymbol{Y}$

$$\boldsymbol{Y} = \begin{bmatrix} y_{11} \\ y_{12} \\ y_{13} \\ \vdots \\ y_{1n} \end{bmatrix} \tag{9.1}$$

The *total sum of squares*, SS_T, is defined as the sum of squares of the measured responses. It may be calculated easily using matrix techniques.

$$SS_T = \boldsymbol{Y}'\boldsymbol{Y} = \sum_{i=1}^{n} y_{1i}^2 \tag{9.2}$$

The total sum of squares has n degrees of freedom associated with it, where n is the *total number of experiments in a set.*

Sum of squares due to the mean

Probably the most interesting and useful aspect of a data set is its ability to reveal how variations in the factor levels result in variations among the responses. The exact values of the responses are not as important as the *variations* among them (see Section 1.4). When this is the case, a β_0 term is usually provided in the model so that the model is not forced to go through the origin, but instead can be offset up or down the response axis by some amount. It is possible to offset the raw data in a similar way by subtracting the mean value of response from each of the individual responses (see Figures 9.1 and 3.3). For this and other purposes, it will be convenient to define a *matrix of mean response*, $\bar{\boldsymbol{Y}}$, of the same form as the response matrix $\boldsymbol{Y}$, but containing for each element the mean response $\bar{y}_1$ (see Section 1.3).

$$\bar{\boldsymbol{Y}} = \begin{bmatrix} \bar{y}_1 \\ \bar{y}_1 \\ \bar{y}_1 \\ \vdots \\ \bar{y}_1 \end{bmatrix} \tag{9.3}$$

This matrix may be used to calculate a useful sum of squares, the *sum of squares due to the mean*, SS_{mean}.

$$SS_{\text{mean}} = \bar{\boldsymbol{Y}}'\bar{\boldsymbol{Y}} = \sum_{i=1}^{n} \bar{y}_1^2 \tag{9.4}$$

The sum of squares due to the mean has one degree of freedom associated with it.

Sum of squares corrected for the mean

The mean response can be subtracted from each of the individual responses to produce the so-called "responses corrected for the mean". This terminology is perhaps unfortunate because it wrongly implies that the original data was somehow "incorrect"; "responses *adjusted* for the mean" might be a better description, but we will use the traditional terminology here. It will be convenient to define a *matrix of responses corrected for the mean*, $\boldsymbol{C}$.

$$\boldsymbol{C} = \boldsymbol{Y} - \bar{\boldsymbol{Y}} = \begin{bmatrix} y_{11} - \bar{y}_1 \\ y_{12} - \bar{y}_1 \\ y_{13} - \bar{y}_1 \\ \vdots \\ y_{1n} - \bar{y}_1 \end{bmatrix} \tag{9.5}$$

This matrix may be used to calculate another useful sum of squares, the *sum of squares corrected for the mean*, SS_{corr}, sometimes called the *sum of squares about the mean* or the *corrected sum of squares*.

$$SS_{corr} = \boldsymbol{C}'\boldsymbol{C} = (\boldsymbol{Y} - \bar{\boldsymbol{Y}})'(\boldsymbol{Y} - \bar{\boldsymbol{Y}}) = \sum_{i=1}^{n} (y_{1i} - \bar{y}_1)^2 \tag{9.6}$$

The sum of squares corrected for the mean has $n - 1$ degrees of freedom associated with it.

It is a characteristic of linear models and least squares parameter estimation that certain sums of squares are additive. One useful relationship is based on the partitioning

$$y_{1i} = \bar{y}_1 + (y_{1i} - \bar{y}_1) \tag{9.7}$$

as shown in Figure 9.1. It is to be emphasized that just because certain quantities are additive (as they are in Equation 9.7), this does not mean that their sums of squares are necessarily additive; it must be shown that a resulting crossproduct is equal to zero before the additivity of sums of squares is proved. As an example, the total sum of squares may be written

$$SS_T = \sum_{i=1}^{n} y_{1i}^2 = \sum_{i=1}^{n} [\bar{y}_1 + (y_{1i} - \bar{y}_1)]^2 \tag{9.8}$$

$$SS_T = \sum_{i=1}^{n} \left[\bar{y}_1^2 + 2\bar{y}_1(y_{1i} - \bar{y}_1) + (y_{1i} - \bar{y}_1)^2\right] \tag{9.9}$$

$$SS_T = \sum_{i=1}^{n} \bar{y}_1^2 + \sum_{i=1}^{n} (y_{1i} - \bar{y}_1)^2 + 2\bar{y}_1 \sum_{i=1}^{n} (y_{1i} - \bar{y}_1) \tag{9.10}$$

By Equations 9.4 and 9.6,

$$SS_T = SS_{mean} + SS_{corr} + 2\bar{y}_1 \sum_{i=1}^{n} (y_{1i} - \bar{y}_1) \tag{9.11}$$

The summation in Equation 9.11 can be shown to be zero. Thus,

$$SS_T = SS_{mean} + SS_{corr} \tag{9.12}$$

The degrees of freedom are partitioned in the same way.

$$DF_T = DF_{mean} + DF_{corr} = 1 + (n - 1) = n \tag{9.13}$$

Although the partitioning of the total sum of squares into a sum of squares due to the mean and a sum of squares corrected for the mean may be carried out for any data set, it is meaningful only for the treatment of models containing a β_0 term. In effect, the β_0 term provides the degree of freedom necessary for offsetting the responses so the mean of the "corrected" responses can be equal to zero.

Models that lack a β_0 term not only force the factors to explain the variation of the responses about the mean, but also require that the factors explain the offset of the mean as well (see Figure 5.9); inclusion of a β_0 term removes this latter requirement.

Sum of squares due to factors

Using matrix least squares techniques (see Section 5.2), the chosen linear model may be fit to the data to obtain a set of parameter estimates, $\hat{\boldsymbol{B}}$, from which predicted values of response, $\hat{y}_{1i}$, may be obtained. It is convenient to define a *matrix of estimated responses*, $\hat{\boldsymbol{Y}}$.

$$\hat{\boldsymbol{Y}} = \boldsymbol{X}\hat{\boldsymbol{B}} = \begin{bmatrix} \hat{y}_{11} \\ \hat{y}_{12} \\ \hat{y}_{13} \\ \vdots \\ \hat{y}_{1n} \end{bmatrix} \tag{9.14}$$

Some of the variation of the responses about their mean is caused by variation of the factors. The effect of the factors as they appear in the model can be measured by the differences between the predicted responses ($\hat{y}_{1i}$) and the mean response ($\bar{y}_1$). For this purpose, it is convenient to define a *matrix of factor contributions*, $\boldsymbol{F}$.

$$\boldsymbol{F} = \hat{\boldsymbol{Y}} - \overline{\boldsymbol{Y}} = \boldsymbol{X}\hat{\boldsymbol{B}} - \overline{\boldsymbol{Y}} = \begin{bmatrix} \hat{y}_{11} - \bar{y}_1 \\ \hat{y}_{12} - \bar{y}_1 \\ \hat{y}_{13} - \bar{y}_1 \\ \vdots \\ \hat{y}_{1n} - \bar{y}_1 \end{bmatrix} \tag{9.15}$$

This matrix may be used to calculate still another useful sum of squares, the *sum of squares due to the factors as they appear in the model*, SS_{fact}, sometimes called the *sum of squares due to regression*.

$$SS_{\text{fact}} = \boldsymbol{F}'\boldsymbol{F} = (\hat{\boldsymbol{Y}} - \overline{\boldsymbol{Y}})'(\hat{\boldsymbol{Y}} - \overline{\boldsymbol{Y}}) = \sum_{i=1}^{n} (\hat{y}_{1i} - \bar{y}_1)^2 \tag{9.16}$$

For models containing a β_0 term, the sum of squares due to the factors has $p - 1$ degrees of freedom associated with it, where p is the *number of parameters in the model*. For models that do not contain a β_0 term, SS_{fact} has p degrees of freedom.

Sum of squares of residuals

We have already defined the *matrix of residuals*, $\boldsymbol{R}$, in Section 5.2. It may be obtained using matrix techniques as

$$\boldsymbol{R} = \boldsymbol{Y} - \hat{\boldsymbol{Y}} = \boldsymbol{Y} - \boldsymbol{X}\hat{\boldsymbol{B}} = \begin{bmatrix} y_{11} - \hat{y}_{11} \\ y_{12} - \hat{y}_{12} \\ y_{13} - \hat{y}_{13} \\ \vdots \\ y_{1n} - \hat{y}_{1n} \end{bmatrix} \tag{9.17}$$

This matrix may be used to calculate the *sum of squares of residuals*, SS_r, sometimes called the *sum of squares about regression*.

$$SS_r = \boldsymbol{R}'\boldsymbol{R} = (\boldsymbol{Y} - \hat{\boldsymbol{Y}})'(\boldsymbol{Y} - \hat{\boldsymbol{Y}}) = \sum_{i=1}^{n} (y_{1i} - \hat{y}_{1i})^2 \tag{9.18}$$

The sum of squares of residuals has $n - p$ degrees of freedom associated with it.

The sum of squares corrected for the mean, SS_{corr}, is equal to the sum of squares due to the factors, SS_{fact}, plus the sum of squares of residuals, SS_r. This result can be obtained from the partitioning

$$(y_{1i} - \bar{y}_1) = (\hat{y}_{1i} - \bar{y}_1) + (y_{1i} - \hat{y}_{1i}) \tag{9.19}$$

(see Figure 9.1). The sum of squares corrected for the mean may be written

$$SS_{corr} = \sum_{i=1}^{n} (y_{1i} - \bar{y}_1)^2 = \sum_{i=1}^{n} [(\hat{y}_{1i} - \bar{y}_1) + (y_{1i} - \hat{y}_{1i})]^2 \tag{9.20}$$

$$SS_{corr} = \sum_{i=1}^{n} \left[(\hat{y}_{1i} - \bar{y}_1)^2 + 2(\hat{y}_{1i} - \bar{y}_1)(y_{1i} - \hat{y}_{1i}) + (y_{1i} - \hat{y}_{1i})^2\right] \tag{9.21}$$

$$SS_{corr} = \sum_{i=1}^{n} (\hat{y}_{1i} - \bar{y}_1)^2 + \sum_{i=1}^{n} (y_{1i} - \hat{y}_{1i})^2 + 2\sum_{i=1}^{n} (\hat{y}_{1i} - \bar{y}_1)(y_{1i} - \hat{y}_{1i}) \tag{9.22}$$

By Equations 9.16 and 9.18,

$$SS_{corr} = SS_{fact} + SS_r + 2\sum_{i=1}^{n} (\hat{y}_{1i} - \bar{y}_1)(y_{1i} - \hat{y}_{1i}) \tag{9.23}$$

It can be shown that the rightmost summation in Equation 9.23 is equal to zero. Thus,

$$SS_{corr} = SS_{fact} + SS_r \tag{9.24}$$

The degrees of freedom are partitioned in the same way.

$$DF_{corr} = DF_{fact} + DF_r = (p - 1) + (n - p) = (n - 1) \tag{9.25}$$

for models containing a β_0 term. If a β_0 term is not included in the model, then the sum of squares due to the mean and the sum of squares corrected for the mean are not permitted, in which case

$$SS_T = SS_{fact} + SS_r \tag{9.26}$$

and

$$DF_T = DF_{fact} + DF_r = p + (n - p) = n \tag{9.27}$$

Sum of squares due to lack of fit

Before discussing the sum of squares due to lack of fit and, later, the sum of squares due to purely experimental uncertainty, it is computationally useful to define a *matrix of mean replicate responses*, $\boldsymbol{J}$, which is structured the same as the $\boldsymbol{Y}$ matrix, *but contains mean values of response from replicates*. For those experiments that were not replicated, the "mean" response is simply the single value of response. The $\boldsymbol{J}$ matrix is of the form

$$\boldsymbol{J} = \begin{bmatrix} \bar{y}_{11} \\ \bar{y}_{12} \\ \bar{y}_{13} \\ \vdots \\ \bar{y}_{1n} \end{bmatrix} \tag{9.28}$$

As an example, suppose that in a set of five experiments, the first and second are replicates, and the third and fifth are replicates. If the matrix of measured responses is

$$\boldsymbol{Y} = \begin{bmatrix} 5 \\ 8 \\ 7 \\ 3 \\ 9 \end{bmatrix} \tag{9.29}$$

then the matrix of mean replicate responses is

$$\boldsymbol{J} = \begin{bmatrix} 6.5 \\ 6.5 \\ 8 \\ 3 \\ 8 \end{bmatrix} \tag{9.30}$$

Note that in the $\boldsymbol{J}$ matrix, the first two elements are the mean of replicate responses one and two; the third and fifth elements are the mean of replicate responses three and five. The fourth elements in the $\boldsymbol{Y}$ and $\boldsymbol{J}$ matrices are the same because the experiment was not replicated.

In a sense, calculating the mean replicate response removes the effect of purely experimental uncertainty from the data. It is not unreasonable, then, to expect that the deviation of these mean replicate responses from the estimated responses is due to a lack of fit of the model to the data. The matrix of lack-of-fit deviations, $\boldsymbol{L}$, is obtained by subtracting $\hat{\boldsymbol{Y}}$ from $\boldsymbol{J}$

$$\boldsymbol{L} = \boldsymbol{J} - \hat{\boldsymbol{Y}} = \boldsymbol{J} - \boldsymbol{X}\hat{\boldsymbol{B}} = \begin{bmatrix} \bar{y}_{11} - \hat{y}_{11} \\ \bar{y}_{12} - \hat{y}_{12} \\ \bar{y}_{13} - \hat{y}_{13} \\ \vdots \\ \bar{y}_{1n} - \hat{y}_{1n} \end{bmatrix} \tag{9.31}$$

This matrix may be used to calculate the *sum of squares due to lack of fit*, SS_{lof}.

$$SS_{\text{lof}} = \boldsymbol{L}'\boldsymbol{L} = (\boldsymbol{J} - \hat{\boldsymbol{Y}})'(\boldsymbol{J} - \hat{\boldsymbol{Y}}) = \sum_{i=1}^{n} (\bar{y}_{1i} - \hat{y}_{1i})^2 \tag{9.32}$$

If f is the number of distinctly different factor combinations at which experiments have been carried out, then the sum of squares due to lack of fit has $f-p$ degrees of freedom associated with it.

Sum of squares due to purely experimental uncertainty

The *matrix of purely experimental deviations*, $\boldsymbol{P}$, is obtained by subtracting $\boldsymbol{J}$ from $\boldsymbol{Y}$.

$$\boldsymbol{P} = \boldsymbol{Y} - \boldsymbol{J} = \begin{bmatrix} y_{11} - \bar{y}_{11} \\ y_{12} - \bar{y}_{12} \\ y_{13} - \bar{y}_{13} \\ \vdots \\ y_{1n} - \bar{y}_{1n} \end{bmatrix} \tag{9.33}$$

Zeros will appear in the $\boldsymbol{P}$ matrix for those experiments that were not replicated ($\bar{y}_{1i} = y_{1i}$ for these experiments). The *sum of squares due to purely experimental uncertainty* is easily calculated.

$$SS_{pe} = \boldsymbol{P}'\boldsymbol{P} = (\boldsymbol{Y} - \boldsymbol{J})'(\boldsymbol{Y} - \boldsymbol{J}) = \sum_{i=1}^{n} (y_{1i} - \bar{y}_{1i})^2 \tag{9.34}$$

The sum of squares due to purely experimental uncertainty has $n-f$ degrees of freedom associated with it.

The sum of squares due to lack of fit, SS_{lof}, and the sum of squares due to purely experimental uncertainty, SS_{pe}, add together to give the sum of squares of residuals, SS_r.

$$SS_r = SS_{lof} + SS_{pe} \tag{9.35}$$

The degrees of freedom are similarly additive.

$$DF_r = DF_{lof} + DF_{pe} = (f-p) + (n-f) = (n-p) \tag{9.36}$$

Table 9.1 summarizes the matrix operations used to calculate the sums of squares discussed in this section for models that contain a β_0 term.

TABLE 9.1

Summary of matrix operations used to calculate sums of squares.

Sum of squares	Matrix operation	Degrees of freedom
SS_T	$\boldsymbol{Y}'\boldsymbol{Y}$	n
SS_{mean}	$\bar{\boldsymbol{Y}}'\bar{\boldsymbol{Y}}$	1
SS_{corr}	$\boldsymbol{C}'\boldsymbol{C} = (\boldsymbol{Y} - \bar{\boldsymbol{Y}})'(\boldsymbol{Y} - \bar{\boldsymbol{Y}})$	$n-1$
SS_{fact}	$\boldsymbol{F}'\boldsymbol{F} = (\hat{\boldsymbol{Y}} - \bar{\boldsymbol{Y}})'(\hat{\boldsymbol{Y}} - \bar{\boldsymbol{Y}}) = (\boldsymbol{X}\hat{\boldsymbol{B}} - \bar{\boldsymbol{Y}})'(\boldsymbol{X}\hat{\boldsymbol{B}} - \bar{\boldsymbol{Y}})$	$p-1$
SS_r	$\boldsymbol{R}'\boldsymbol{R} = (\boldsymbol{Y} - \hat{\boldsymbol{Y}})'(\boldsymbol{Y} - \hat{\boldsymbol{Y}}) = (\boldsymbol{Y} - \boldsymbol{X}\hat{\boldsymbol{B}})'(\boldsymbol{Y} - \boldsymbol{X}\hat{\boldsymbol{B}})$	$n-p$
SS_{lof}	$\boldsymbol{L}'\boldsymbol{L} = (\boldsymbol{J} - \hat{\boldsymbol{Y}})'(\boldsymbol{J} - \hat{\boldsymbol{Y}}) = (\boldsymbol{J} - \boldsymbol{X}\hat{\boldsymbol{B}})'(\boldsymbol{J} - \boldsymbol{X}\hat{\boldsymbol{B}})$	$f-p$
SS_{pe}	$\boldsymbol{P}'\boldsymbol{P} = (\boldsymbol{Y} - \boldsymbol{J})'(\boldsymbol{Y} - \boldsymbol{J})$	$n-f$

9.2 Additivity of sums of squares and degrees of freedom

Figure 9.2 summarizes the additivity of sums of squares and degrees of freedom for models containing a β_0 term. For models that do not include a β_0 term, the partitioning is shown in Figure 9.3.

The information contained in Figures 9.2 and 9.3 is commonly presented in an analysis of variance (ANOVA) table similar to that shown in Table 9.2 for models containing a β_0 term; Table 9.3 is an ANOVA table for models that do not contain a β_0 term. The column farthest to the right in each of these tables contains the values that are obtained by dividing a sum of squares by its associated degrees of freedom. These values are called "mean squares". Statistically, the *mean squares are estimates of variances*. Although ANOVA tables are often used to present least squares results for linear models, the diagrammatic tree structure of Figures 9.2 and

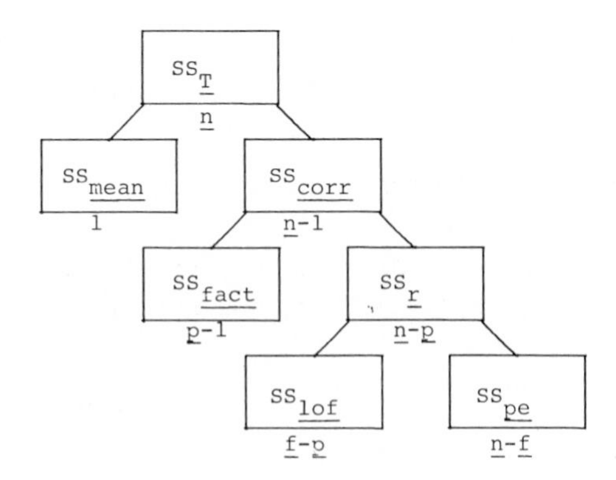

Figure 9.2. Sums of squares and degrees of freedom tree illustrating additive relationships for linear models that contain a β_0 term.

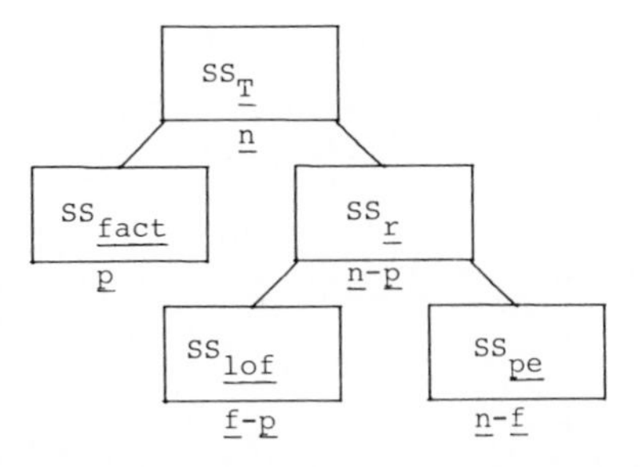

Figure 9.3. Sums of squares and degrees of freedom tree illustrating additive relationships for linear models that do not contain a β_0 term.

TABLE 9.2
ANOVA table for linear models containing a β_0 term.

Source	Degrees of freedom	Sum of squares	Mean square
Total	n	SS_T	SS_T/n
Mean	1	SS_{mean}	$SS_{mean}/1$
Corrected	$n-1$	SS_{corr}	$SS_{corr}/(n-1)$
Factor effects	$p-1$	SS_{fact}	$SS_{fact}/(p-1)$
Residuals	$n-p$	SS_r	$SS_r/(n-p)$
Lack of fit	$f-p$	SS_{lof}	$SS_{lof}/(f-p)$
Purely experimental uncertainty	$n-f$	SS_{pe}	$SS_{pe}/(n-f)$

TABLE 9.3
ANOVA table for linear models lacking a β_0 term.

Source	Degrees of freedom	Sum of squares	Mean square
Total	n	SS_T	SS_T/n
Factor effects	p	SS_{fact}	SS_{fact}/p
Residuals	$n-p$	SS_r	$SS_r/(n-p)$
Lack of fit	$f-p$	SS_{lof}	$SS_{lof}/(f-p)$
Purely experimental uncertainty	$n-f$	SS_{pe}	$SS_{pe}/(n-f)$

9.3 better illustrates several important statistical concepts. We will continue to use these diagrams throughout the remainder of the text.

9.3 Coefficients of determination and correlation

Figure 9.4 emphasizes the relationship among three sums of squares in the ANOVA tree – the sum of squares due to the factors as they appear in the model, SS_{fact}; the sum of squares of residuals, SS_r; and the sum of squares corrected for the mean, SS_{corr} (or the total sum of squares, SS_T, if there is no β_0 term in the model).

If the factors have very little effect on the response, we would expect that the sum

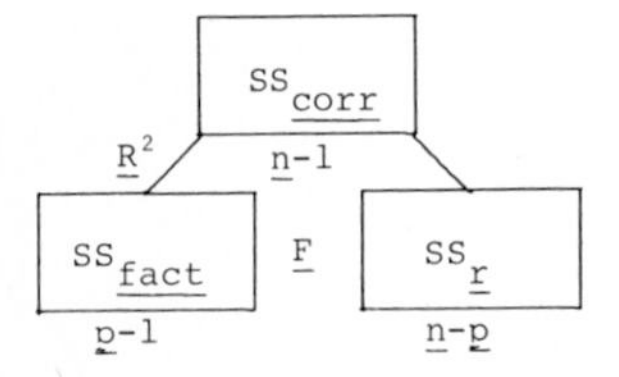

Figure 9.4. Relationships among SS_{fact}, SS_r, and SS_{corr} for calculating both the coefficient of multiple determination, R^2, and the variance-ratio for the significance of the factor effects, $F_{(p-1,n-p)}$.

of squares removed from SS_{corr} by SS_{fact} would be small, and therefore SS_r would be large – about the same size as SS_{corr}. Conversely, if the factors do a very good job of explaining the responses, we would expect the residuals to be very small and SS_{fact} to be relatively large – about the same size as SS_{corr}.

The *coefficient of multiple determination*, R^2, is a measure of how much of SS_{corr} is accounted for by the factor effects.

$$R^2 = SS_{fact}/SS_{corr} \tag{9.37}$$

The coefficient of multiple determination ranges from 0 (indicating that the factors, as they appear in the model, have no effect upon the response) to 1 (indicating that the factors, as they appear in the model, "explain" the data "perfectly"). The square root of the coefficient of multiple determination is the *coefficient of multiple correlation*, R.

If the model is the two-parameter (β_0 and β_1) single-factor straight-line relationship $y_{1i} = \beta_0 + \beta_1 x_{1i} + r_{1i}$, then R^2 is given the symbol r^2 and is called the *coefficient of determination*. It is defined the same way R^2 is defined.

$$r^2 = SS_{fact}/SS_{corr} \tag{9.38}$$

Again, r^2 ranges from 0 to 1. The square root of the coefficient of determination is the *coefficient of correlation*, r, often called the *correlation coefficient*.

$$r = \mathrm{SGN}(b_1)\ \mathrm{SQR}(r^2) \tag{9.39}$$

where $\mathrm{SGN}(b_1)$ is the sign of the slope of the straight-line relationship. Although r ranges from -1 to $+1$, only the absolute value is indicative of how much the factor explains the data; the sign of r simply indicates the sign of b_1.

It is important to realize that R might give a false sense of how well the factors explain the data. For example, the R value of 0.956 arises because the factors explain 91.4% of the sum of squares corrected for the mean. An R value of 0.60 indicates that only 36% of SS_{corr} has been explained by the factors.

Although the coefficients of determination and the correlation coefficients are conceptually simple and attractive, and are frequently used as a measure of how well a model fits a set of data, they are not, by themselves, a good measure of the effectiveness of the factors as they appear in the model, primarily because they do not take into account the degrees of freedom. Thus, the value of R^2 can usually be increased by adding another parameter to the model (until $p = f$), but this increased R^2 value does not necessarily mean that the expanded model offers a *significantly* better fit. It should also be noted that *the coefficient of determination gives no indication of whether the lack of perfect prediction is caused by an inadequate model or by purely experimental uncertainty*.

9.4 Statistical test for the effectiveness of the factors

A statistically valid measure of the effectiveness of the factors in fitting a model to the data is given by the Fisher variance ratio

$$F_{(p-1,n-p)} = s^2_{\text{fact}}/s^2_{\text{r}} = [SS_{\text{fact}}/(p-1)]/[SS_{\text{r}}/(n-p)] \tag{9.40}$$

for linear models containing a β_0 term. (For models that do not contain a β_0 term, the ratio is still $s^2_{\text{fact}}/s^2_{\text{r}}$, but there are p degrees of freedom in the numerator, not $p - 1$.) The F-test for the significance of the factor effects is usually called the *test for the significance of the regression*.

Although it is beyond the scope of this presentation, it can be shown that s^2_{fact}, s^2_{r}, and s^2_{corr} would all be expected to have the same value if the factors had no effect upon the response. However, if the factors do have an effect upon the response, then s^2_{fact} will become larger than s^2_{r}.

The statistical test for the effectiveness of the factors asks the question, "Has a significant amount of variance in the data set been accounted for by the factors as they appear in the model"? Another way of asking this is to question whether or not one or more of the parameters associated with the factor effects is significant. For models containing a β_0 term, the null hypothesis to be tested is

$$\text{H}_0: \quad \beta_1 = \ldots = \beta_{p-1} = 0 \tag{9.41}$$

with an alternative hypothesis that *one* (*or more*) of the parameters is not equal to zero. We will designate

$$y_{1i} = \beta_0 + \beta_1(x) + \ldots + \beta_{p-1}(x) + r_{1i} \tag{9.42}$$

as the *expanded model* and

$$y_{1i} = \beta_0 + r_{1i} \tag{9.43}$$

as the *reduced model*. If one or more of the parameters in the set (β_1, ..., β_{p-1}) is significant, then the variance of residuals for the expanded model should be significantly less than the variance of residuals for the reduced model. The difference, s^2_{fact}, must be due to one or more of the factor effects (i.e., the β_1, ..., β_{p-1} in the expanded model). The more significant the factor effects, the larger s^2_{fact} will be with respect to s^2_{r}.

9.5 Statistical test for the lack of fit

Figure 9.5 emphasizes the relationships among three other sums of squares in the ANOVA tree – the sum of squares due to lack of fit, SS_{lof}; the sum of squares due to purely experimental uncertainty, SS_{pe}; and the sum of squares of residuals, SS_{r}. Two of the resulting variances, s^2_{lof} and s^2_{pe}, were used in Section 6.5 where a statistical test was developed for estimating the significance of the lack of fit of a model to a set of data. The null hypothesis

$$\text{H}_0: \quad s^2_{\text{lof}} - s^2_{\text{pe}} = 0 \tag{9.44}$$

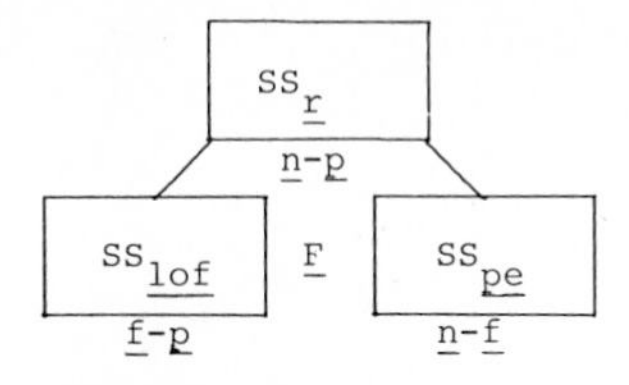

Figure 9.5. Relationship between SS_{lof} and SS_{pe} for calculating the variance-ratio for the significance of the lack of fit, $F_{(f-p,n-f)}$.

is tested with the Fisher variance ratio

$$F_{(f-p,n-f)} = s_{lof}^2/s_{pe}^2 = [SS_{lof}/(f-p)]/[SS_{pe}/(n-f)] \tag{9.45}$$

The value of F calculated in this manner can be tested against a tabular critical value of F_{crit} at a given level of confidence (see Section 6.5), or the level of confidence at which the calculated value of F_{calc} is critical may be determined (see Section 6.6).

If the null hypothesis of Equation 9.44 is disproved (or is seriously questioned), the conclusion is that there is still a significant amount of the variation in the measured responses that is not explained by the model. That is, there is a significant lack of fit between the model and the data.

We emphasize that if the lack of fit of a model is to be tested, $f-p$ (the degrees of freedom associated with SS_{lof}) and $n-p$ (the degrees of freedom associated with SS_{pe}) must each be greater than zero; that is, the number of factor combinations must be greater than the number of parameters in the model, and there should be replication to provide an estimate of the variance due to purely experimental uncertainty.

9.6 Statistical test for a set of parameters

Let us now consider the more general case in which the reduced model contains more parameters than the single parameter β_0. We assume that we have an expanded model

$$y_{1i} = \beta_0 + \beta_1(x) + \ldots + \beta_{g-1}(x) + \beta_g(x) + \ldots + \beta_{p-1}(x) + r_{1i} \tag{9.46}$$

containing p parameters, and wish to test the significance of the *set* of parameters β_g through β_{p-1}. The reduced model is then

$$y_{1i} = \beta_0 + \beta_1(x) + \ldots + \beta_{g-1}(x) + r_{1i} \tag{9.47}$$

and contains g parameters. An appropriate partitioning of the sums of squares for these models is shown in Figure 9.6 and Table 9.4, where SS_{rr} is the sum of squares of residuals for the reduced model, and SS_{re} is the sum of squares of residuals for the expanded model. The difference in these sums of squares (SS_{exp}) must be due to

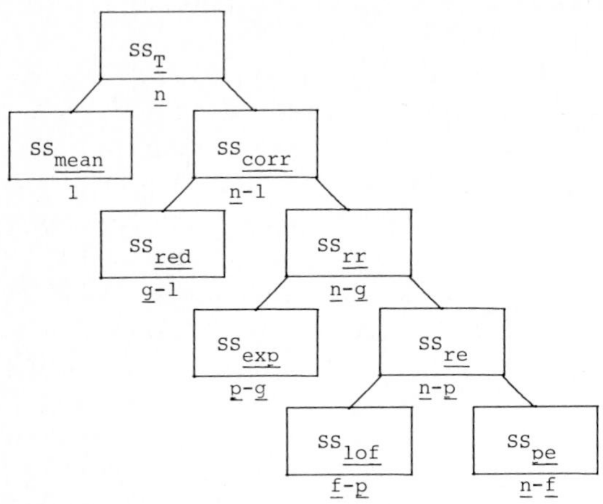

Figure 9.6. Sums of squares and degrees of freedom tree illustrating additive relationships for testing a subset of the parameters in a model.

the presence of the additional $p - g$ parameters in the expanded model. (In Figure 9.6, the notation SS_{red} refers to the sum of squares due to the factor effects in the reduced model only; SS_{exp} refers to the sum of squares due to the *additional* factor effects in the expanded model.) An appropriate test of the hypothesis that

$$H_0: \quad \beta_g = ... = \beta_{p-1} = 0 \tag{9.48}$$

is

$$F_{(p-g,n-p)} = s^2_{exp}/s^2_{re} = \left[SS_{exp}/(p-g)\right]/\left[SS_{re}/(n-p)\right] \tag{9.49}$$

If $F_{calc} > F_{crit}$, then the null hypothesis is disproved at the given level of confidence, and one or more of the parameters β_g, ..., β_{p-1} offers a significant reduction in the variance of the data. Alternatively, the level of confidence (or risk) at which one or more of the parameter estimates is significantly different from zero may be

TABLE 9.4
ANOVA table for testing the significance of a set of parameters.

Source	Degrees of freedom	Sum of squares	Mean square
Total	n	SS_T	SS_T/n
Mean	1	SS_{mean}	$SS_{mean}/1$
Corrected	$n-1$	SS_{corr}	$SS_{corr}/(n-1)$
Factors, reduced model	$g-1$	SS_{red}	$SS_{red}/(g-1)$
Residuals, reduced model	$n-g$	SS_{rr}	$SS_{rr}/(n-g)$
Factors, *additional*	$p-g$	SS_{exp}	$SS_{exp}/(p-g)$
Residuals	$n-p$	SS_{re}	$SS_{re}/(n-p)$
Lack of fit	$f-p$	SS_{lof}	$SS_{lof}/(f-p)$
Purely experimental uncertainty	$n-f$	SS_{pe}	$SS_{pe}/(n-f)$

calculated (see Section 6.6). Equation 9.40 is seen to be a special case of Equation 9.49 for which $g = 1$.

Why bother with an F-test for the significance of a *set* of parameters? Why not simply use a single-parameter test to determine the significance of each parameter estimate individually? The answer is that the risk of falsely rejecting the null hypothesis that $\beta_j = 0$ for *at least* one of the parameters is no longer α: the probability of falsely rejecting at least one of the null hypotheses when all k null hypotheses are true is $[1 - (1 - \alpha)^k]$. If there are two parameters in the model, the overall risk becomes 0.0975 (if $\alpha = 0.05$); for three parameters, the risk is 0.1426; and so on. The use of the F-test developed in this section allows the simultaneous testing of a set of parameters with a specified risk of falsely rejecting the overall null hypothesis when it is true.

It should be pointed out that if the null hypothesis is disproved, it offers no indication of which parameter(s), either individually or jointly, are significantly different from zero. In addition, if the null hypothesis cannot be rejected, it does not mean that the parameter estimates in question are *insignificant*; it means only that they are *not significant* at the given level of probability (see Chapter 6). Again, determining the level of confidence at which F is significant is useful.

9.7 Statistical significance and practical significance

The F-tests for the effectiveness of the factors and for the lack of fit sometimes give seemingly conflicting results: with some sets of data, it will happen that each of the F-tests will be highly significant. The question then arises, "How can a model exhibit both highly significant factor effects and a highly significant lack of fit"?

Such a situation will often arise if the model does indeed fit the data well, and if the measurement process is highly precise. Recall that the F-test for lack of fit compares the variance due to lack of fit with the variance due to purely experimental uncertainty. The reference point of this comparison is the precision with which measurements can be made. Thus, although the lack of fit might be so small as to be of no practical importance, the F-test for lack of fit will show that it is statistically significant if the estimated variance due to purely experimental uncertainty is relatively very small.

It is important in this case to keep in mind the distinction between "statistical significance" and "practical significance". If, *in a practical sense*, the residuals are small enough to be considered acceptable for the particular application, it is not necessary to test for lack of fit.

References

Arkin, H. and Colton, R.R. (1970). *Statistical Methods*, 5th ed. Barnes and Noble, New York.

Box, G.E.P., Hunter, W.G., and Hunter, J.S. (1978). *Statistics for Experimenters. An Introduction to Design, Data Analysis, and Model Building*. Wiley, New York.

Cochran, W.G., and Cox, G.M. (1957). *Experimental Designs*, 2nd. ed. Wiley, New York.
Daniel, C., and Wood, F.S. (1971). *Fitting Equations to Data*. Wiley, New York.
Davies, O.L., Ed. (1956). *The Design and Analysis of Industrial Experiments*. Hafner, New York.
Draper, N.R., and Smith, H. (1966). *Applied Regression Analysis*. Wiley, New York.
Fisher, R.A. (1966). *The Design of Experiments*, 8th ed. Hafner, New York.
Fisher, R.A. (1970). *Statistical Methods for Research Workers*, 14th ed. Hafner, New York.
Fisher, R.A., and Yates, F. (1963). *Statistical Tables for Biological, Agricultural and Medical Research*. Hafner, New York.
Himmelblau, D.M. (1970). *Process Analysis by Statistical Methods*. Wiley, New York.
Kempthorne, O. (1952). *The Design and Analysis of Experiments*. Wiley, New York.
Mandel, J. (1964). *The Statistical Analysis of Experimental Data*. Wiley, New York.
Mendenhall, W. (1968). *Introduction to Linear Models and the Design and Analysis of Experiments*. Duxbury Press, Belmont, California.
Meyer, S.L. (1975). *Data Analysis for Scientists and Engineers*. Wiley, New York.
Moore, D.S. (1979). *Statistics. Concepts and Controversies*. Freeman, San Francisco, California.
Natrella, M.G. (1963). *Experimental Statistics* (Nat. Bur. of Stand. Handbook 91). US Govt. Printing Office, Washington, DC.
Neter, J., and Wasserman, W. (1974). *Applied Linear Statistical Models. Regression, Analysis of Variance, and Experimental Design*. Irwin, Homewood, Illinois.
Scheffé, H. (1953). *Analysis of Variance*. Wiley, New York.
Wilson, E.B., Jr. (1952). *An Introduction to Scientific Research*. McGraw-Hill, New York.
Youden, W.J. (1951). *Statistical Methods for Chemists*. Wiley, New York.

Exercises

9.1 Total sum of squares

Calculate the total sum of squares, SS_T, for the nine responses in Section 3.1 (see Equation 9.2). How many degrees of freedom are associated with this sum of squares?

9.2 Sum of squares due to the mean

Calculate the sum of squares due to the mean, SS_{mean}, for the nine responses in Section 3.1 (see Equation 9.4). How many degrees of freedom are associated with this sum of squares?

9.3 Sum of squares corrected for the mean

Use the $\boldsymbol{C}$ matrix of Equation 9.5 to calculate the sum of squares corrected for the mean, SS_{corr}, for the nine responses in Section 3.1 (see Equation 9.6). How many degrees of freedom are associated with this sum of squares?

9.4 Variance

Calculation of SS_T is allowed if there is a β_0 term in the model. A sum of squares divided by its associated degrees of freedom gives an estimated variance. Comment on the model underlying the calculations in Sections 3.1 and 3.3. Comment on the fact that the numerical value of Equation 3.5 is equal to SS_{corr}/DF_{corr} from Exercise 9.3.

9.5 Additivity of sums of squares and degrees of freedom

Use the results of Exercises 9.1-9.3 to illustrate that $SS_T = SS_{mean} + SS_{corr}$ and that $DF_T = DF_{mean} + DF_{corr}$.

9.6 Sum of squares due to factors

Fit the model $y_{1i} = \beta_0 + \beta_1 x_{1i} + r_{1i}$ to the following data:

$$\boldsymbol{D} = \begin{bmatrix} 1 \\ 3 \\ 2 \\ 2 \end{bmatrix} \qquad \boldsymbol{Y} = \begin{bmatrix} 2 \\ 10 \\ 6 \\ 8 \end{bmatrix}$$

Calculate SS_T, SS_{mean}, and SS_{corr}. Calculate $\hat{\boldsymbol{Y}}$ (Equation 9.14), $\boldsymbol{F}$ (Equation 9.15), and SS_{fact} (Equation 9.16). How many degrees of freedom are associated with SS_{fact}?

9.7 Sum of squares of residuals

Use Equations 9.17 and 9.18 to calculate the sum of squares of residuals, SS_r, for the data and model of Exercise 9.6. How many degrees of freedom are associated with this sum of squares? Use numerical values to show that $SS_{corr} = SS_{fact} + SS_r$ and that $DF_{corr} = DF_{fact} + DF_r$.

9.8 Matrix of mean replicate responses and sum of squares due to lack of fit

Calculate the $\boldsymbol{J}$ matrix for Exercise 9.6. Calculate the corresponding $\boldsymbol{L}$ matrix and SS_{lof}. How many degrees of freedom are associated with this sum of squares?

9.9 Sum of squares due to purely experimental uncertainty

Calculate the sum of squares due to purely experimental uncertainty, SS_{pe}, for the model and data of Exercise 9.6. How many degrees of freedom are associated with SS_{pe}? Use numerical values to show that $SS_r = SS_{lof} + SS_{pe}$ and that $DF_r = DF_{lof} + DF_{pe}$.

9.10 Sums of squares and degrees of freedom tree

Use the data in Exercises 9.6-9.9 to construct a sums of squares and degrees of freedom tree (Figure 9.2).

9.11 Matrix operations for sums of squares

Note the additivities that exist between each odd-numbered line (except the last) and the two lines below it in Table 9.1. What are the relationships between Table 9.1 and Figure 9.1?

9.12 Coefficients of determination and correlation

Calculate the coefficient of determination, r^2, and the coefficient of correlation, r, for the model and data of Exercise 9.6. What is the difference between the coefficient of determination and the coefficient of multiple determination?

9.13 Correlation coefficient

What information is contained in $r = -0.70$? What information is contained in $r = +1.36$?

9.14 Significance of the regression

Calculate the Fisher F-ratio for the significance of the factor effects (Equation 9.40) for the model and data of Exercise 9.6. At approximately what level of confidence is the factor x_1 significant?

9.15 Significance of the regression

In Section 9.4 it is stated that the statistical test for the effectiveness of the factors asks the question, "Has a significant amount of variance in the data set been accounted for by the factors *as they appear in the model*"? Why is it necessary to qualify this with the phrase, "...as they appear in the model"?

9.16 Significance of the lack of fit

Calculate the Fisher F-ratio for the significance of the lack of fit (Equation 9.45) for the model and data of Exercise 9.6. Is the lack of fit very significant?

9.17 Degrees of freedom

The following data were obtained by an experimenter who believed a straight line relationship would be a good model. How many degrees of freedom are there for lack of fit?

x_{1i}	y_{1i}
3.5	2.8
7.6	5.9
3.5	3.1
3.5	2.6
7.6	5.5
3.5	2.7
7.6	6.3

9.18 Degrees of freedom

The following data were obtained by an experimenter who believed a straight line relationship would be a good model. How many degrees of freedom are there for lack of fit? Can the experimenter test the significance of the lack of fit?

x_{1i}	y_{1i}
2.6	3.5
2.7	3.5
2.8	3.5
3.1	3.5
5.5	7.6
5.9	7.6
6.3	7.6

9.19 Degrees of freedom

An experimenter wanted to fit the model $y_{1i} = \beta_0 + \beta_1 x_{1i} + \beta_{11} x_{1i}^2 + r_{1i}$ to the data of Exercise 9.17. Comment.

9.20 Hypothesis testing

Why is Equation 9.49 an appropriate test of the hypothesis H_0: $\beta_g = \ldots = \beta_{p-1} = 0$?

9.21 Statistical test for a set of parameter

Section 9.6 shows how to test the significance of a set of parameters in a model. This "set" could contain just one parameter. Is it possible to fit a large, multi-parameter model, and then test each parameter in turn, eliminating those parameters that do not have a highly significant effect, until a concise, "best" model is obtained? Is it possible to start with a small, single-parameter model, and then add parameters, testing each in turn, keeping only those that do have a highly significant effect, until a "best" model is obtained? Would the order of elimination or addition of parameters make a difference? [See, for example, Draper, N.R., and Smith, H. (1966). *Applied Regression Analysis*, Chapter 6. Wiley, New York.]

9.22 Hypothesis testing

Suppose a food manufacturer uses the following procedure to modify popular recipes. When first developed, the recipe is a gourmet delight! Before long, however, someone asks the question, "I wonder if it would taste *different* if we used only half as many eggs"? The recipe is prepared with only half as many eggs, and is presented with the original product to a panel of tasters. The null hypothesis is usually H_0: taste of original recipe − taste of modified recipe = 0. The alternative hypothesis is that the difference in taste is significant, and the level of confidence is usually set at about 95%. Most of the time the null hypothesis cannot be disproved, so the modified recipe is used from then on.

Soon, a second question is asked: "I wonder if the modified recipe would taste different if we used only half as much butter"? And a third question, "I wonder if the modified-modified recipe would taste different if we used artificial flavoring"? And so on. The result is a recipe that "tastes like cardboard".

If the "cardboard" recipe were tested against the original recipe, do you think the difference in taste would be significant at a high level of probability? What does all this say about testing parameters one at a time, rather than as a group?

Is, in fact, the correct question being asked? [See Section 1.2, Table 1.1, and Figure 6.3.]

9.23 Lack of fit

A good method of detecting lack of fit is to plot residuals as a function of the factor levels. The technique works best for single-factor systems, but often can be used effectively for multifactor systems. If a straight line model were fit to the data in Figure 2.3, what would the plot of *residuals* vs. the level of factor x_1 look like? What would this suggest about the adequacy of the model? Would the pattern of residuals suggest a form for an additional term in the model? What terms might be suggested if the residuals were distributed in an "N"-type pattern? An "M"-type pattern?

9.24 Statistical significance and practical significance

Look up an account of the discovery of the planet Neptune. What can be said about the effectiveness of the factors in the models that described the motions of the planets other than Neptune? What can be said about the lack of fit of these models to the available astronomical data? How was the lack of fit accounted for?

9.25 *Experiment vs. observation*

Are the data in Exercise 9.24 the results of experimentation or observation (see Section 1.2)?

9.26 *Effectiveness of factors and lack of fit*

Science is often described as an iterative process that involves an interplay between hypothesis and experimentation. [See, for example, Box, G.E.P. (1976). "Science and Statistics," *J. Am. Statist. Assoc.*, 71, 791.] How are the concepts of "goodness of fit" and "lack of fit" used in this iterative process?

9.27 *Experimental design*

A system is thought to be described by the model $y_{1i} = \beta_0 + \beta_1 x_{1i} + r_{1i}$; in fact, it can be better described by the model $y_{1i} = \beta_0 + \beta_1 x_{1i} + \beta_{11} x_{1i}^2 + r_{1i}$. Can you design a set of experiments that would tend to show that the factor effects in the first model are highly significant, and that the model exhibits very little lack of fit? Can you design a set of experiments that would tend to show that the factor effects in the first model are not very significant, and that the model exhibits a highly significant lack of fit? Can experimental design be used (or misused) to strengthen or otherwise influence the conclusions drawn from experimental data?

9.28 *Significance of regression*

It has been suggested that it is "unfair" to judge the effectiveness of the factors (i.e., the significance of the regression) on the basis of $s_{\text{fact}}^2/s_{\text{r}}^2$: one of the components of s_{r}^2 is s_{pe}^2, and the factor effects should not be asked (or expected) to account for imprecision. An alternative *F*-test might be $s_{\text{fact}}^2/s_{\text{lof}}^2$. Comment.

9.29 *ANOVA table*

Write an ANOVA table for the model $y_{1i} = \beta_1 x_{1i} + r_{1i}$ fit to the following data:

x_{1i}	y_{1i}
3.0	2.6
3.0	3.5
3.0	2.9
6.0	5.9
6.0	6.3
6.0	6.1
9.0	9.1
9.0	9.3
9.0	9.5

9.30 Mean squares

A "mean square residuals" is equal to 1.395. If the model contained β_0 and five additional parameters, and if the model was fit to data from twelve experiments, what is the variance of residuals? The sum of squares of residuals?

9.31 Sums of squares

The calculation of the mean and standard deviation in Chapter 3 can be viewed in terms of the linear model $y_{1i} = \beta_0 + r_{1i}$. Prepare a sums of squares and degrees of freedom tree (Figure 9.2) for this model and the nine data points of Section 3.1. How do you interpret the fact that $SS_{corr} = SS_r = SS_{pe}$? Why are DF_{fact} and DF_{lof} equal to zero?

CHAPTER 10

A Ten-Experiment Example

Enzymes are large biological molecules that act as catalysts for chemical reactions in living organisms. As catalysts, they allow chemical reactions to proceed at a rapid rate, a rate much faster than would occur if the enzymes were not present. The action of an enzyme is often represented as

$$A \xrightarrow{E} B \tag{10.1}$$

where A represents one or more "reactants" and B represents one or more "products" of the chemical reaction. The enzyme E is assumed to take part in the reaction, but to be unchanged by it. Alternative expressions for enzyme catalyzed reactions are

$$A + E \rightarrow B + E \tag{10.2}$$

and

$$\begin{matrix} A \rightarrow & \rightarrow B \\ E \rightarrow & \rightarrow E \end{matrix} \tag{10.3}$$

which emphasize the "recyclable" nature of the enzyme.

A systems view of enzyme catalyzed reactions is given in Figure 10.1. In this figure, an additional input to the system is shown – the pH, or "negative logarithm base ten of the hydrogen ion concentration". If the pH of a chemical system is low (e.g., pH = 1), the concentration of hydrogen ions is large and the system is said to be "acidic". If the pH of a chemical system is high (e.g., pH = 13), the concentration of hydrogen ions is small and the system is said to be "basic".

The reasons for the effect of pH upon the catalytic properties of enzymes are numerous and will not be discussed here. For most enzymes, however, there is a pH at which they are optimally effective: changing the pH to lower (more acidic) levels or to higher (more basic) levels will decrease the overall rate at which the associated chemical reaction occurs. In the region of the optimum pH, the reaction rate vs. pH response surface can usually be approximated reasonably well by a second-order, parabolic relationship.

Let us assume that we are primarily interested in estimating the optimum pH for a particular enzyme catalyzed reaction. We might also be interested in determining if a parabolic relationship provides a significantly better fit than is provided by a straight line relationship over the domain of pH studied. If it does provide a better

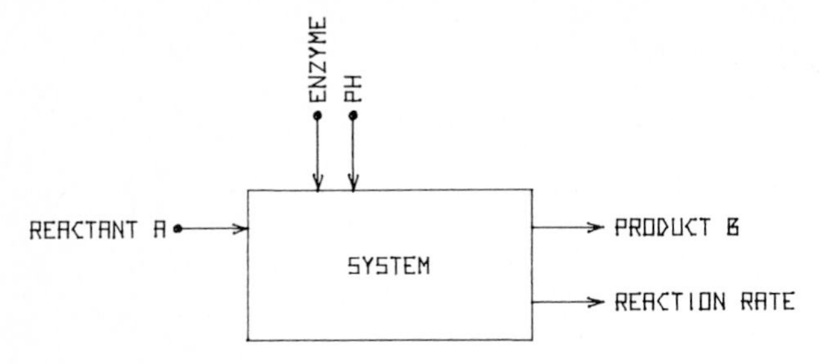

Figure 10.1. General system theory view of an enzyme catalyzed reaction.

fit, then we might ask if the parabolic relationship is an adequate model or if it exhibits a highly significant lack of fit.

We assume a limited budget that allows resources for only ten experiments. How can these experiments be distributed among factor combinations and replicates? At what levels of pH should the experiments be carried out? How can the results be interpreted?

10.1 Allocation of degrees of freedom

A set of n measured responses has a total of n degrees of freedom. Of these, $n-f$ degrees of freedom are given to the estimation of variance due to purely experimental uncertainty (s_{pe}^2), $f-p$ degrees of freedom are used to estimate the variance due to lack of fit (s_{lof}^2), and p degrees of freedom are used to estimate the parameters of the model (see Table 9.2).

In the present example, the total degrees of freedom is ten. The parabolic model that is to be fit,

$$\text{rate}_i = \beta_0 + \beta_1 \text{pH}_1 + \beta_{11} \text{pH}_1^2 + r_i \tag{10.4}$$

or

$$y_{1i} = \beta_0 + \beta_1 x_{1i} + \beta_{11} x_{1i}^2 + r_{1i} \tag{10.5}$$

contains three parameters, β_0, β_1, and β_{11}. The number of degrees of freedom associated with the residuals is ten minus three ($n-p$), or seven. Thus, the number of degrees of freedom allocated to lack of fit and purely experimental uncertainty must together total seven.

What information could be obtained if all ten experiments were carried out at only three levels of pH? Three levels of x_1 ($f=3$) provide sufficient factor combinations for being able to fit a three-parameter model, but leave no degrees of freedom for estimating lack of fit: $f-p=3-3=0$. Because one of our objectives was to determine if a parabolic relationship provides an adequate model for the observed rate, we must be able to estimate the variance due to lack of fit of the model; the number of factor combinations (levels of x_1 in this example) must therefore be greater than three.

TABLE 10.1
Possible allocation of seven residual degrees of freedom.

Number of factor combinations	Degrees of freedom allocated to	
	s^2_{lof}	s^2_{pe}
4	1	6
5	2	5
6	3	4
7	4	3
8	5	2
9	6	1

What would be the consequences of carrying out experiments at ten factor combinations ($f = 10$)? There would certainly be an adequate number of degrees of freedom available for estimating the variance due to lack of fit of the model: $f - p = 10 - 3 = 7$. However, the test of significance for the lack of fit is based upon an F-ratio, the denominator of which is the estimated variance due to purely experimental uncertainty. Unfortunately, ten factor combinations leave no degrees of freedom for the estimation of s^2_{pe} ($n - f = 10 - 10 = 0$) and the significance of the lack of fit could not be tested. Clearly, then, the number of factor combinations must be less than ten. This restriction and the previous one for lack of fit require that the number of factor combinations (f) be greater than or equal to four and less than or equal to nine: $4 \leqslant f \leqslant 9$. Thus, the number of degrees of freedom given to lack of fit can range from one to six $[1 \leqslant (f - 3) \leqslant 6]$ and the number of degrees of freedom given to purely experimental uncertainty can range from six to one $[6 \geqslant (10 - f) \geqslant 1]$.

Let us consider some of the possible allocations of degrees of freedom shown in Table 10.1, keeping in mind that the confidence of an estimated variance improves as the number of degrees of freedom associated with that estimate is increased. The effect of s^2_{pe} on the values of the V matrix was shown in Equation 7.1. Thus, we might allocate most of the available degrees of freedom to the estimation of s^2_{pe} to make it a more precise estimator of σ^2_{pe} and decrease the uncertainties in the parameter estimates. The price to be paid, of course, is that s^2_{lof} will be a less precise estimator of σ^2_{lof}. On the basis of this reasoning, let us allocate five degrees of freedom to the estimation of σ^2_{pe} and two degrees of freedom to the estimation of σ^2_{lof}. The number of factor combinations will therefore be five.

10.2 Placement of experiments

There remain two questions to be answered concerning the design of the experiments: what five levels of pH should be chosen, and how should the replicate experiments be allocated among these five levels?

If the factor combinations are chosen too close together, the variances and covariances of the parameter estimates will be large (see Sections 7.2–7.4). Further, it might happen that the chosen levels of pH do not enclose the optimal pH and the extrapolated location of the optimum might be very imprecise.

If the factor combinations are chosen far apart, the variances and covariances of the parameter estimates will be smaller, and the probability of bracketing the optimal pH will be greater. However, the assumed second-order model might not be as good an approximation to the true response surface over such a large domain of the factor as it would be over a smaller domain.

In this as in all other problems of experimental design, prior information is helpful. For example, if the enzyme we are dealing with is naturally found in a neutral environment, then it would probably be most active at a neutral pH, somewhere near pH = 7. If it were found in an acidic environment, say in the stomach, it would be expected to exhibit its optimal activity at a low (acidic) pH. When information such as this is available, it is appropriate to center the experimental design about the "best guess" of where the desired region might be. In the absence of prior information, factor combinations might be centered about the midpoint of the factor domain.

Let us assume that the enzyme of interest is naturally found in a basic environment for which the pH ≈ 10; the midpoint of the factor combinations will therefore be taken to be 10.

From the known pH dependence of the activity of other enzymes, a domain of approximately 4 pH units should be sufficient to bracket the optimum, yet not be so wide as to seriously invalidate a parabolic approximation to the true behavior of the system. The chosen treatment combinations will therefore be pH = 8, 9, 10, 11, and 12. We will code them as −2, −1, 0, +1, and +2 ($c_{x_1} = 10$, $d_{x_1} = 1$; see Section 8.5).

Given these five levels of pH, how can the five replicate experiments be allocated? One way is to place all of the replicates at the center factor level. Doing so would give a good estimate of σ^2_{pe} at the center of the experimental design, but it would give no information about the heteroscedasticity of the response surface. Also, in a coded data system such as this, the $(\boldsymbol{X}^{*\prime}\boldsymbol{X}^*)^{-1}$ matrix element associated with the variance of the parameter estimate b_1^* is no better with this design than for a five-experiment design with one experiment at each factor level. For the five-experiment design,

$$\boldsymbol{X}^* = \begin{bmatrix} 1 & -2 & +4 \\ 1 & -1 & +1 \\ 1 & 0 & 0 \\ 1 & +1 & +1 \\ 1 & +2 & +4 \end{bmatrix} \tag{10.6}$$

The corresponding $(\boldsymbol{X}^{*\prime}\boldsymbol{X}^*)$ matrix is

$$(\boldsymbol{X}^{*\prime}\boldsymbol{X}^*) = \begin{bmatrix} 5 & 0 & 10 \\ 0 & 10 & 0 \\ 10 & 0 & 34 \end{bmatrix} \tag{10.7}$$

and the inverse is

$$(\boldsymbol{X}^{*\prime}\boldsymbol{X}^{*})^{-1} = \begin{bmatrix} 0.486 & 0.000 & -0.143 \\ 0.000 & 0.100 & 0.000 \\ -0.143 & 0.000 & 0.071 \end{bmatrix} \tag{10.8}$$

For the ten-experiment design with six replicates at the center point,

$$\boldsymbol{X}^{*} = \begin{bmatrix} 1 & -2 & +4 \\ 1 & -1 & +1 \\ 1 & 0 & 0 \\ 1 & 0 & 0 \\ 1 & 0 & 0 \\ 1 & 0 & 0 \\ 1 & 0 & 0 \\ 1 & 0 & 0 \\ 1 & +1 & +1 \\ 1 & +2 & +4 \end{bmatrix} \tag{10.9}$$

$$(\boldsymbol{X}^{*\prime}\boldsymbol{X}^{*}) = \begin{bmatrix} 10 & 0 & 10 \\ 0 & 10 & 0 \\ 10 & 0 & 34 \end{bmatrix} \tag{10.10}$$

and the inverse is

$$(\boldsymbol{X}^{*\prime}\boldsymbol{X}^{*})^{-1} = \begin{bmatrix} 0.142 & 0.000 & -0.042 \\ 0.000 & 0.100 & 0.000 \\ -0.042 & 0.000 & 0.042 \end{bmatrix} \tag{10.11}$$

Let us try instead an experimental design in which two replicates are carried out at the center point and three replicates are carried out at each of the extreme points (-2 and $+2$). Then

$$\boldsymbol{X}^{*} = \begin{bmatrix} 1 & -2 & +4 \\ 1 & -2 & +4 \\ 1 & -2 & +4 \\ 1 & -1 & +1 \\ 1 & 0 & 0 \\ 1 & 0 & 0 \\ 1 & +1 & +1 \\ 1 & +2 & +4 \\ 1 & +2 & +4 \\ 1 & +2 & +4 \end{bmatrix} \tag{10.12}$$

The corresponding $(\boldsymbol{X}^{*\prime}\boldsymbol{X}^{*})$ matrix is

$$(\boldsymbol{X}^{*\prime}\boldsymbol{X}^{*}) = \begin{bmatrix} 10 & 0 & 26 \\ 0 & 26 & 0 \\ 26 & 0 & 98 \end{bmatrix} \tag{10.13}$$

and the inverse is

$$(X^{*\prime}X^{*})^{-1} = \begin{bmatrix} 0.322 & 0.000 & -0.086 \\ 0.000 & 0.038 & 0.000 \\ -0.086 & 0.000 & 0.033 \end{bmatrix} \quad (10.14)$$

The parameter estimates with this design will be more precise than they would be with either a five-experiment design (see Equation 10.8) or, with the exception of the estimate of β_0^*, a ten-experiment design with six replicates at the center point (see Equation 10.11). We will use as our design here, the allocation represented in Equation 10.12.

The corresponding coded *experimental design matrix*, $\boldsymbol{D}^*$, contains the coded factor combinations at which experiments are to be carried out. The coded experimental design matrix is usually not the same as the matrix of coded parameter coefficients, $\boldsymbol{X}^*$; the design matrix is determined by the chosen factor combinations only, while the matrix of parameter coefficients is determined also by the model to be fit. Each *row* of the experimental design matrix corresponds to a given experiment, and each *column* of the experimental design matrix indicates a particular factor. The experimental design matrix for this study of the effect of pH on reaction rate is

$$\boldsymbol{D}^* = \begin{bmatrix} -2 \\ -2 \\ -2 \\ -1 \\ 0 \\ 0 \\ +1 \\ +2 \\ +2 \\ +2 \end{bmatrix} \quad (10.15)$$

10.3 Results for the reduced model

Let us assume that we have carried out the indicated experiments and have obtained the following values for the measured reaction rates:

$$\boldsymbol{Y} = \begin{bmatrix} 33 \\ 50 \\ 37 \\ 53 \\ 89 \\ 87 \\ 69 \\ 69 \\ 67 \\ 80 \end{bmatrix} \quad (10.16)$$

We will first fit the reduced model

$$y_{1i} = \beta_0^* + \beta_1^* x_{1i}^* + r_{1i} \tag{10.17}$$

The matrix of parameter coefficients is then

$$X^* = \begin{bmatrix} 1 & -2 \\ 1 & -2 \\ 1 & -2 \\ 1 & -1 \\ 1 & 0 \\ 1 & 0 \\ 1 & +1 \\ 1 & +2 \\ 1 & +2 \\ 1 & +2 \end{bmatrix} \tag{10.18}$$

$$(X^{*\prime}X^*) = \begin{bmatrix} 10 & 0 \\ 0 & 26 \end{bmatrix} \tag{10.19}$$

$$(X^{*\prime}X^*)^{-1} = \begin{bmatrix} 1/10 & 0 \\ 0 & 1/26 \end{bmatrix} \tag{10.20}$$

$$(X^{*\prime}Y) = \begin{bmatrix} 634 \\ 208 \end{bmatrix} \tag{10.21}$$

$$\hat{B}^* = \begin{bmatrix} b_0^* \\ b_1^* \end{bmatrix} = \begin{bmatrix} 63.4 \\ 8.00 \end{bmatrix} \tag{10.22}$$

The data and the least squares straight line relationship are shown in Figure 10.2. It is to be remembered that the parameter estimates are those for the coded factor levels (see Section 10.2) and refer to the model

$$y_{1i} = 63.4 + 8.00(\text{pH}_i - 10.00) + r_{1i} \tag{10.23}$$

Thus, the parameter estimate b_0^* is the estimated response at pH = 10.00.

The overall mean response is

$$\bar{y}_1 = 63.4 \tag{10.24}$$

The matrix $\bar{Y}$ expressing the contribution of the overall mean response to each experiment is

$$\bar{Y} = \begin{bmatrix} 63.4 \\ 63.4 \\ 63.4 \\ 63.4 \\ 63.4 \\ 63.4 \\ 63.4 \\ 63.4 \\ 63.4 \\ 63.4 \end{bmatrix} \tag{10.25}$$

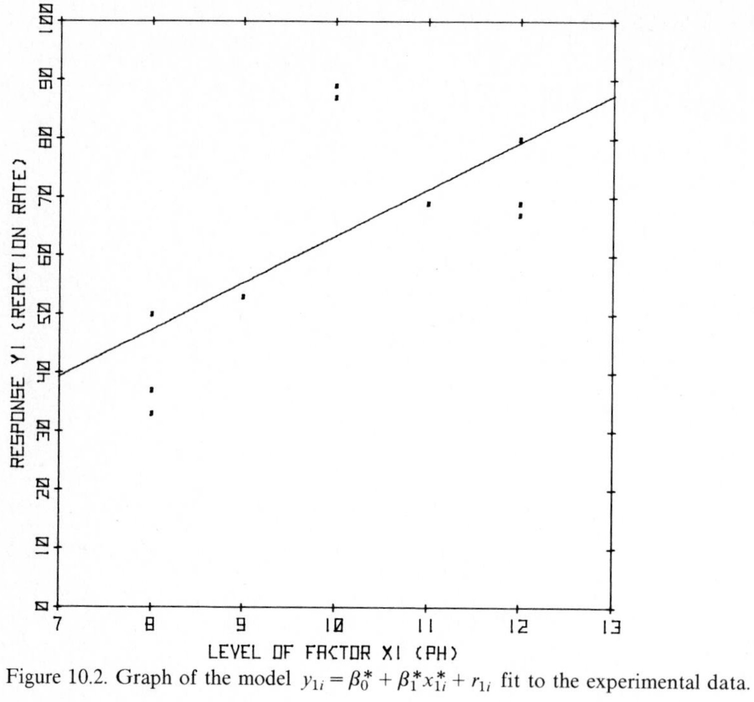

Figure 10.2. Graph of the model $y_{1i} = \beta_0^* + \beta_1^* x_{1i}^* + r_{1i}$ fit to the experimental data.

The matrix of predicted responses, $\hat{\boldsymbol{Y}}$, is given by

$$\hat{\boldsymbol{Y}} = \boldsymbol{X}^* \hat{\boldsymbol{B}}^* = \begin{bmatrix} 47.4 \\ 47.4 \\ 47.4 \\ 55.4 \\ 63.4 \\ 63.4 \\ 71.4 \\ 79.4 \\ 79.4 \\ 79.4 \end{bmatrix} \tag{10.26}$$

The matrix of residuals, $\boldsymbol{R}$, is given by

$$\boldsymbol{R} = \boldsymbol{Y} - \hat{\boldsymbol{Y}} = \begin{bmatrix} -14.4 \\ 2.6 \\ -10.4 \\ -2.4 \\ 25.6 \\ 23.6 \\ -2.4 \\ -10.4 \\ -12.4 \\ 0.6 \end{bmatrix} \tag{10.27}$$

The matrix of mean replicate responses, $\boldsymbol{J}$, is

$$\boldsymbol{J} = \begin{bmatrix} 40 \\ 40 \\ 40 \\ 53 \\ 88 \\ 88 \\ 69 \\ 72 \\ 72 \\ 72 \end{bmatrix} \tag{10.28}$$

Using these matrices, the following sums of squares are easily calculated (see Section 9.1).

$$SS_T = \boldsymbol{Y}'\boldsymbol{Y} = 43\,668 \tag{10.29}$$

$$SS_{\text{mean}} = \bar{\boldsymbol{Y}}'\bar{\boldsymbol{Y}} = 40\,195.6 \tag{10.30}$$

$$SS_{\text{corr}} = \boldsymbol{C}'\boldsymbol{C} = 3472.4 \tag{10.31}$$

$$SS_{\text{fact}} = \boldsymbol{F}'\boldsymbol{F} = 1664 \tag{10.32}$$

$$SS_r = \boldsymbol{R}'\boldsymbol{R} = 1808.4 \tag{10.33}$$

$$SS_{\text{lof}} = \boldsymbol{L}'\boldsymbol{L} = 1550.4 \tag{10.34}$$

$$SS_{\text{pe}} = \boldsymbol{P}'\boldsymbol{P} = 258 \tag{10.35}$$

The relationships of the sums of squares and degrees of freedom are given in Figure 10.3 and Table 10.2.

The coefficient of determination is

$$R^2 = (SS_{\text{fact}})/(SS_{\text{corr}}) = 1664/3472.4 = 0.479 \tag{10.36}$$

and indicates in this case that the term $\beta_1^* x_{1i}^*$ removes 47.9% of the corrected sum of

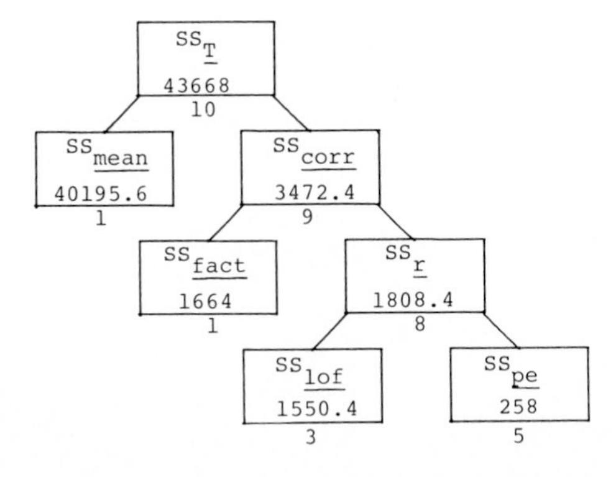

Figure 10.3. Sums of squares and degrees of freedom tree for Figure 10.2.

TABLE 10.2
ANOVA table for reduced model.

Source	Degrees of freedom	Sum of squares	Mean square
Total	10	43668.0	4366.8
Mean	1	40195.6	40195.6
Corrected	9	3472.4	385.82
Factor effects	1	1664.0	1664.0
Residuals	8	1808.4	226.05
Lack of fit	3	1550.4	516.8
Purely experimental uncertainty	5	258.0	51.6

squares; 52.1% remains as residuals. The significance of the coefficient of determination is contained in

$$F_{(p-1,n-p)} = F_{1,8} = (SS_{\text{fact}}/DF_{\text{fact}})/(SS_{\text{r}}/DF_{\text{r}}) = (1664/1)/(1808.4/8) = 7.36 \quad (10.37)$$

which is significant at the 97.3% level of confidence. The significance of the lack of fit is determined by

$$F_{(f-p,n-f)} = F_{3,5} = (SS_{\text{lof}}/DF_{\text{lof}})/(SS_{\text{pe}}/DF_{\text{pe}}) = (1550.4/3)/(258/5) = 10.02 \quad (10.38)$$

which is significant at the 98.5% level of confidence.

10.4 Results for the expanded model

Fitting the expanded (parabolic) model

$$y_{1i} = \beta_0^* + \beta_1^* x_{1i}^* + \beta_{11}^* x_{1i}^{*2} + r_{1i} \quad (10.39)$$

to the data in Equation 10.16 gives the parameter estimates

$$\hat{\boldsymbol{B}}^* = \begin{bmatrix} b_0^* \\ b_1^* \\ b_{11}^* \end{bmatrix} = \begin{bmatrix} 79.0 \\ 8.00 \\ -6.00 \end{bmatrix} \quad (10.40)$$

The estimated response surface and the data are shown in Figure 10.4. The allocation of the sums of squares and degrees of freedom are given in Figure 10.5 and Table 10.3.

The coefficient of multiple determination is

$$R^2 = 2758.4/3472.4 = 0.794 \quad (10.41)$$

and indicates in this case that the terms $\beta_1^* x_{1i}^*$ and $\beta_{11}^* x_{1i}^{*2}$ remove 79.4% of the corrected sum of squares; only 20.6% remains as residuals. The significance of the coefficient of multiple determination is contained in

$$F_{2,7} = (2758.4/2)/(714/7) = 13.52 \quad (10.42)$$

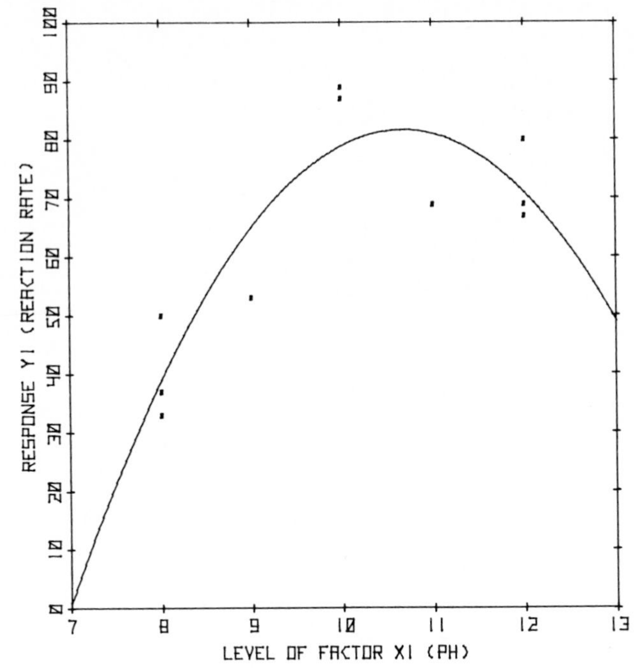

Figure 10.4. Graph of the model $y_{1i} = \beta_0^* + \beta_1^* x_{1i}^* + \beta_{11}^* x_{1i}^{*2} + r_{1i}$ fit to the experimental data.

which is significant at the 99.6% level of confidence. The significance of the lack of fit is determined by

$$F_{2,5} = (456/2)/(258/5) = 4.42 \tag{10.43}$$

which is significant at the 92.2% level of confidence.

The partitioning of sums of squares and degrees of freedom shown in Figure 10.6 and Table 10.4 allows us to determine the individual significance of the "set of

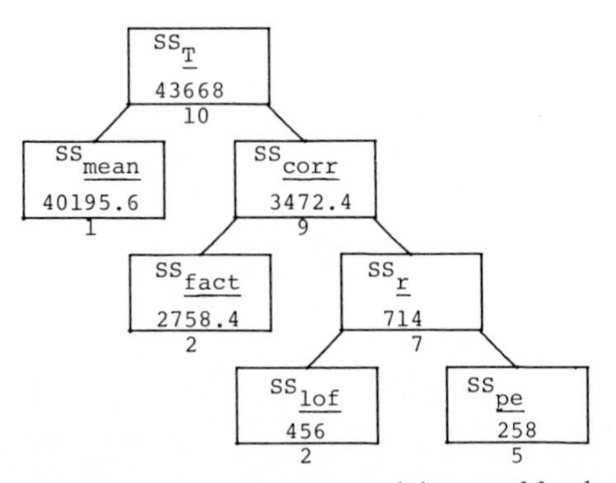

Figure 10.5. Sums of squares and degrees of freedom tree for Figure 10.4.

TABLE 10.3
ANOVA table for expanded model.

Source	Degrees of freedom	Sum of squares	Mean square
Total	10	43668.0	4366.8
Mean	1	40195.6	40195.6
Corrected	9	3472.4	385.82
Factor effects	2	2758.4	1379.2
Residuals	7	714.0	102.0
Lack of fit	2	456.0	228.0
Purely experimental uncertainty	5	258.0	51.6

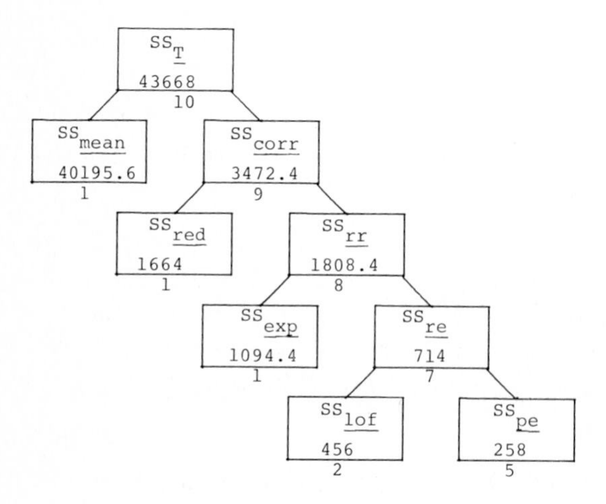

Figure 10.6. Sums of squares and degrees of freedom tree for estimating the significance of the additional parameter β_{11}^{*}.

TABLE 10.4
ANOVA table for significance of $\beta_{11}^{*} x_{1i}^{*2}$.

Source	Degrees of freedom	Sum of squares	Mean square
Total	10	43668.0	4366.8
Mean	1	40195.6	40195.6
Corrected	9	3472.4	385.82
Factors, reduced model	1	1664.0	1664.0
Residuals, reduced model	8	1808.4	226.05
Factors, *additional*	1	1094.4	1094.4
Residuals	7	714.0	102.0
Lack of fit	2	456.0	228.0
Purely experimental uncertainty	5	258.0	51.6

parameters" containing only β_{11}^* (that is, the significance of adding the single term $\beta_{11}^* x_{1i}^{*2}$).

$$F_{(p-g,n-p)} = F_{1,7} = (SS_{\text{exp}}/DF_{\text{exp}})/(SS_{\text{re}}/DF_{\text{re}}) = 10.73 \tag{10.44}$$

which is significant at the 98.6% level of confidence.

10.5 Coding transformations of parameter estimates

The parabolic relationship between reaction rate and uncoded pH is obtained by expansion of the coded model.

$$\text{rate}_i = 79.0 + 8.00(\text{pH}_i - 10.00) - 6.00(\text{pH}_i - 10.00)^2 \tag{10.45}$$

$$\text{rate}_i = 79.0 + 8.00\text{pH}_i - 80.0 - 6.00\text{pH}_i^2 + 120\text{pH}_i - 600 \tag{10.46}$$

$$\text{rate}_i = -601 + 128\text{pH}_i - 6.00\text{pH}_i^2 \tag{10.47}$$

(see Equation 10.4). The same transformation may be accomplished using general matrix techniques. It can be shown that

$$\hat{\boldsymbol{B}} = \boldsymbol{A}^{-1}\hat{\boldsymbol{B}}^* \tag{10.48}$$

where $\hat{\boldsymbol{B}}$ is the matrix of parameter estimates in the uncoded coordinate system, $\hat{\boldsymbol{B}}^*$ is the matrix of parameter estimates in the coded coordinate system, and the transformation matrix, $\boldsymbol{A}$, is obtained from the matrix of uncoded parameter coefficients, $\boldsymbol{X}$, and from the matrix of coded parameter coefficients, $\boldsymbol{X}^*$.

$$\boldsymbol{A} = (\boldsymbol{X}^{*\prime}\boldsymbol{X}^*)^{-1}(\boldsymbol{X}^{*\prime}\boldsymbol{X}) \tag{10.49}$$

In the present example, $(\boldsymbol{X}^{*\prime}\boldsymbol{X}^*)$ is given by Equation 10.14. Other arrays are

$$\boldsymbol{X}^{*\prime} = \begin{bmatrix} 1 & 1 & 1 & 1 & 1 & 1 & 1 & 1 & 1 & 1 \\ -2 & -2 & -2 & -1 & 0 & 0 & +1 & +2 & +2 & +2 \\ +4 & +4 & +4 & +1 & 0 & 0 & +1 & +4 & +4 & +4 \end{bmatrix} \tag{10.50}$$

$$\boldsymbol{X} = \begin{bmatrix} 1 & 8 & 64 \\ 1 & 8 & 64 \\ 1 & 8 & 64 \\ 1 & 9 & 81 \\ 1 & 10 & 100 \\ 1 & 10 & 100 \\ 1 & 11 & 121 \\ 1 & 12 & 144 \\ 1 & 12 & 144 \\ 1 & 12 & 144 \end{bmatrix} \tag{10.51}$$

$$(\boldsymbol{X}^{*\prime}\boldsymbol{X}) = \begin{bmatrix} 10 & 100 & 1026 \\ 0 & 26 & 520 \\ 26 & 260 & 2698 \end{bmatrix} \tag{10.52}$$

$$A = (X^{*\prime}X^*)^{-1}(X^{*\prime}X) = \begin{bmatrix} 49/152 & 0 & -13/152 \\ 0 & 1/26 & 0 \\ -13/152 & 0 & 5/152 \end{bmatrix} \begin{bmatrix} 10 & 100 & 1026 \\ 0 & 26 & 520 \\ 26 & 260 & 2698 \end{bmatrix}$$

$$= \begin{bmatrix} 1 & 10 & 100 \\ 0 & 1 & 20 \\ 0 & 0 & 1 \end{bmatrix} \tag{10.53}$$

$$A^{-1} = \begin{bmatrix} 1 & -10 & 100 \\ 0 & 1 & -20 \\ 0 & 0 & 1 \end{bmatrix} \tag{10.54}$$

Finally,

$$\hat{B} = \begin{bmatrix} b_0 \\ b_1 \\ b_{11} \end{bmatrix} = A^{-1}\hat{B}^* = \begin{bmatrix} -601 \\ 128 \\ -6 \end{bmatrix} \tag{10.55}$$

which may be compared with the coefficients obtained in Equation 10.47.

The location of the predicted optimum pH may be found by differentiating the fitted model with respect to pH. Setting the derivative of Equation 10.47 equal to zero gives the location of the stationary point (in this case, a maximum).

$$0 = 128 - 12\text{pH} \tag{10.56}$$

$$\text{pH}_{max} = 128/12 = 10\tfrac{2}{3} \tag{10.57}$$

Anticipating a later section on canonical analysis of second-order polynomial models, we will show that the first-order term can be made to equal zero if we code the model using the stationary point as the center of the symmetrical design. For this new system of coding, $c_{x_1} = 10\tfrac{2}{3}$ and $d_{x_1} = 1$ (see Section 8.5).

$$X^* = \begin{bmatrix} 1 & -8/3 & 64/9 \\ 1 & -8/3 & 64/9 \\ 1 & -8/3 & 64/9 \\ 1 & -5/3 & 25/9 \\ 1 & -2/3 & 4/9 \\ 1 & -2/3 & 4/9 \\ 1 & 1/3 & 1/9 \\ 1 & 4/3 & 16/9 \\ 1 & 4/3 & 16/9 \\ 1 & 4/3 & 16/9 \end{bmatrix} \tag{10.58}$$

$$(X^{*\prime}X^*) = \begin{bmatrix} 10 & -20/3 & 274/9 \\ -20/3 & 279/9 & -1484/27 \\ 274/9 & -1484/27 & 13\,714/81 \end{bmatrix} \tag{10.59}$$

$$(X^{*\prime}X^*)^{-1} = \begin{bmatrix} 0.2699 & -0.06890 & -0.07091 \\ -0.06890 & 0.09694 & 0.04386 \\ -0.07091 & 0.04386 & 0.03289 \end{bmatrix} \tag{10.60}$$

$$(X^{*\prime}X) = \begin{bmatrix} 10 & 100 & 1026 \\ -20/3 & -122/3 & -492/3 \\ 274/9 & 2428/9 & 22146/9 \end{bmatrix} \quad (10.61)$$

$$A = (X^{*\prime}X^{*})^{-1}(X^{*\prime}X) = \begin{bmatrix} 1 & 10.67 & 113.8 \\ 0 & 1 & 21.33 \\ 0 & 0 & 1 \end{bmatrix} \quad (10.62)$$

$$\hat{B}^{*} = \begin{bmatrix} b_0^{*} \\ b_1^{*} \\ b_{11}^{*} \end{bmatrix} = A\hat{B} = \begin{bmatrix} 81.67 \\ 0 \\ -6.00 \end{bmatrix} \quad (10.63)$$

Thus, in this new system of coding, $b_1^{*} = 0$; the first-order parameter is zero.

10.6 Confidence intervals for response surfaces

In Section 6.1, the concept of confidence intervals of parameter estimates was presented. In this section, we consider a general approach to the estimation of confidence intervals for parameter estimates and response surfaces based upon models that have been shown to be adequate (i.e., the lack of fit is not highly significant, either in a statistical or in a practical sense).

If a model does not show a serious lack of fit, then

$$s_r^2 = SS_r/(n-p) \quad (10.64)$$

is a valid estimate of σ_{pe}^2 (see Section 6.4) and the equation

$$V = s_r^2(X'X)^{-1} \quad (10.65)$$

is often used to calculate the variance-covariance matrix. It is important to realize that this is a valid estimate only if the model is adequate.

The variances of the parameter estimates can be used to set confidence intervals that would include the true value of the parameter a certain percentage of the time. In general, the confidence interval for a parameter β, based on s_r^2, is given by

$$b \pm \mathrm{SQR}\left(F_{(1,n-p)} \times s_b^2\right) \quad (10.66)$$

where β is the true value of the parameter, b is its estimated value, s_b^2 is the corresponding variance from the diagonal of the variance-covariance matrix, and $F_{(1,n-p)}$ is the tabular value of F at the desired level of confidence.

Although the derivation is beyond the scope of this presentation, it can be shown that the estimated variance of predicting a *single new value of response at a given point in factor space*, $s_{y_{10}}^2$, is equal to the purely experimental uncertainty variance, s_{pe}^2, plus the variance of estimating the mean response at that point, $s_{\hat{y}_{10}}^2$; that is,

$$s_{y_{10}}^2 = s_{pe}^2 + s_{\hat{y}_{10}}^2 \quad (10.67)$$

where the subscript "0" is used to indicate that the factor combination of interest does not necessarily correspond to one of the experiments that was previously carried out. Let a $1 \times f$ matrix $\boldsymbol{X}_0$ contain only one row; let $\boldsymbol{X}_0$ have columns that correspond to the columns of the $\boldsymbol{X}$ matrix (the matrix of parameter coefficients); and let the elements of $\boldsymbol{X}_0$ correspond to the factor combination of interest. For example, if we are interested in the point represented by pH = 8.5, then in the uncoded factor space given in Section 10.2, $x_{10} = -1.5$, and for the second-order model of Equation 10.39,

$$\boldsymbol{X}_0 = [1 \quad x_{10} \quad x_{10}^2] = [1 \quad -1.5 \quad 2.25] \tag{10.68}$$

For a given experimental design (such as that of Equation 10.15), the variance of predicting the mean response at a point in factor space is

$$s_{\hat{y}_{10}}^2 = s_{\text{pe}}^2 \left[\boldsymbol{X}_0 (\boldsymbol{X}'\boldsymbol{X})^{-1} \boldsymbol{X}_0' \right] \tag{10.69}$$

Thus,

$$s_{y_{10}}^2 = s_{\text{pe}}^2 + s_{\text{pe}}^2 \left[\boldsymbol{X}_0 (\boldsymbol{X}'\boldsymbol{X})^{-1} \boldsymbol{X}_0' \right] = s_{\text{pe}}^2 \left\{ 1 + \left[\boldsymbol{X}_0 (\boldsymbol{X}'\boldsymbol{X})^{-1} \boldsymbol{X}_0' \right] \right\} \tag{10.70}$$

If the model is adequate but still not perfectly correct, then the estimate $s_{y_{10}}^2$ based on s_{pe}^2 (Equation 10.70) will be too low because it does not take into account the lack of fit of the model. To partially compensate for the possibility of a slight lack of fit between the model and the data, it is customary to use s_{r}^2 to estimate $s_{y_{10}}^2$ in setting confidence intervals for response surfaces.

$$s_{y_{10}}^2 = s_{\text{r}}^2 \left\{ 1 + \left[\boldsymbol{X}_0 (\boldsymbol{X}'\boldsymbol{X})^{-1} \boldsymbol{X}_0' \right] \right\} \tag{10.71}$$

It is to be stressed that if the model is grossly incorrect, it is of little practical use to estimate confidence intervals for response surfaces.

For the example at pH = 8.5, by Equation 10.14 and Table 10.3,

$$s_{y_{10}}^2 = (102.0)(1.191) = 121.5 \tag{10.72}$$

The standard uncertainty in the single new value of response is

$$s_{y_{10}} = \text{SQR}\left(s_{y_{10}}^2 \right) = 11.02 \tag{10.73}$$

It is evident that Equation 10.71 can be used to plot the variance (or uncertainty) of predicting a single new value of response if $\boldsymbol{X}_0$ is made to vary across the domain of factor space. Such a plot of standard deviation of predicting a single new value of response as a function of pH is shown in Figure 10.7 for the experimental design of Equation 10.15, the data of Equation 10.16, and the second-order model of Equation 10.39.

It is possible to use $s_{y_{10}}^2$ to obtain confidence limits for predicting a single new value of response. The interval is given by

$$y_{10} = \hat{y}_{10} \pm \text{SQR}\left(F_{(1,n-p)} \times s_{y_{10}}^2 \right) \tag{10.74}$$

where F has one degree of freedom in the numerator and $n - p$ degrees of freedom in the denominator because $s_{y_{10}}^2$ is based upon s_{r}^2. Because

$$y_{10} = \boldsymbol{X}_0 \hat{\boldsymbol{B}} \tag{10.75}$$

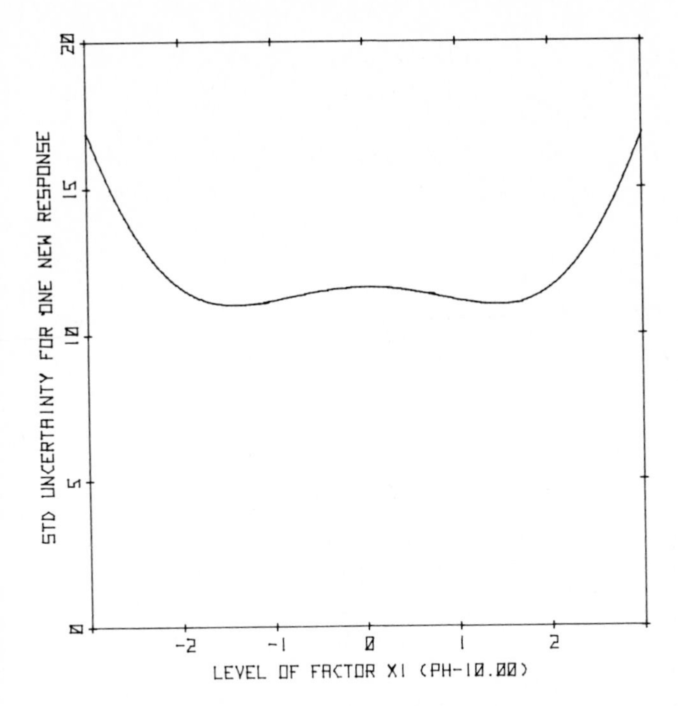

Figure 10.7. Standard uncertainty for estimating one new response as a function of the factor x_1^* for the model $y_{1i} = \beta_0^* + \beta_1^* x_{1i}^* + \beta_{11}^* x_{1i}^{*2} + r_{1i}$.

an equivalent expression for the confidence interval is obtained by substituting Equations 10.71 and 10.75 in Equation 10.74.

$$y_{10} = \boldsymbol{X}_0 \hat{\boldsymbol{B}} \pm \text{SQR}\left(F_{(1,n-p)} \times s_r^2 \left\{ 1 + \left[\boldsymbol{X}_0 (\boldsymbol{X}'\boldsymbol{X})^{-1} \boldsymbol{X}_0' \right] \right\} \right) \tag{10.76}$$

At pH = 8.5, the 95% confidence interval is

$$y_{10} = 53.5 \pm \text{SQR}(5.59 \times 121.5) = 53.5 \pm 26.06 \tag{10.77}$$

Equation 10.76 can be used to plot the confidence limits for predicting a single new value of response if $\boldsymbol{X}_0$ is made to vary across the domain of factor space. Figure 10.8 gives the 95% confidence limits for the data and model of Section 10.4.

A related confidence interval is used for estimating a *single mean of several new values of response at a given point in factor space*. It can be shown that the estimated variance of predicting the mean of m new values of response at a given point in factor space, $s_{\bar{y}_{10}}^2$, is

$$s_{\bar{y}_{10}}^2 = s_r^2 \left\{ (1/m) + \left[\boldsymbol{X}_0 (\boldsymbol{X}'\boldsymbol{X})^{-1} \boldsymbol{X}_0' \right] \right\} \tag{10.78}$$

A plot of the corresponding confidence limits can be obtained from

$$\bar{y}_{10} = \boldsymbol{X}_0 \hat{\boldsymbol{B}} \pm \text{SQR}\left(F_{(1,n-p)} \times s_r^2 \left\{ (1/m) + \left[\boldsymbol{X}_0 (\boldsymbol{X}'\boldsymbol{X})^{-1} \boldsymbol{X}_0' \right] \right\} \right) \tag{10.79}$$

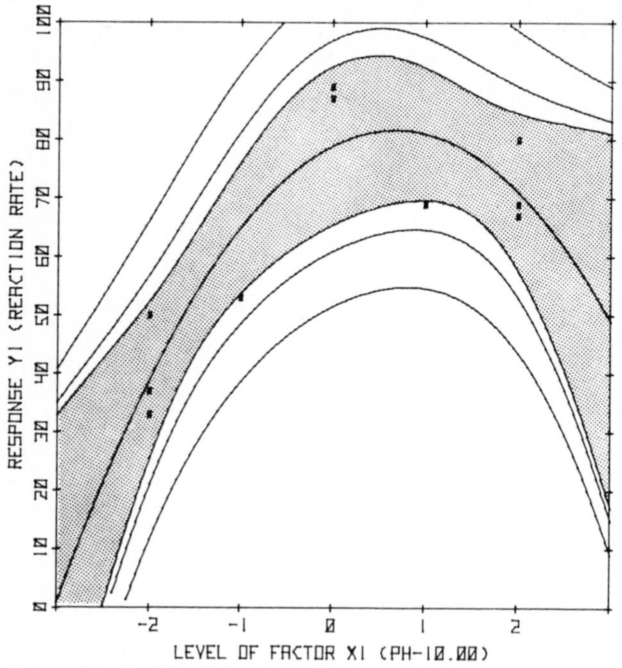

Figure 10.8. Confidence bands (95% level) for predicting a single new value of response (outer band), a single mean of four new values of response (middle band), and a single estimate of the true mean response (inner band).

Figure 10.8 gives the 95% confidence limits for predicting a single mean of four responses at any point in factor space for the data and model of Section 10.4.

If m is large, Equation 10.79 reduces to

$$\hat{y}_{10} = \boldsymbol{X}_0\hat{\boldsymbol{B}} \pm \mathrm{SQR}\left(F_{(1,n-p)} \times s_r^2\left[\boldsymbol{X}_0(\boldsymbol{X}'\boldsymbol{X})^{-1}\boldsymbol{X}_0'\right]\right) \quad (10.80)$$

which gives the confidence interval for a *single estimate of the true mean*, $\hat{y}_{10}$, *at a given point in factor space*. Figure 10.8 plots the 95% confidence limits for predicting the true mean at any single point in factor space for the data and model of Section 10.4.

Finally, we turn to an entirely different question involving confidence limits. Suppose we were to carry out the experiments indicated by the design matrix of Equation 10.15 a second time. We would probably not obtain the same set of responses we did the first time (Equation 10.16), but instead would have a different $\boldsymbol{Y}$ matrix. This would lead to a different set of parameter estimates, $\hat{\boldsymbol{B}}$, and a predicted response surface that in general would not be the same as that shown in Figure 10.4. A third repetition of the experiments would lead to a third predicted response surface, and so on. The question, then, is what limits can we construct about these response surfaces so that in a given percentage of cases, those limits will include the entire true response surface?

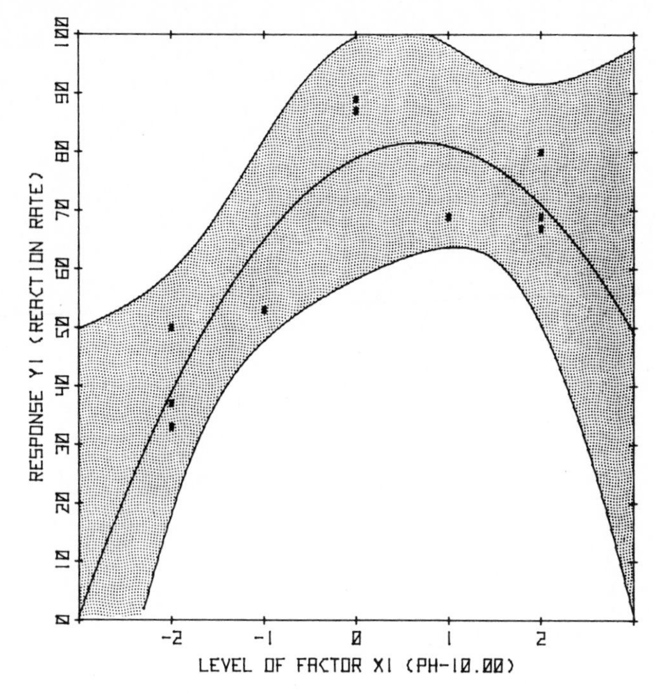

Figure 10.9. Confidence band (95% level) for predicting the entire true response surface.

The answer was provided by Working and Hotelling and is of the same form as Equation 10.80, but the statistic W^2 (named after Working) replaces F.

$$W^2 = p \times F_{(p,n-p)} \tag{10.81}$$

$$\hat{y}_{10} = \boldsymbol{X}_0\hat{\boldsymbol{B}} \pm \mathrm{SQR}\left(W^2 \times s_r^2\left[\boldsymbol{X}_0(\boldsymbol{X}'\boldsymbol{X})^{-1}\boldsymbol{X}_0'\right]\right) \tag{10.82}$$

The 95% confidence limits are plotted in Figure 10.9. The Working-Hotelling

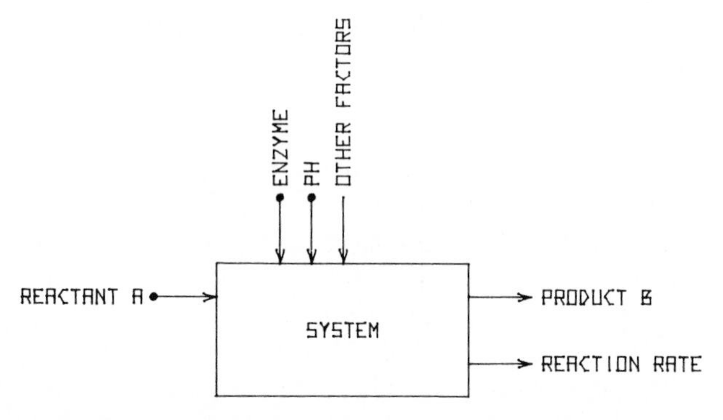

Figure 10.10. General system theory view emphasizing the effect of uncontrolled factors on the noise associated with responses.

confidence limits are used when it is necessary to estimate *true mean responses*, $\hat{y}_{10}$, *at several points in factor space from the same set of experimental data.*

Before leaving this example, we would point out that there is an excessive amount of purely experimental uncertainty in the system under study. The range of values obtained for replicate determinations is rather large, and suggests the existence of important factors that are not being controlled (see Figure 10.10). If steps are taken to bring the system under better experimental control, the parameters of the model can be estimated with better precision (see Equation 7.1).

An old statement is very true – statistical treatment is no substitute for good data.

References

Box, G.E.P., Hunter, W.G., and Hunter, J.S. (1978). *Statistics for Experimenters. An Introduction to Design, Data Analysis, and Model Building*. Wiley, New York.

Cochran, W.G., and Cox, G.M. (1957). *Experimental Designs*, 2nd ed. Wiley, New York.

Daniel, C., and Wood, F.S. (1971). *Fitting Equations to Data*. Wiley, New York.

Draper, N.R., and Smith, H. (1966). *Applied Regression Analysis*. Wiley, New York.

Fisher, R.A. (1966). *The Design of Experiments*, 8th ed. Hafner, New York.

Himmelblau, D.M. (1970). *Process Analysis by Statistical Methods*. Wiley, New York.

Mendenhall, W. (1968). *Introduction to Linear Models and the Design and Analysis of Experiments*. Duxbury Press, Belmont, California.

Moore, D.S. (1979). *Statistics, Concepts and Controversies*. Freeman, San Francisco, California.

Natrella, M.G. (1963). *Experimental Statistics* (Nat. Bur. of Stand. Handbook 91). US Govt. Printing Office, Washington, DC.

Neter, J., and Wasserman, W. (1974). *Applied Linear Statistical Models. Regression, Analysis of Variance, and Experimental Designs*. Irwin, Homewood, Illinois.

Shewhart, W.A. (1939). *Statistical Method from the Viewpoint of Quality Control*. Lancaster Press, Pennsylvania.

Wilde, D.J., and Beightler, C.S. (1967). *Foundations of Optimization*. Prentice-Hall, Englewood Cliffs, New Jersey.

Youden, W.J. (1951). *Statistical Methods for Chemists*. Wiley, New York.

Exercises

10.1 Placement of experiments

Rework Sections 10.3 and 10.4 with $x_{12}^* = -1$ and $x_{1(10)}^* = 1$. Compare $(\boldsymbol{X}^{*\prime}\boldsymbol{X}^*)^{-1}$ with Equations 10.11 and 10.14. Which experimental design gives the most precise estimate of β_1^*?

10.2 Effect of coding

Rework Sections 10.3 and 10.4 using $c_{x_1} = 9$, $d_{x_1} = 0.5$. Does coding appear to affect the significance of the statistical tests?

10.3 Residual plot

Plot the residuals of Equation 10.27 against the factor levels of Equation 10.15. Is there a probable pattern in these residuals? What additional term might be added to the model to improve the fit?

10.4 Residual plot

Plot the residuals of Figure 10.4 against the factor levels of Equation 10.15. Is there a probable pattern in these residuals? Would an additional term in the model be helpful?

10.5 Matrix inversion

Verify that Equation 10.8 is the correct inverse of Equation 10.7 (see Appendix A).

10.6 Coding transformations

Calculate $\hat{\boldsymbol{B}}$ from $\hat{\boldsymbol{B}}^*$ in Section 10.3 (see Section 10.5 and Equation 10.48). What is the interpretation of b_0? What is the relationship between b_1^* and b_1? Why?

10.7 Experimental design

A system is thought to be described between $x_1 = 0$ and $x_1 = 10$ by the model $y_{1i} = \beta_0 + \beta_1 x_{1i} + r_{1i}$. What is the minimum number of experiments that will allow the model to be fit, the significance of regression to be tested, and the lack of fit to be tested? How should these experiments be distributed along x_1? Why?

10.8 Experimental design

A system is thought to be described between $x_1 = 0$ and $x = 10$ by the model $y_{1i} = \beta_0 + \beta_1 x_{1i} + \beta_1(\log x_{1i}) + r_{1i}$. If resources are available for ten experiments, how would you place the experiments along x_1? Why?

10.9 Experimental design

A *two-factor* system is thought to be described by the model $y_{1i} = \beta_0 + \beta_1 x_{1i} + \beta_2 x_{2i} + r_{1i}$ over the domain $0 \leqslant x_1 \leqslant 10$, $0 \leqslant x_2 \leqslant 10$ (see Figure 2.16). What is the minimum number of experiments that will allow the model to be fit? What are some of the ways you would *not* place these experiments in the domain of x_1 and x_2? [Hint: calculate $(\boldsymbol{X}'\boldsymbol{X})^{-1}$ for questionable designs.]

10.10 Confidence intervals

Compare Equations 10.76, 10.79, and 10.80. In what ways are they similar? In what way are they different?

10.11 Confidence intervals

Rewrite Equation 10.82 substituting $p \times F_{(p,n-p)}$ for W^2. In what ways is it similar to Equation 10.80? In what way is it different?

10.12 System theory

What other factors might affect the reaction rate of the enzyme catalyzed system shown in Figure 10.10? Which of these can be easily controlled? Which would be very difficult to control? Which would be expected to exert a large effect on the system? Which would be expected to exert only a small effect on the system? [See Exercise 1.1.]

10.13 Measurement systems

Measurement systems are used to obtain information from systems under study:

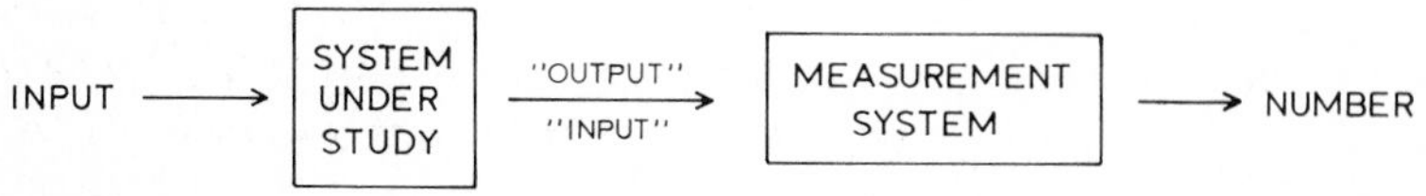

INPUT → SYSTEM UNDER STUDY → "OUTPUT" / "INPUT" → MEASUREMENT SYSTEM → NUMBER

Comment on known and unknown, controlled and uncontrolled *measurement system factors* and their influence on the quality of information about the system under study.

10.14 Measurement systems

An ideal measurement system does not perturb the system under study. Give three examples of ideal measurement systems. Give three examples of non-ideal measurement systems.

10.15 Experimental design

Draw a systems diagram for a process with which you are familiar. Choose one input for investigation. Over what domain might this input be varied and controlled? What linear mathematical model might be used to approximate the effect of this factor on one of the responses from the system? Design a set of experiments to test the validity of this model.

10.16 Confidence intervals

Calculate the 99% confidence interval for predicting a single new value of response at pH = 7.0 for the data of Equation 10.16 and the second-order model of Equation 10.39. Calculate the 99%, confidence interval for predicting the mean of seven new values of response for these conditions. Calculate the 99% confidence interval for predicting the true mean for these conditions. What confidence interval would be used if it were necessary to predict the true mean at several points in factor space?

10.17 Number of experiments

The experimental design represented by Equation 10.6 requires only half as many experiments as the design represented by Equation 10.12. Each design can be used to fit the same parabolic model. What are the advantages and disadvantages of each design?

10.18 Coefficients of correlation

In Sections 10.3 and 10.4 the R^2 value increased from 0.479 to 0.794 and the F-ratio for the significance of regression increased from the 97.3% to the 99.6% level of confidence when the term $\beta_1^* x_{1i}^{*2}$ was added to the model. Is it possible that in some instances R^2 will increase, but the significance of regression will decrease? If so, why? [See, for example, Neter, J., and Wasserman, W. (1974). *Applied Linear Statistical Models. Regression, Analysis of Variance, and Experimental Designs*, p. 229. Irwin, Homewood, Illinois.]

10.19 Confidence intervals

Comment on the practice of writing results in the following form: 5.63 ± 0.16.

10.20 Variance-covariance matrix

In Section 10.3, is s_r^2 a "valid estimate" of σ_{pe}^2? Why or why not? In Section 10.4, is s_r^2 a "valid estimate" of σ_{pe}^2? Why or why not?

CHAPTER 11

Approximating a Region of a Multifactor Response Surface

In previous chapters, many of the fundamental concepts of experimental design have been presented for single-factor systems. Several of these concepts are now expanded and new ones are introduced to begin the treatment of multifactor systems.

Although the complexity of multifactor systems increases roughly exponentially as the number of factors being investigated increases, most multifactor concepts can be introduced using the relatively simple two-factor case. Thus, in most of this chapter we will consider the system shown in Figure 11.1, a system having two inputs designated factor x_1 and factor x_2 (see Section 1.2), and a single output designated response y_1 (see Section 1.3).

11.1. Elementary concepts

In Chapter 2 it was seen that a response surface for a one-factor system can be represented by a *line*, either straight or curved, existing in the *plane* of two-dimensional experiment space (one factor dimension and one response dimension). In two-factor systems, a response surface can be represented by a true *surface*, either flat or curved, existing in the *volume* of three-dimensional experiment space (two factor dimensions and one response dimension). By extension, a response surface associated with three- or higher-dimensional factor space can be thought of as a *hypersurface* existing in the *hypervolume* of four- or higher-dimensional experiment space.

Figure 11.2 is a graphic representation of a portion of two-dimensional factor space associated with the system shown in Figure 11.1. In this illustration, the

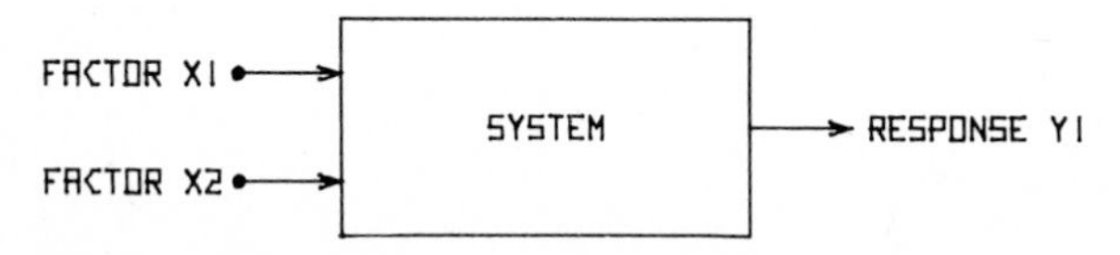

Figure 11.1. Two-factor, single-response system for discussion of multifactor experimentation.

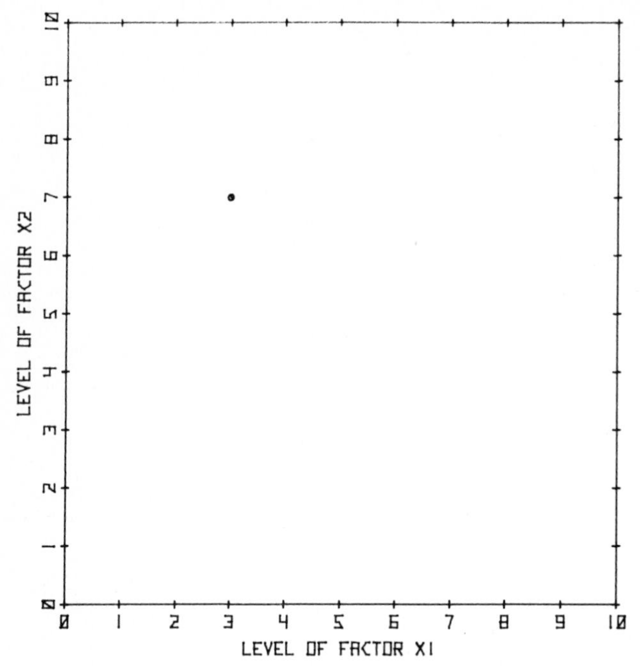

Figure 11.2. Location of a single experiment in two-dimensional factor space.

domain of factor x_1 (the "horizontal axis") lies between 0 and +10; similarly, the domain of factor x_2 (the "vertical axis") lies between 0 and +10. The response axis is not shown in this representation, although it might be imagined to rise perpendicularly from the intersection of the factor axes (at $x_1 = 0$, $x_2 = 0$). Figure 11.2 shows the location in factor space of a single experiment at $x_{11} = +3$, $x_{21} = +7$.

Figure 11.3 is a pseudo-three-dimensional representation of a portion of the three-dimensional experiment space associated with the system shown in Figure 11.1. The two-dimensional factor subspace is shaded with a one-unit grid. The factor domains are again $0 \leqslant x_1 \leqslant +10$ and $0 \leqslant x_2 \leqslant +10$. The response axis ranges from 0 to +8. The location in factor space of the single experiment at $x_{11} = +3$, $x_{21} = +7$ is shown as a point in the plane of factor space. The response ($y_{11} = +4.00$) associated with this experiment is shown as a point above the plane of factor space, and is "connected" to the factor space by a dotted vertical line.

A two-factor response surface is the graph of a system output or objective function plotted against the system's two inputs. It is assumed that all other controllable factors are held constant, each at a specified level. Again, it is important that this assumption be true; otherwise, as will be seen in Section 11.2, the response surface might appear to change shape or to be excessively noisy.

Figure 11.4 is a pseudo-three-dimensional representation of a response surface showing a system response, y_1, plotted against the two system factors, x_1 and x_2.

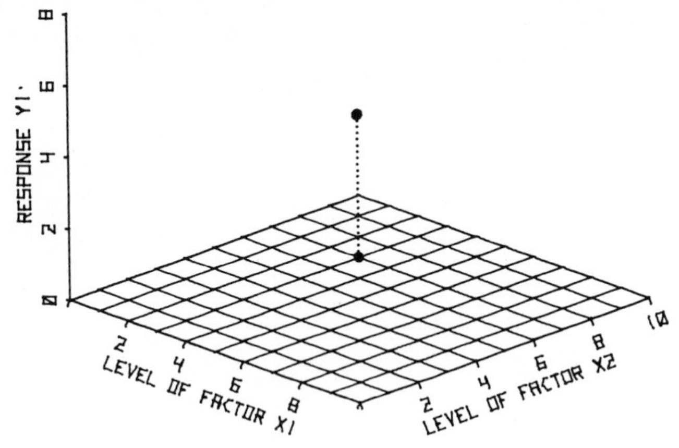

Figure 11.3. Location of a single experiment in three-dimensional experiment space.

The response surface might be described by some mathematical function T that relates the response y_1 to the factors x_1 and x_2.

$$y_1 = T(x_1, x_2) \tag{11.1}$$

For Figure 11.4, the exact relationship is

$$y_1 = 1.68 + 0.24x_1 + 0.56x_2 - 0.04x_1^2 - 0.04x_2^2 \tag{11.2}$$

Such a relationship (and corresponding response surface) might represent magnetic field strength as a function of position in a plane parallel to the pole face of a magnet, or reaction yield as a function of reactor pressure and temperature for a chemical process.

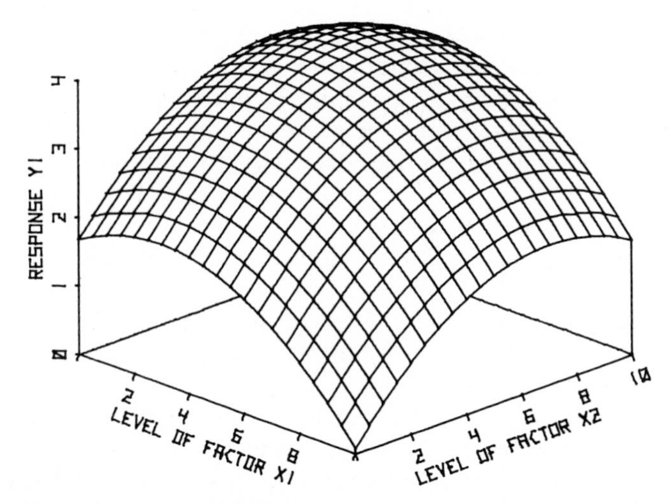

Figure 11.4. A two-factor response surface.

11.2. Factor interaction

The concept of "self interaction" was introduced in Section 8.6 where it was shown that for a single-factor, second-order model, "the (first-order) effect of the factor depends upon the level of the factor". It was shown that the model

$$y_{1i} = \beta_0 + \beta_1 x_{1i} + \beta_{11} x_{1i}^2 + r_{1i} \tag{11.3}$$

can be written

$$y_{1i} = \beta_0 + (\beta_1 + \beta_{11} x_{1i}) x_{1i} + r_{1i} \tag{11.4}$$

from which the "slope" with respect to x_1 (the first-order effect of x_1) can be obtained:

$$\text{slope of factor } x_1 = (\beta_1 + \beta_{11} x_{1i}) \tag{11.5}$$

In multifactor systems, it is possible that the effect of one factor will depend upon the level of a *second* factor. For example, the slope of factor x_1 might depend upon the level of factor x_2 in the following way:

$$\text{slope of factor } x_1 = (\beta_1 + \beta_{12} x_{2i}) \tag{11.6}$$

When incorporated into the complete second-order model, we obtain

$$y_{1i} = \beta_0 + (\beta_1 + \beta_{12} x_{2i}) x_{1i} + r_{1i} \tag{11.7}$$

and finally

$$y_{1i} = \beta_0 + \beta_1 x_{1i} + \beta_{12} x_{1i} x_{2i} + r_{1i} \tag{11.8}$$

The term $\beta_{12} x_{1i} x_{2i}$ is said to be an *interaction term*. The subscripts on β_{12} indicate that the parameter is used to assess the interaction between the two factors x_1 and x_2.

According to Equation 11.8 (and seen more easily in Equation 11.7), the response y_1 should be offset from the origin an amount equal to β_0 and should change according to a first-order relationship with the factor x_1. However, the sign and magnitude of that change depend not only upon the parameters β_1 and β_{12}, but also upon the value of the factor x_2. If we set $\beta_0 = +4.0$, $\beta_1 = -0.4$, and $\beta_{12} = +0.08$, then Equation 11.7 becomes

$$y_{1i} = 4.0 + (-0.4 + 0.08 x_{2i}) x_{1i} + r_{1i} \tag{11.9}$$

or

$$y_{1i} = 4.0 - 0.4 x_{1i} + 0.08 x_{1i} x_{2i} + r_{1i} \tag{11.10}$$

When $x_2 = 0$, the effect of x_1 will be to *decrease* the response y_1 by 0.4 units for every unit increase in x_1. When $x_2 = 5$, y_1 will not be influenced by x_1; i.e., the slope of y_1 with respect to x_1 will be zero. When $x_2 = 10$, the effect of x_1 will be to *increase* the response y_1 by 0.4 units for every unit increase in x_1. Figure 11.5 graphs the response surface of Equation 11.10.

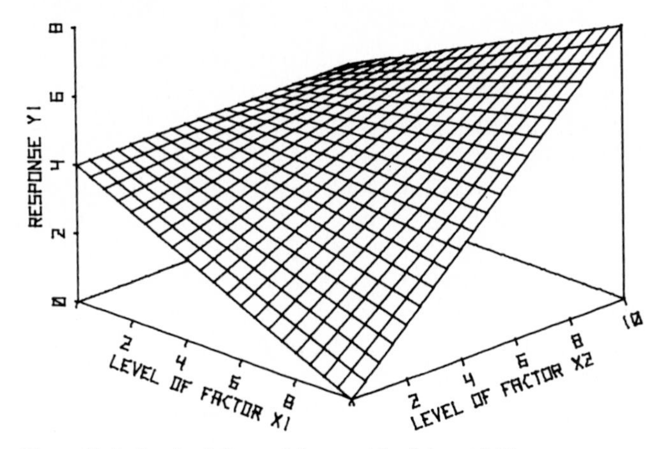

Figure 11.5. Graph of the model $y_{1i} = 4.0 - 0.4x_{1i} + 0.08x_{1i}x_{2i}$.

We could also look at this interaction from the point of view of the second factor x_2 and say that its effect depends upon the level of the first factor x_1. Rewriting Equation 11.8 gives

$$y_{1i} = \beta_0 + \beta_1 x_{1i} + (\beta_{12} x_{1i}) x_{2i} + r_{1i} \tag{11.11}$$

That is, the

$$\text{slope of factor } x_2 = (\beta_{12} x_{1i}) \tag{11.12}$$

This interpretation is confirmed in Figure 11.5 where it can be seen that when $x_1 = 0$, the response does not depend upon the factor x_2. However, the dependence of y_1 upon x_2 gets larger as the level of the factor x_1 increases.

The concept of interaction is fundamental to an understanding of multifactor systems. Much time can be lost and many serious mistakes can be made if interaction is not considered.

As a simple example, let us suppose that a research director wants to know the effect of temperature upon yield in an industrial process. The yield is very low, so in an attempt to increase the yield he asks one of his research groups to find out the effect of temperature (T) on yield over the temperature domain from 0°C to 10°C. The group reports that yield *decreases* with temperature according to the equation

$$\text{yield} = (4.0 - 0.4T)\% \tag{11.13}$$

Not certain of this result, he asks another of his research groups to repeat the work. They report that yield *increases* with temperature according to the equation

$$\text{yield} = (4.0 + 0.4T)\% \tag{11.14}$$

Many research directors would be upset at this seemingly conflicting information – how can one group report that yield decreases with temperature and the other group report that yield increases with temperature?

In many instances, including this example, there could be a second factor in the system that is interacting with the factor of interest. It is entirely possible that the first research group understood their director to imply that catalyst should not be included in the process when they studied the effect of temperature. The second group, however, might have thought the research director intended them to add the usual 10 milligrams of catalyst. Catalyst could then be the second, interacting factor: if catalyst were absent, the yield would decrease; with catalyst present in the amount of 10 milligrams, the yield would increase. Presumably, intermediate amounts of catalyst would cause intermediate effects. The results of this example are illustrated in Figure 11.5 if temperature in °C is represented by the factor x_1, milligrams of catalyst is represented by x_2, and yield is represented by y_1. Equations 11.13 and 11.14 are obtained from Equation 11.10.

At the beginning of Section 2.1 where single-factor response surfaces were first discussed, the following caution was presented: "It is assumed that all other controllable factors are held constant, each at a specified level.... It is important that this assumption be true; otherwise, the single-factor response surface might appear to change shape or to be excessively noisy". This caution was stated again for multifactor response surfaces in Section 11.1. We have just seen how controlling a second factor (catalyst) at *different* levels causes the single-factor response surface

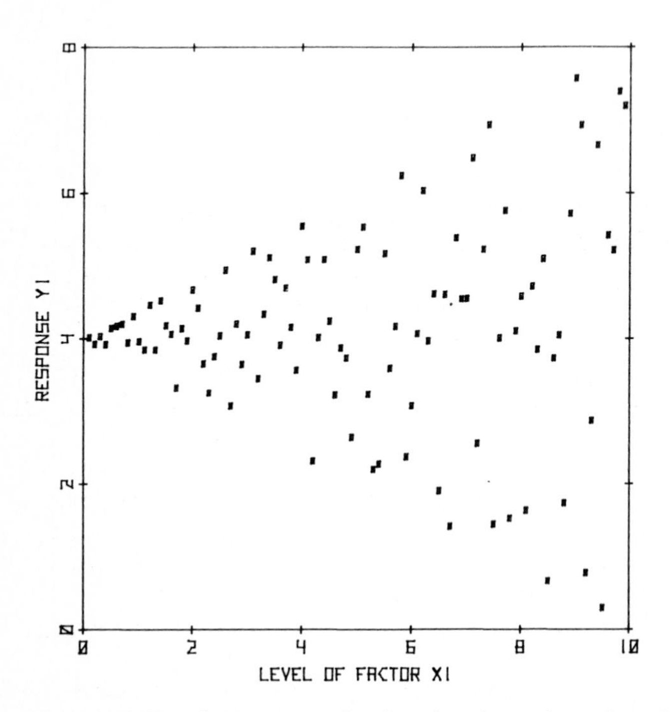

Figure 11.6. Plot of response as a function of x_1 for random values of x_2 ($0 \leqslant x_2 \leqslant 10$) in the model $y_{1i} = 4.0 - 0.4x_{1i} + 0.08x_{1i}x_{2i}$.

to change shape. Let us now see why an *uncontrolled* second factor often produces a single-factor response surface that appears to be excessively noisy.

Suppose a third research group were to attempt to determine the effect of temperature on yield. If they were not aware of the importance of catalyst and took no precautions to control the catalyst level at a specified concentration, then their results would depend not only upon the known and controlled factor temperature (x_1), but also upon the unknown and uncontrolled factor catalyst (x_2). A series of four experiments might give the results

x_1, °C	x_2, mg	y_1, %
1	3.65	3.89
2	9.00	4.64
3	6.33	4.32
4	5.31	4.10

Plotting this data and the results of several other similar experiments in which catalyst level was not controlled might give the results shown in Figure 11.6. The relationship between y_1 and x_1 is seen to be "fuzzy"; the data is noisy because catalyst (x_2) has not been controlled and interacts with the factor x_1.

A similar effect is observed when non-interacting factors are not controlled. However, uncontrolled non-interacting factors usually produce homoscedastic noise; uncontrolled interacting factors often produce heteroscedastic noise, as they do in the present example.

11.3. Factorial designs

Factorial designs are an enormously popular class of experimental designs that are often used to investigate multifactor response surfaces. The word "factorial" does not have its usual mathematical meaning of an integer multiplied by all integers smaller than itself (e.g., $3! = 3 \times 2 \times 1$), but instead indicates that many factors are varied simultaneously in a systematic way. One of the major advantages of factorial designs is that they can be used to reveal the existence of factor interaction when it is present in a system.

Historically, factorial designs were introduced by Sir R. A. Fisher to counter the then prevalent idea that if one were to discover the effect of a factor, all other factors must be held constant and only the factor of interest could be varied. Fisher showed that all factors of interest could be varied simultaneously, and the individual factor effects and their interactions could be estimated by proper mathematical treatment. The Yates algorithm and its variations are often used to obtain these estimates, but the use of least squares fitting of linear models gives essentially identical results.

Important descriptors of factorial designs are the number of factors involved in the design and the number of levels of each factor. For example, if a factorial design has three levels (low, middle, and high) of each of two factors (x_1 and x_2), it is said

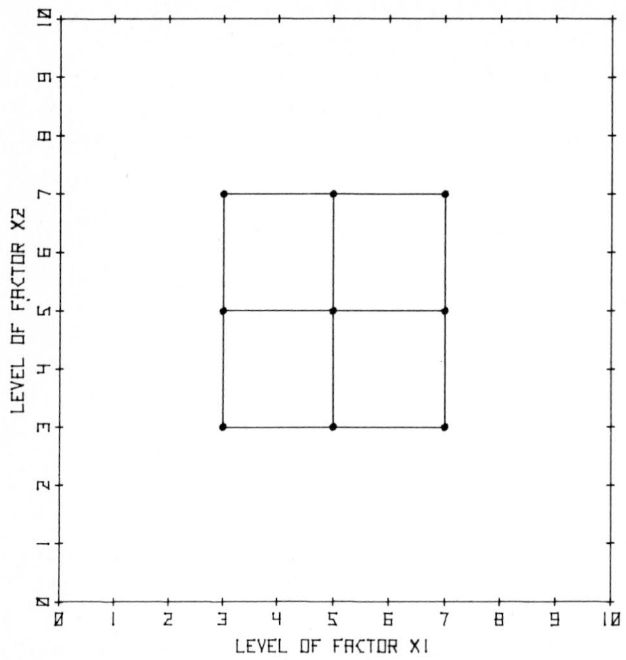

Figure 11.7. Factor combinations for a 3^2 factorial experimental design in two-dimensional factor space.

to be a 3×3 or 3^2 design. A factorial design involving three factors with each factor at two levels (low and high) would be called a $2 \times 2 \times 2$ or 2^3 factorial. Figure 11.7 illustrates possible factor combinations for a 3^2 factorial and Figure 11.8 shows factor combinations for a 2^3 design. Note that all possible combinations of the chosen factor levels are present in the experimental design; thus, the description of a factorial design gives the number of factor combinations (f) contained in the design: $3^2 = 9$, and $2^3 = 8$. Note also that as the number of factors increases, the number of factor combinations required by a full factorial design increases exponentially (see Table 11.1). In general, if k is the number of factors being investigated, and m is the number of levels for each factor, then m^k factor combinations are generated by a full factorial design. We will restrict our discussion here to the two-level, two-factor design shown in Figure 11.9.

The linear model most commonly fit to the data from 2^2 factorial designs is

$$y_{1i} = \beta_0 + \beta_1 x_{1i} + \beta_2 x_{2i} + \beta_{12} x_{1i} x_{2i} + r_{1i} \tag{11.15}$$

This model gives estimates of an offset term (β_0), a first-order effect (β_1) of the first factor x_1, a first-order effect (β_2) of the second factor x_2, and a second-order interaction effect (β_{12}) between the two factors. When the model of Equation 11.15 is fit to data from a 2^2 factorial design, the number of factor combinations ($f = 4$) is equal to the number of parameters ($p = 4$), and the number of degrees of freedom

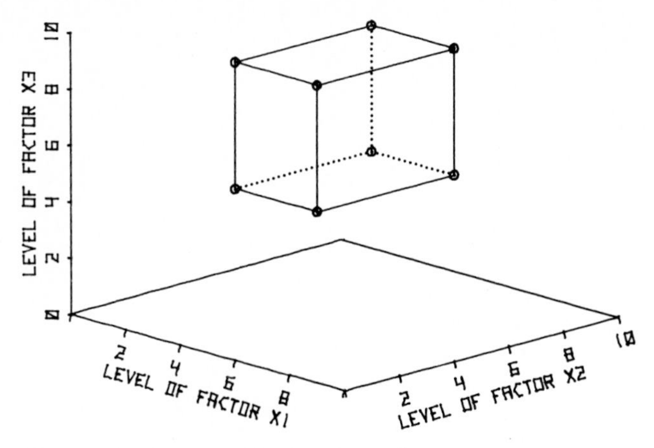

Figure 11.8. Factor combinations for a 2^3 factorial experimental design in three-dimensional factor space.

for estimating lack of fit is zero $(f - p = 4 - 4 = 0)$. Further, if there is no replication, the number of degrees of freedom for residuals is zero $(n - p = 4 - 4 = 0)$; under these conditions, the number of degrees of freedom for purely experimental uncertainty is also zero $(n - f = 4 - 4 = 0)$. If replication is carried out, purely experimental uncertainty can be estimated, but an estimate of lack of fit is still not possible.

Let us suppose the results of the experimental design shown in Figure 11.9 are obtained from the response surface shown in Figure 11.5. These results are

i	x_{1i}	x_{2i}	y_{1i}
1	3	3	3.52
2	3	7	4.48
3	7	3	2.88
4	7	7	5.12

Then

$$\boldsymbol{D} = \begin{bmatrix} 3 & 3 \\ 3 & 7 \\ 7 & 3 \\ 7 & 7 \end{bmatrix} \tag{11.16}$$

$$\boldsymbol{Y} = \begin{bmatrix} 3.52 \\ 4.48 \\ 2.88 \\ 5.12 \end{bmatrix} \tag{11.17}$$

$$\boldsymbol{X} = \begin{bmatrix} 1 & 3 & 3 & 9 \\ 1 & 3 & 7 & 21 \\ 1 & 7 & 3 & 21 \\ 1 & 7 & 7 & 49 \end{bmatrix} \tag{11.18}$$

TABLE 11.1
Number of factor combinations required by full factorial designs as a function of the number of factors.

Number of factors, k	Number of factor combinations required for full factorial		
	2-level	3-level	4-level
1	2	3	4
2	4	9	16
3	8	27	64
4	16	81	256
5	32	243	1,024
6	64	729	4,096
7	128	2,187	16,384
8	256	6,561	65,536
9	512	19,683	262,144
10	1,024	59,049	1,048,576

Using the method of least squares, we obtain

$$(\boldsymbol{X}'\boldsymbol{X}) = \begin{bmatrix} 4 & 20 & 20 & 100 \\ 20 & 116 & 100 & 580 \\ 20 & 100 & 116 & 580 \\ 100 & 580 & 580 & 3364 \end{bmatrix} \tag{11.19}$$

$$(\boldsymbol{X}'\boldsymbol{X})^{-1} = \begin{bmatrix} 13.1 & -2.27 & -2.27 & 0.391 \\ -2.27 & 0.453 & 0.391 & -0.0781 \\ -2.27 & 0.391 & 0.453 & -0.0781 \\ 0.391 & -0.0781 & -0.0781 & 0.0156 \end{bmatrix} \tag{11.20}$$

$$(\boldsymbol{X}'\boldsymbol{Y}) = \begin{bmatrix} 1 & 1 & 1 & 1 \\ 3 & 3 & 7 & 7 \\ 3 & 7 & 3 & 7 \\ 9 & 21 & 21 & 49 \end{bmatrix} \begin{bmatrix} 3.52 \\ 4.48 \\ 2.88 \\ 5.12 \end{bmatrix} = \begin{bmatrix} 16.0 \\ 80 \\ 86.4 \\ 437.12 \end{bmatrix} \tag{11.21}$$

$$\hat{\boldsymbol{B}} = \begin{bmatrix} b_0 \\ b_1 \\ b_2 \\ b_{12} \end{bmatrix} = (\boldsymbol{X}'\boldsymbol{X})^{-1}(\boldsymbol{X}'\boldsymbol{Y}) = \begin{bmatrix} 4.00 \\ -0.40 \\ 0.00 \\ 0.08 \end{bmatrix} \tag{11.22}$$

Because four parameters were estimated from data obtained at four factor combinations, there are no degrees of freedom for lack of fit; further, there was no replication in this example, so there are no degrees of freedom for purely experimental uncertainty. Thus, there can be no degrees of freedom for residuals, and the estimated model will appear to fit the data "perfectly". This is verified by estimating the responses using the fitted model parameters.

$$\hat{\boldsymbol{Y}} = \boldsymbol{X}\hat{\boldsymbol{B}} = \begin{bmatrix} 1 & 3 & 3 & 9 \\ 1 & 3 & 7 & 21 \\ 1 & 7 & 3 & 21 \\ 1 & 7 & 7 & 49 \end{bmatrix} \begin{bmatrix} 4.00 \\ -0.40 \\ 0.00 \\ 0.08 \end{bmatrix} = \begin{bmatrix} 3.52 \\ 4.48 \\ 2.88 \\ 5.12 \end{bmatrix} \tag{11.23}$$

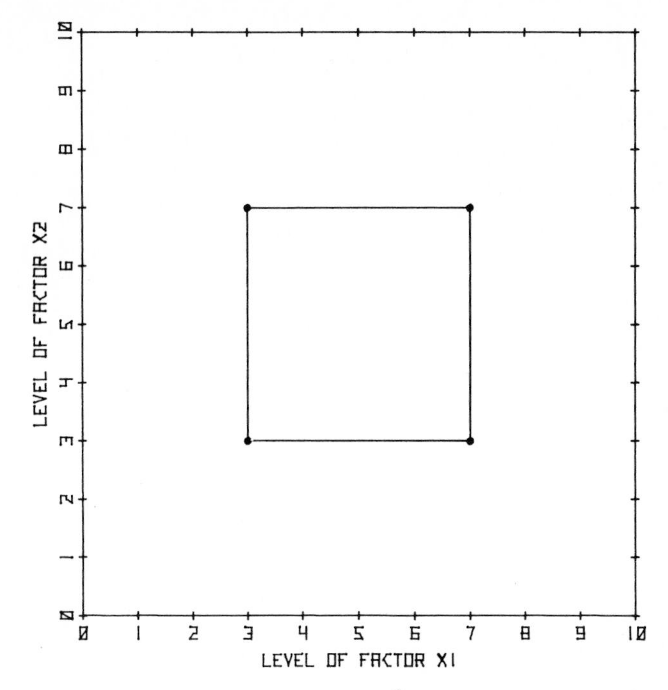

Figure 11.9. Factor combinations for a 2^2 factorial experimental design in two-dimensional factor space.

This reproduces the original data exactly (see Equation 11.17).

Because the data for this example were taken from Figure 11.5, it is not surprising that these parameter estimates are the same as the values in Equation 11.10 from which Figure 11.5 was drawn. Note that β_2 does not appear in Equation 11.10; thus, the estimated value of $\beta_2 = 0$ in Equation 11.22 is reasonable and expected for this example.

One might be tempted to conclude from this fitted model that the factor x_2 is not important – after all, the coefficient β_2 is equal to zero. This would be an incorrect conclusion. It is true that the $\beta_2 x_{2i}$ term in the fitted model does not influence the response, but x_2 is still an important factor because of its effect in the interaction term $\beta_{12} x_{1i} x_{2i}$.

Other models can be fit to data from two-level factorial designs. For example, fitting the model expressed by Equation 11.8 to the data used in this section will produce the fitted model given by Equation 11.10. Some models cannot be fit to data from two-level factorial experiments; for example, the model

$$y_{1i} = \beta_0 + \beta_1 x_{1i} + \beta_2 x_{2i} + \beta_{22} x_{2i}^2 + r_{1i} \quad (11.24)$$

produces an $(\boldsymbol{X}'\boldsymbol{X})$ matrix that cannot be inverted.

11.4. Coding of factorial designs

Calculations for factorial designs are often greatly simplified if coding of factor levels is employed (see Section 8.5). For the present example, setting $c_{x_1} = c_{x_2} = 5$ and $d_{x_1} = d_{x_2} = 2$,

$$\boldsymbol{X}^* = \begin{bmatrix} 1 & -1 & -1 & 1 \\ 1 & -1 & 1 & -1 \\ 1 & 1 & -1 & -1 \\ 1 & 1 & 1 & 1 \end{bmatrix} \tag{11.25}$$

$$(\boldsymbol{X}^{*\prime}\boldsymbol{X}^*) = \begin{bmatrix} 4 & 0 & 0 & 0 \\ 0 & 4 & 0 & 0 \\ 0 & 0 & 4 & 0 \\ 0 & 0 & 0 & 4 \end{bmatrix} \tag{11.26}$$

Inversion of this matrix is trivial.

$$(\boldsymbol{X}^{*\prime}\boldsymbol{X}^*)^{-1} = \begin{bmatrix} 1/4 & 0 & 0 & 0 \\ 0 & 1/4 & 0 & 0 \\ 0 & 0 & 1/4 & 0 \\ 0 & 0 & 0 & 1/4 \end{bmatrix} \tag{11.27}$$

$$(\boldsymbol{X}^{*\prime}\boldsymbol{Y}) = \begin{bmatrix} 1 & 1 & 1 & 1 \\ -1 & -1 & 1 & 1 \\ -1 & 1 & -1 & 1 \\ 1 & -1 & -1 & 1 \end{bmatrix} \begin{bmatrix} 3.52 \\ 4.48 \\ 2.88 \\ 5.12 \end{bmatrix} = \begin{bmatrix} 16.00 \\ 0.00 \\ 3.20 \\ 1.28 \end{bmatrix} \tag{11.28}$$

$$\hat{\boldsymbol{B}}^* = \begin{bmatrix} b_0 \\ b_1 \\ b_2 \\ b_{12} \end{bmatrix} = (\boldsymbol{X}^{*\prime}\boldsymbol{X}^*)^{-1}(\boldsymbol{X}^{*\prime}\boldsymbol{Y}) = \begin{bmatrix} 4.00 \\ 0.00 \\ 0.80 \\ 0.32 \end{bmatrix} \tag{11.29}$$

Transformation back to the original coordinate system is accomplished as described in Section 10.5.

$$(\boldsymbol{X}^{*\prime}\boldsymbol{X}) = \begin{bmatrix} 4 & 20 & 20 & 100 \\ 0 & 8 & 0 & 40 \\ 0 & 0 & 8 & 40 \\ 0 & 0 & 0 & 16 \end{bmatrix} \tag{11.30}$$

$$\boldsymbol{A} = (\boldsymbol{X}^{*\prime}\boldsymbol{X}^*)^{-1}(\boldsymbol{X}^{*\prime}\boldsymbol{X}) = \begin{bmatrix} 1 & 5 & 5 & 25 \\ 0 & 2 & 0 & 10 \\ 0 & 0 & 2 & 10 \\ 0 & 0 & 0 & 4 \end{bmatrix} \tag{11.31}$$

$$\hat{\boldsymbol{B}} = \boldsymbol{A}^{-1}\hat{\boldsymbol{B}}^* = \begin{bmatrix} 1 & -2.5 & -2.5 & 6.25 \\ 0 & 0.5 & 0 & -1.25 \\ 0 & 0 & 0.5 & -1.25 \\ 0 & 0 & 0 & 0.25 \end{bmatrix} \begin{bmatrix} 4.00 \\ 0.00 \\ 0.80 \\ 0.32 \end{bmatrix} = \begin{bmatrix} 4.00 \\ -0.40 \\ 0.00 \\ 0.08 \end{bmatrix} \tag{11.32}$$

which is the same result as was obtained in the previous section (see Equation 11.22).

Many workers do not transform the parameter estimates back to the original coordinate system, but instead work with the parameter estimates obtained in the coded factor space. This can often lead to surprising and seemingly contradictory results. As an example, the fitted model in the coded factor space was found to be

$$y_{1i} = 4.00 + 0.00x_{1i}^* + 0.80x_{2i}^* + 0.32x_{1i}^*x_{2i}^* \tag{11.33}$$

This suggests that the β_1^* term can be omitted and leads to a simpler model of the form

$$y_{1i} = \beta_0^* + \beta_2^*x_{2i}^* + \beta_{12}^*x_{1i}^*x_{2i}^* + r_{1i} \tag{11.34}$$

which is clearly not the form of the equation from which the data were generated (see Equations 11.8 and 11.10). The coded equation has replaced the $\beta_1 x_{1i}$ term with a $\beta_2^* x_{2i}^*$ term! This is not an uncommon phenomenon: the mathematical form of the model can often appear to change when factor levels are coded. The reason for the present change of model is seen in the algebra of coding.

$$y_{1i} = \beta_0 + \beta_1 x_{1i} + \beta_2 x_{2i} + \beta_{12} x_{1i} x_{2i} \tag{11.35}$$

Substituting coded values gives

$$y_{1i} = \beta_0 + \beta_1(2x_{1i}^* + 5) + \beta_2(2x_{2i}^* + 5) + \beta_{12}(2x_{1i}^* + 5)(2x_{2i}^* + 5) \tag{11.36}$$

$$y_{1i} = \beta_0 + 2\beta_1 x_{1i}^* + 5\beta_1 + 2\beta_2 x_{2i}^* + 5\beta_2 + \beta_{12}(4x_{1i}^*x_{2i}^* + 10x_{1i}^* + 10x_{2i}^* + 25) \tag{11.37}$$

$$y_{1i} = \beta_0 + 2\beta_1 x_{1i}^* + 5\beta_1 + 2\beta_2 x_{2i}^* + 5\beta_2 + 4\beta_{12}x_{1i}^*x_{2i}^* + 10\beta_{12}x_{1i}^* + 10\beta_{12}x_{2i}^* + 25\beta_{12} \tag{11.38}$$

$$y_{1i} = (\beta_0 + 5\beta_1 + 5\beta_2 + 25\beta_{12}) + (2\beta_1 + 10\beta_{12})x_{1i}^* + (2\beta_2 + 10\beta_{12})x_{2i}^* + (4\beta_{12})x_{1i}^*x_{2i}^* \tag{11.39}$$

From this it is seen that

$$\beta_0^* = (\beta_0 + 5\beta_1 + 5\beta_2 + 25\beta_{12}) = (4.00 - 2.00 + 0.00 + 2.00) = 4.00 \tag{11.40}$$

$$\beta_1^* = (2\beta_1 + 10\beta_{12}) = (-0.80 + 0.80) = 0.00 \tag{11.41}$$

$$\beta_2^* = (2\beta_2 + 10\beta_{12}) = (0.00 + 0.80) = 0.80 \tag{11.42}$$

$$\beta_{12}^* = (4\beta_{12}) = 0.32 \tag{11.43}$$

Thus, it is the algebraic effect of coding and the numerical values of the estimated parameters that cause some terms to be added to the model (e.g., β_2) and other terms to disappear from the model (e.g., β_1). Transformation of parameter estimates back to the original factor space usually avoids possible misinterpretation of results.

11.5. Star designs

Two-level factorial designs are useful for estimating first-order factor effects and interaction effects between factors, but they cannot be used to estimate additional second-order curvature effects such as those represented by the terms $\beta_{11}x_{1i}^2$ and $\beta_{22}x_{2i}^2$ in the model

$$y_{1i} = \beta_0 + \beta_1 x_{1i} + \beta_2 x_{2i} + \beta_{11} x_{1i}^2 + \beta_{22} x_{2i}^2 + r_{1i} \tag{11.44}$$

A different class of experimental designs, the *star designs*, provide information that can be used to fit models of the general form described by Equation 11.44. Models of this class contain $2k + 1$ parameters, where k is the number of factors included in the model.

Star designs are located by a center point from which other factor combinations are generated by moving a positive and a negative distance in each factor dimension, one factor dimension at a time. Star designs thus generate $2k + 1$ factor combinations, and are sufficient for estimating the $2k + 1$ parameters of models such as Equation 11.44. Star designs for two- and three-dimensional factor spaces are shown in Figures 11.10 and 11.11, respectively.

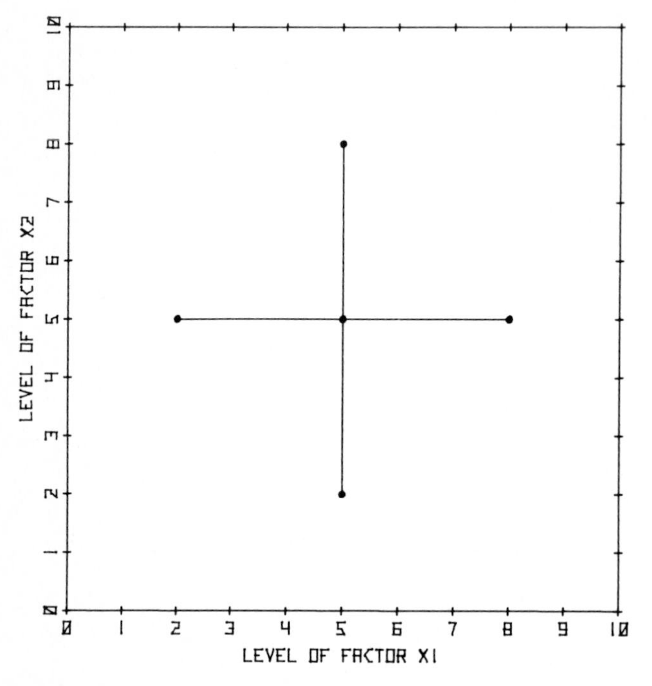

Figure 11.10. Factor combinations for a star experimental design in two-dimensional factor space.

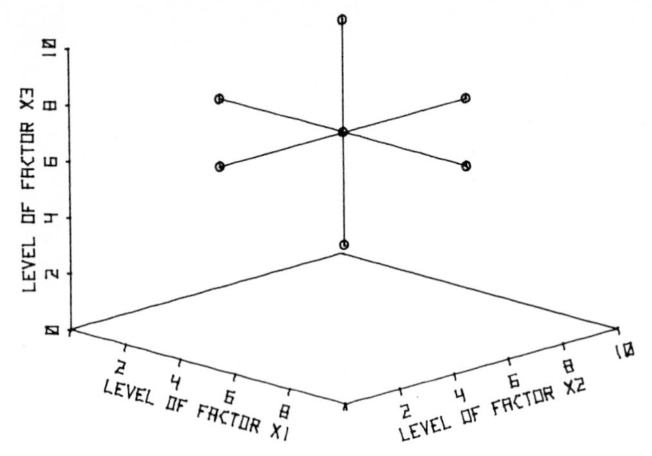

Figure 11.11. Factor combinations for a star experimental design in three-dimensional factor space.

As an example of the use of star designs, let us fit the model of Equation 11.44 to data obtained at locations specified by the star design in Figure 11.10.

i	x_{1i}	x_{2i}	y_{1i}
1	2	5	3.50
2	5	2	5.75
3	5	5	8.00
4	5	8	5.75
5	8	5	3.50

Let us code the factor levels using $c_{x_1} = c_{x_2} = 5$ and $d_{x_1} = d_{x_2} = 3$. Then,

$$(\boldsymbol{X}^{*\prime}\boldsymbol{X}^{*}) = \begin{bmatrix} 1 & 1 & 1 & 1 & 1 \\ -1 & 0 & 0 & 0 & 1 \\ 0 & -1 & 0 & 1 & 0 \\ 1 & 0 & 0 & 0 & 1 \\ 0 & 1 & 0 & 1 & 0 \end{bmatrix} \begin{bmatrix} 1 & -1 & 0 & 1 & 0 \\ 1 & 0 & -1 & 0 & 1 \\ 1 & 0 & 0 & 0 & 0 \\ 1 & 0 & 1 & 0 & 1 \\ 1 & 1 & 0 & 1 & 0 \end{bmatrix}$$

$$= \begin{bmatrix} 5 & 0 & 0 & 2 & 2 \\ 0 & 2 & 0 & 0 & 0 \\ 0 & 0 & 2 & 0 & 0 \\ 2 & 0 & 0 & 2 & 0 \\ 2 & 0 & 0 & 0 & 2 \end{bmatrix} \tag{11.45}$$

$$(\boldsymbol{X}^{*\prime}\boldsymbol{X}^{*})^{-1} = \begin{bmatrix} 1 & 0 & 0 & -1 & -1 \\ 0 & 0.5 & 0 & 0 & 0 \\ 0 & 0 & 0.5 & 0 & 0 \\ -1 & 0 & 0 & 1.5 & 1 \\ -1 & 0 & 0 & 1 & 1.5 \end{bmatrix} \tag{11.46}$$

$$(\boldsymbol{X}^{*\prime}\boldsymbol{Y}) = \begin{bmatrix} 1 & 1 & 1 & 1 & 1 \\ -1 & 0 & 0 & 0 & 1 \\ 0 & -1 & 0 & 1 & 0 \\ 1 & 0 & 0 & 0 & 1 \\ 0 & 1 & 0 & 1 & 0 \end{bmatrix} \begin{bmatrix} 3.50 \\ 5.75 \\ 8.00 \\ 5.75 \\ 3.50 \end{bmatrix} = \begin{bmatrix} 26.5 \\ 0.0 \\ 0.0 \\ 7.0 \\ 11.5 \end{bmatrix} \tag{11.47}$$

$$\hat{\boldsymbol{B}}^* = \begin{bmatrix} b_0^* \\ b_1^* \\ b_2^* \\ b_{11}^* \\ b_{22}^* \end{bmatrix} = (\boldsymbol{X}^{*\prime}\boldsymbol{X}^{\prime})^{-1}(\boldsymbol{X}^{*\prime}\boldsymbol{Y}) = \begin{bmatrix} 8.0 \\ 0.0 \\ 0.0 \\ -4.5 \\ -2.25 \end{bmatrix} \tag{11.48}$$

$$(\boldsymbol{X}^{*\prime}\boldsymbol{X}) = \begin{bmatrix} 5 & 25 & 25 & 143 & 143 \\ 0 & 6 & 0 & 60 & 0 \\ 0 & 0 & 6 & 0 & 60 \\ 2 & 10 & 10 & 68 & 50 \\ 2 & 10 & 10 & 50 & 68 \end{bmatrix} \tag{11.49}$$

$$\boldsymbol{A} = (\boldsymbol{X}^{*\prime}\boldsymbol{X}^*)^{-1}(\boldsymbol{X}^{*\prime}\boldsymbol{X}) = \begin{bmatrix} 1 & 5 & 5 & 25 & 25 \\ 0 & 3 & 0 & 30 & 0 \\ 0 & 0 & 3 & 0 & 30 \\ 0 & 0 & 0 & 9 & 0 \\ 0 & 0 & 0 & 0 & 9 \end{bmatrix} \tag{11.50}$$

$$\hat{\boldsymbol{B}} = \begin{bmatrix} b_0 \\ b_1 \\ b_2 \\ b_{11} \\ b_{22} \end{bmatrix} = \boldsymbol{A}^{-1}\hat{\boldsymbol{B}}^* = \begin{bmatrix} -10.75 \\ 5.00 \\ 2.50 \\ -0.50 \\ -0.25 \end{bmatrix} \tag{11.51}$$

Because $2k+1$ parameters were estimated from data obtained at $2k+1$ factor combinations, there are no degrees of freedom for lack of fit ($f-p=5-5=0$); again, there was no replication in this example, so the estimated model should give a "perfect" fit to the data. This is verified by estimating the responses using the fitted model parameters.

$$\hat{\boldsymbol{Y}} = \boldsymbol{X}\hat{\boldsymbol{B}} = \begin{bmatrix} 1 & 2 & 5 & 4 & 25 \\ 1 & 5 & 2 & 25 & 2 \\ 1 & 5 & 5 & 25 & 25 \\ 1 & 5 & 8 & 25 & 64 \\ 1 & 8 & 5 & 64 & 25 \end{bmatrix} \begin{bmatrix} -10.75 \\ 5.00 \\ 2.50 \\ -0.50 \\ -0.25 \end{bmatrix} = \begin{bmatrix} 3.50 \\ 5.75 \\ 8.00 \\ 5.75 \\ 3.50 \end{bmatrix} \tag{11.52}$$

This reproduces the original data exactly, as expected.

Other models can be fit to data from star designs. For example, the model

$$y_{1i} = \beta_0 + \beta_1 x_{1i} + \beta_{22} x_{2i}^2 + r_{1i} \tag{11.53}$$

can be fit to the data used in this section. Some models cannot be fit to data obtained from star designs; in particular, many models possessing factor interaction terms produce ($\boldsymbol{X}'\boldsymbol{X}$) matrices that cannot be inverted.

11.6. Central composite designs

One of the most useful models for approximating a region of a multifactor response surface is the *full second-order polynomial model*. For two factors, the model is of the form

$$y_{1i} = \beta_0 + \beta_1 x_{1i} + \beta_2 x_{2i} + \beta_{11} x_{1i}^2 + \beta_{22} x_{2i}^2 + \beta_{12} x_{1i} x_{2i} + r_{1i} \tag{11.54}$$

In general, if k is the number of factors being investigated, the full second-order polynomial model contains $\frac{1}{2}(k+1)(k+2)$ parameters. A rationalization for the widespread use of full second-order polynomial models is that they represent a truncated Taylor series expansion of any continuous function, and such models would therefore be expected to provide a reasonably good approximation of the true response surface over a local region of experiment space.

If we choose to use a full second-order polynomial model, then we must use an appropriate experimental design for estimating the $\frac{1}{2}(k+1)(k+2)$ parameters of the model. Two-level factorial designs are an attractive possibility because they allow estimation of the β_0, β_1, β_2, and β_{12} parameters; however, they do not allow estimation of the second-order parameters β_{11} and β_{22}, and for situations involving fewer than four factors, there are too few factor combinations (2^k) to estimate all $\frac{1}{2}(k+1)(k+2)$ parameters of the model. Star designs are also an attractive possibility: they allow the estimation of the β_{11} and β_{22} parameters along with the β_0, β_1, and β_2 parameters; unfortunately, star designs do not allow the estimation of the β_{12} interaction parameter, and for all situations involving more than one factor, there are too few factor combinations ($2k+1$) to estimate all parameters of the full second-order polynomial model.

The juxtaposition of a two-level factorial design with a star design gives a *composite design* that can be used to estimate all parameters in a full second-order polynomial model. If the centers of the two separate experimental designs coincide, the resulting design is said to be a *central composite design*. If the centers do not coincide, the result is a *non-central composite design*. Figures 11.12 and 11.13 illustrate central composite designs for the two- and three-dimensional factor spaces.

Central composite designs are relatively efficient for small numbers of factors. "Efficiency" in this case means obtaining the required parameter estimates with little wasted effort. One measure of efficiency is the *efficiency value*, E.

$$E = p/f \tag{11.55}$$

where p is the number of parameters in the model to be fit, and f is the number of

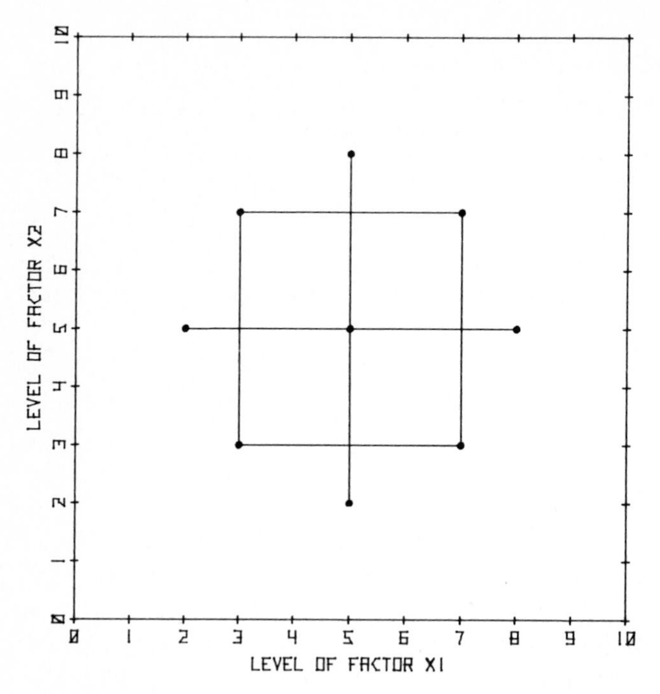

Figure 11.12. Factor combinations for a central composite experimental design in two-dimensional factor space.

factor combinations in the experimental design. The efficiency value ranges from unity to zero; if it is greater than unity, the model cannot be fit. In general, it is not desirable that the efficiency value be unity – that is, the design should not be

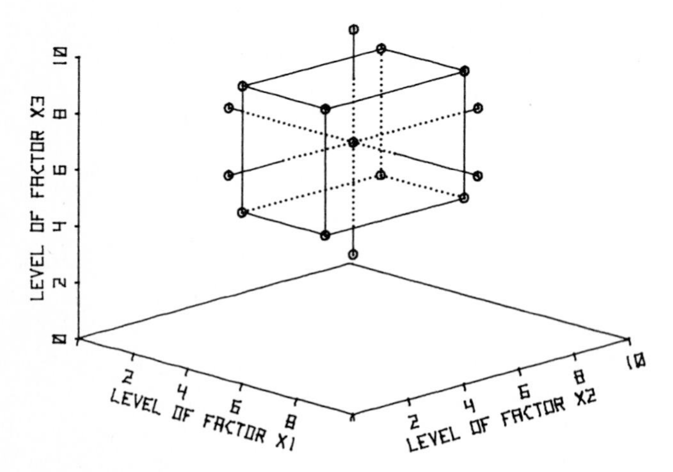

Figure 11.13. Factor combinations for a central composite experimental design in three-dimensional factor space.

TABLE 11.2
Efficiency of full second-order polynomial models fit to data from central composite experimental designs without replication.

Experimental factors, k	Parameters, $p=(k+1)(k+2)/2$	Factor combinations $f=2^k+2k+1$	Efficiency $E=p/f$
1	3	5	0.60
2	6	9	0.67
3	10	15	0.67
4	15	25	0.60
5	21	43	0.49
6	28	77	0.36
7	36	143	0.25
8	45	273	0.16
9	55	533	0.10
10	66	1045	0.06

perfectly efficient. Some inefficiency should usually be included to estimate lack of fit. Table 11.2 shows efficiency as functions of the number of experimental factors for fitting a full second-order polynomial model using a central composite experimental design. For up to about five factors the efficiency is very good, but decreases rapidly for six or more factors.

Replication is often included in central composite designs. If the response surface is thought to be reasonably homoscedastic, only one of the factor combinations (commonly the center point) need be replicated, usually three or four times to provide sufficient degrees of freedom for s_{pe}^2. If the response surface is thought to be heteroscedastic, the replicates can be spread over the response surface to obtain an "average" purely experimental uncertainty.

As an illustration of the use of central composite experimental designs, let us fit the model of Equation 11.54 to the following responses $\boldsymbol{Y}$ obtained at the coded factor combinations $\boldsymbol{D}^*$. We will assume that $c_{x_1}=c_{x_2}=5$, and $d_{x_1}=d_{x_2}=2$.

$$\boldsymbol{Y}=\begin{bmatrix} 7.43 \\ 7.93 \\ 6.84 \\ 6.40 \\ 6.50 \\ 4.42 \\ 7.45 \\ 7.41 \\ 7.95 \\ 7.81 \\ 7.86 \\ 7.85 \end{bmatrix} \tag{11.56}$$

$$D^* = \begin{bmatrix} -1 & -1 \\ -1 & 1 \\ 1 & -1 \\ 1 & 1 \\ -2 & 0 \\ 2 & 0 \\ 0 & -2 \\ 0 & 2 \\ 0 & 0 \\ 0 & 0 \\ 0 & 0 \\ 0 & 0 \end{bmatrix} \tag{11.57}$$

Inspection of the coded experimental design matrix shows that the first four experiments belong to the two-level two-factor factorial part of the design, the next four experiments are the extreme points of the star design, and the last four experiments are replicates of the center point. The corresponding $\boldsymbol{X}^*$ matrix for the six-parameter model of Equation 11.54 is

$$X^* = \begin{bmatrix} 1 & -1 & -1 & 1 & 1 & 1 \\ 1 & -1 & 1 & 1 & 1 & -1 \\ 1 & 1 & -1 & 1 & 1 & -1 \\ 1 & 1 & 1 & 1 & 1 & 1 \\ 1 & -2 & 0 & 4 & 0 & 0 \\ 1 & 2 & 0 & 4 & 0 & 0 \\ 1 & 0 & -2 & 0 & 4 & 0 \\ 1 & 0 & 2 & 0 & 4 & 0 \\ 1 & 0 & 0 & 0 & 0 & 0 \\ 1 & 0 & 0 & 0 & 0 & 0 \\ 1 & 0 & 0 & 0 & 0 & 0 \\ 1 & 0 & 0 & 0 & 0 & 0 \end{bmatrix} \tag{11.58}$$

$$(X^{*\prime}X^*) = \begin{bmatrix} 12 & 0 & 0 & 12 & 12 & 0 \\ 0 & 12 & 0 & 0 & 0 & 0 \\ 0 & 0 & 12 & 0 & 0 & 0 \\ 12 & 0 & 0 & 36 & 4 & 0 \\ 12 & 0 & 0 & 4 & 36 & 0 \\ 0 & 0 & 0 & 0 & 0 & 4 \end{bmatrix} \tag{11.59}$$

$$(X^{*\prime}X^*)^{-1} = \begin{bmatrix} 5/24 & 0 & 0 & -1/16 & -1/16 & 0 \\ 0 & 1/12 & 0 & 0 & 0 & 0 \\ 0 & 0 & 1/12 & 0 & 0 & 0 \\ -1/16 & 0 & 0 & 3/64 & 1/64 & 0 \\ -1/16 & 0 & 0 & 1/64 & 3/64 & 0 \\ 0 & 0 & 0 & 0 & 0 & 1/4 \end{bmatrix} \tag{11.60}$$

Note that with this coded experimental design, the estimates of b_1^*, b_2^*, and b_{12}^* will be independent of the other estimated parameters; the estimates of b_0^*, b_{11}^*, and b_{22}^* will be interdependent.

$$(\boldsymbol{X}^{*\prime}\boldsymbol{Y}) = \begin{bmatrix} 85.85 \\ -6.280 \\ -0.02000 \\ 72.28 \\ 88.04 \\ -0.9400 \end{bmatrix} \tag{11.61}$$

$$\hat{\boldsymbol{B}}^* = (\boldsymbol{X}^{*\prime}\boldsymbol{X}^*)^{-1}(\boldsymbol{X}^{*\prime}\boldsymbol{Y}) = \begin{bmatrix} 7.865 \\ -0.5233 \\ -0.001667 \\ -0.6019 \\ -0.1094 \\ -0.2350 \end{bmatrix} \tag{11.62}$$

Thus, the fitted model in coded factor space is

$$y_1 = 7.865 - 0.5233x_1^* - 0.001667x_2^* - 0.6019x_1^{*2} - 0.1094x_2^{*2} - 0.2350x_1^*x_2^* \tag{11.63}$$

In uncoded factor space the parameter estimates are

$$y_1 = 3.264 + 1.537x_1 + 0.5664x_2 - 0.1505x_1^2 - 0.02734x_2^2 - 0.05875x_1x_2 \tag{11.64}$$

The fitted model of Equation 11.64 is drawn in Figure 11.14. The negative estimates

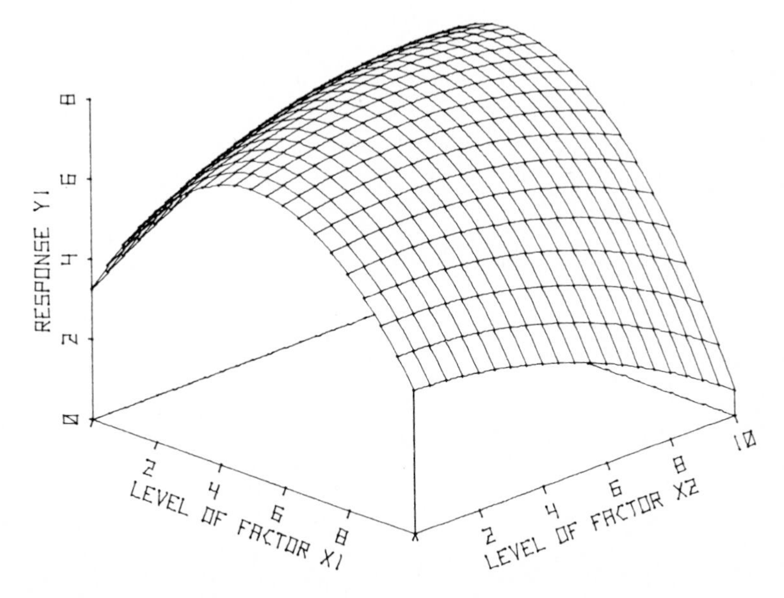

Figure 11.14. Graph of the fitted full second-order polynomial model $y_{1i} = 3.264 + 1.537x_{1i} + 0.5664x_{2i} - 0.1505x_{1i}^2 - 0.02734x_{2i}^2 - 0.0578x_{1i}x_{2i}$.

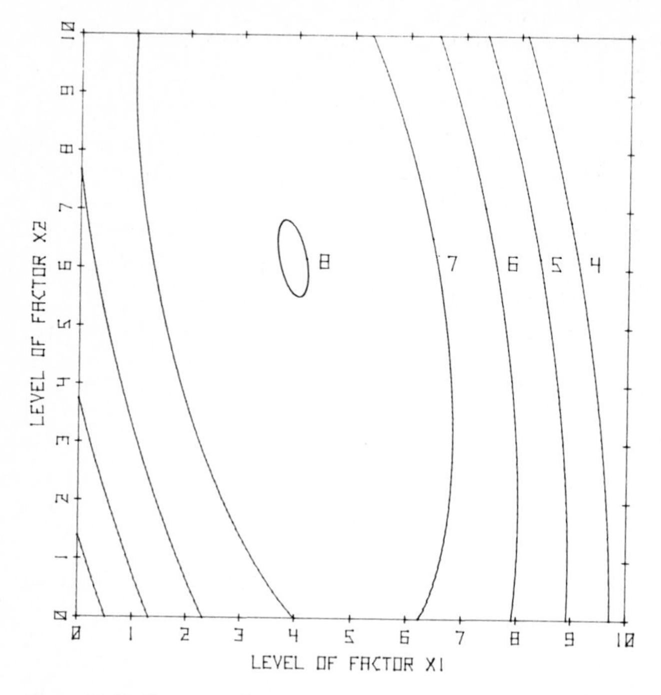

Figure 11.15. Contours of constant response as functions of x_1 and x_2 for the response surface of Figure 11.14.

of b_{11} and b_{22} cause the estimated response surface to fold downward quadratically in both factors, x_1 and x_2, although less rapidly in factor x_2 than in factor x_1. Careful inspection of Figure 11.14 reveals that the "ridge" of the surface is tilted with respect to the factor axes – the ridge runs obliquely from the middle of the left front side toward the far back corner of Figure 11.14. The "rotation" of the

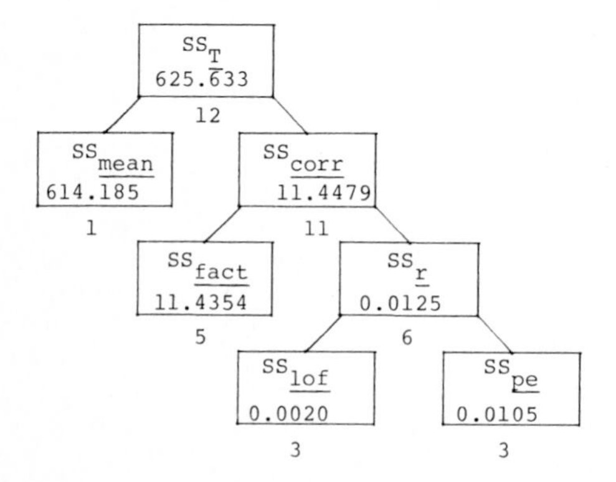

Figure 11.16. Sums of squares and degrees of freedom tree for Figure 11.14.

response surface with respect to the factor axes is caused by the interaction term $b_{12}x_1x_2$. This ridge is seen more clearly in Figure 11.15 which shows "contours of constant response" as a function of the factors x_1 and x_2 for Equation 11.64. Each contour in Figure 11.15 represents the intersection of the response surface in Figure 11.14 with a plane parallel to the x_1–x_2 plane at a given response level. Response contours for $y_1 = 8, 7, 6, 5,$ and 4 are shown in Figure 11.15.

The sum of squares and degrees of freedom tree for the fitted model is given in Figure 11.16. The R^2 value is 0.9989. The Fisher F-ratio for the significance of the factor effects is $F_{5,6} = 1096.70$ which is significant at the 100.0000% level of confidence. The F-ratio for the lack of fit is $F_{3,3} = 0.19$ which is not very significant. As expected, the residuals are small:

$$\boldsymbol{R} = \begin{bmatrix} -0.014 \\ 0.019 \\ -0.028 \\ 0.006 \\ -0.005 \\ 0.009 \\ 0.019 \\ -0.015 \\ 0.085 \\ -0.055 \\ -0.005 \\ -0.015 \end{bmatrix} \tag{11.65}$$

11.7. Canonical analysis

Full second-order polynomial models used with central composite experimental designs are very powerful tools for approximating the true behavior of many systems. However, the interpretation of the large number of estimated parameters in multifactor systems is not always straightforward. As an example, the parameter estimates of the coded and uncoded models in the previous section are quite different, even though the two models describe essentially the same response surface (see Equations 11.63 and 11.64). It is difficult to see this similarity by simple inspection of the two equations. Fortunately, *canonical analysis* is a mathematical technique that can be applied to full second-order polynomial models to reveal the essential features of the response surface and allow a simpler understanding of the factor effects and their interactions.

Canonical analysis achieves this geometric interpretation of the response surface by transforming the estimated polynomial model into a simpler form. The origin of the factor space is first *translated* to the *stationary point* of the estimated response surface, the point at which the partial derivatives of the response with respect to all of the factors are simultaneously equal to zero (see Section 10.5). The new factor

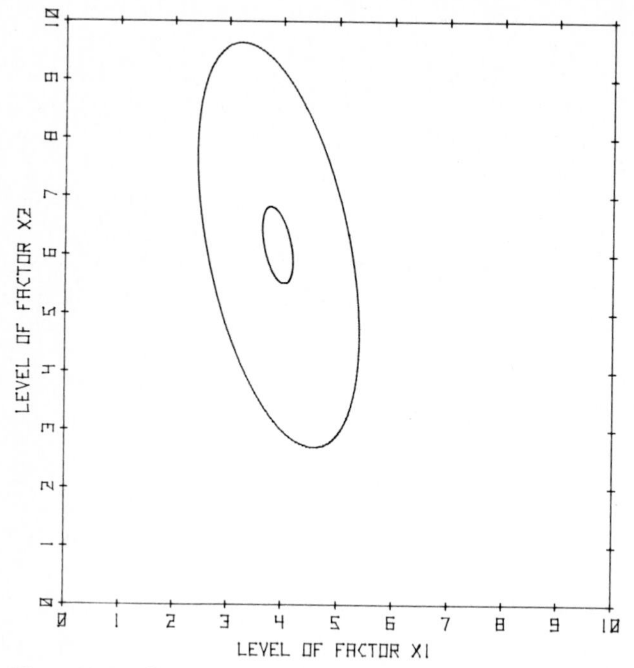

Figure 11.17a. Contours of constant response in two-dimensional factor space.

axes are then *rotated* to coincide with the principal axes of the second-order response surface. The process of canonical analysis is illustrated in Figure 11.17 for a two-factor response surface with elliptical contours of constant response. Translation has the effect of removing the first-order terms from the polynomial model; rotation has the effect of removing the interaction terms. It is the signs and magnitudes of the remaining second-order terms that reveal the essential features of the response surface.

To find the coordinates of the stationary point, we first differentiate the full second-order polynomial model with respect to each of the factors and set each derivative equal to zero. For two-factor models we obtain

$$\begin{aligned} \partial y_1/\partial x_1 &= b_1 + 2b_{11}x_1 + b_{12}x_2 = 0 \\ \partial y_1/\partial x_2 &= b_2 + 2b_{22}x_2 + b_{12}x_1 = 0 \end{aligned} \tag{11.66}$$

The coordinates of the stationary point (s_{x_1} and s_{x_2}) are those values of x_1 and x_2 that simultaneously satisfy both of these partial derivatives. Equation 11.66 may be rewritten as

$$\begin{aligned} 2b_{11}s_{x_1} + b_{12}s_{x_2} &= -b_1 \\ b_{12}s_{x_1} + 2b_{22}s_{x_2} &= -b_2 \end{aligned} \tag{11.67}$$

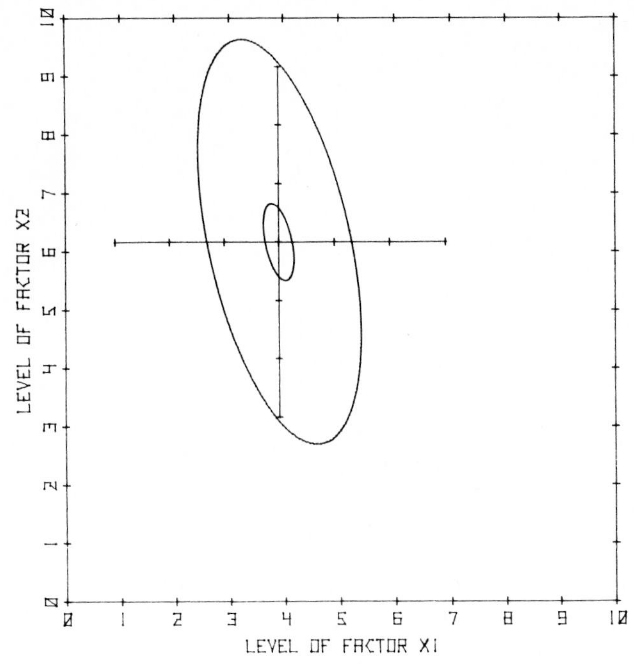

Figure 11.17b. Canonical axes translated to stationary point of response surface.

or in matrix notation

$$\begin{bmatrix} 2b_{11} & b_{12} \\ b_{12} & 2b_{22} \end{bmatrix} \begin{bmatrix} s_{x_1} \\ s_{x_2} \end{bmatrix} = \begin{bmatrix} b_1 \\ b_2 \end{bmatrix} \tag{11.68}$$

Let us define a $1 \times k$ *matrix of stationary point coordinates*, $\boldsymbol{s}$;

$$\boldsymbol{s} = [s_{x_1} \quad s_{x_2}] \tag{11.69}$$

Let us also define a $k \times 1$ *matrix of first-order parameter estimates*, $\boldsymbol{f}$;

$$\boldsymbol{f} = \begin{bmatrix} b_1 \\ b_2 \end{bmatrix} \tag{11.70}$$

Finally, we define a $k \times k$ *matrix of second-order parameter estimates*, $\boldsymbol{S}$;

$$\boldsymbol{S} = \begin{bmatrix} b_{11} & b_{12}/2 \\ b_{12}/2 & b_{22} \end{bmatrix} \tag{11.71}$$

The single-factor second-order parameter estimates lie along the diagonal of the $\boldsymbol{S}$ matrix, and the two-factor interaction parameter estimates are divided in half on either side of the diagonal.

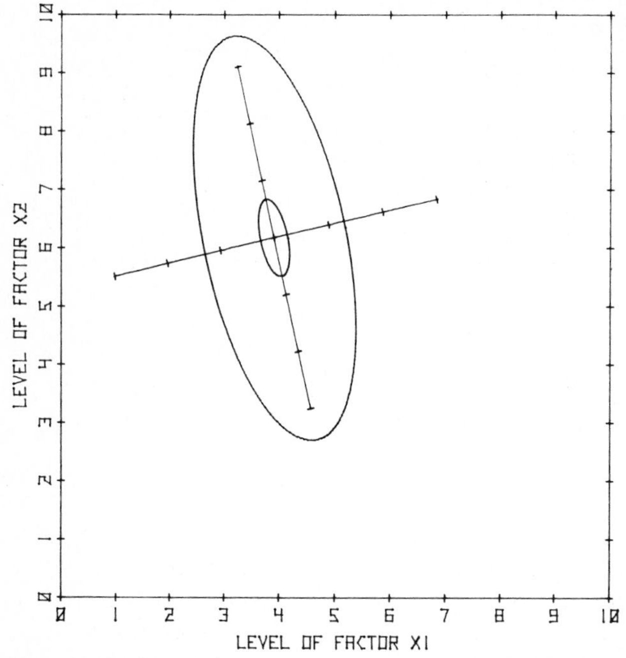

Figure 11.17c. Canonical axes rotated to coincide with principal axes of response surface.

By Equations 11.69–11.71, Equation 11.68 may be rewritten

$$2\boldsymbol{S}\boldsymbol{s}' = -\boldsymbol{f} \tag{11.72}$$

or

$$\boldsymbol{S}\boldsymbol{s}' = -0.5\boldsymbol{f} \tag{11.73}$$

The solution to this set of simultaneous equations gives the transpose of the stationary point coordinate matrix.

$$\boldsymbol{s}' = -0.5\boldsymbol{S}^{-1}\boldsymbol{f} \tag{11.74}$$

For the parameter estimates of Equation 11.64,

$$\boldsymbol{s}' = -0.5\begin{bmatrix} -0.1505 & -0.05875/2 \\ -0.05875/2 & -0.02734 \end{bmatrix}^{-1}\begin{bmatrix} 1.537 \\ 0.5664 \end{bmatrix} = \begin{bmatrix} 3.903 \\ 6.165 \end{bmatrix} \tag{11.75}$$

That is, the coordinates of the stationary point are $x_1 = 3.903$, $x_2 = 6.165$. The corresponding response at the stationary point is 8.009. These results may be verified qualitatively by inspection of Figures 11.14 and 11.15.

Translation of the origin of factor space to the location specified by $\boldsymbol{s}$ (i.e., coding in which $c_{x_1} = s_{x_1}$, $c_{x_2} = s_{x_2}$, and $d_{x_1} = d_{x_2} = 1$) has the effect of making $b_1^t = 0$ and $b_2^t = 0$, where the superscript t indicates coding involving translation only. The parameters b_{11}^t, b_{22}^t, and b_{12}^t have the same values as their uncoded estimates. If the

TABLE 11.3
Interpretation of two-factor canonical parameter estimates.

Sign of $\tilde{b}^{t}_{11}$	Relationship	Sign of $\tilde{b}^{t}_{22}$	Interpretation	Illustrated in figure
+	$\simeq$	+	Parabolic bowl opening upward	11.18
−	$\simeq$	−	Parabolic bowl opening downward	11.19
−	$<$	−	Flattened parabolic bowl opening downward	11.20
−	$\ll$	−	Ridge	11.21
+	$\simeq$	−	Saddle region or col	11.22

b^{t}_{0} term is subtracted from both sides of the equation, a simpler model for the response surface is obtained, a form containing second-order terms only.

$$y_1 - b^t_0 = b^t_{11}x^{t2}_1 + b^t_{22}x^{t2}_2 + b^t_{12}x^t_1x^t_2 \tag{11.76}$$

Rotation of the translated factor axes is an eigenvalue-eigenvector problem, the complete discussion of which is beyond the scope of this presentation. It may be shown that there exists a set of rotated factor axes such that the off-diagonal terms of the resulting $\tilde{\boldsymbol{S}}$ matrix are equal to zero (the ˜ indicates rotation); that is, in the translated and rotated coordinate system, *there are no interaction terms*. The

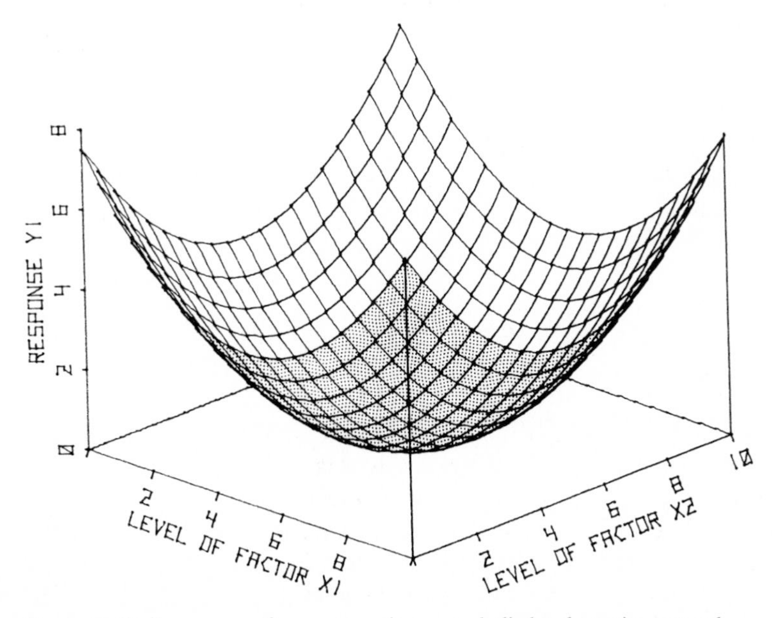

Figure 11.18. Response surface representing a parabolic bowl opening upward.

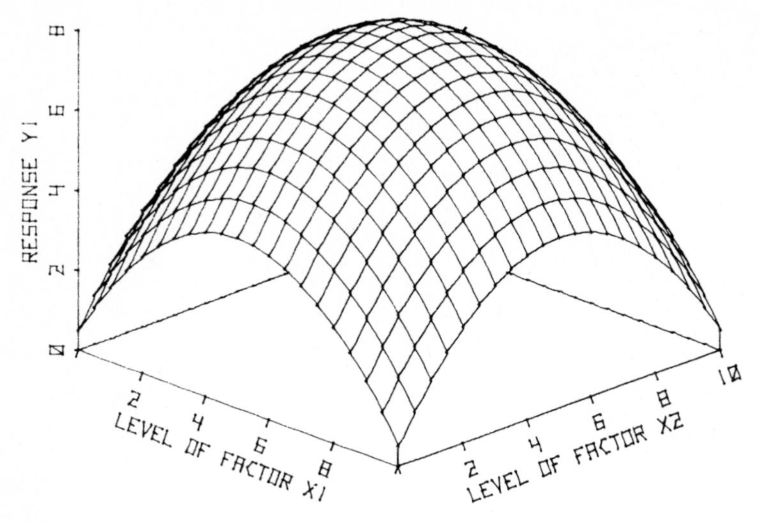

Figure 11.19. Response surface representing a parabolic bowl opening downward.

relationship between the rotated coordinate system and the translated coordinate system centered at the stationary point is given by

$$\begin{aligned} \tilde{x}^t_{1i} &= e_{11}x^t_{1i} + e_{12}x^t_{2i} = e_{11}(x_{1i} - s_{x_1}) + e_{12}(x_{2i} - s_{x_2}) \\ \tilde{x}^t_{2i} &= e_{21}x^t_{1i} + e_{22}x^t_{2i} = e_{21}(x_{1i} - s_{x_1}) + e_{22}(x_{2i} - s_{x_2}) \end{aligned} \quad (11.77)$$

where $\tilde{x}^t_{1i}$ and $\tilde{x}^t_{2i}$ are the coordinates in the translated and rotated factor space, and

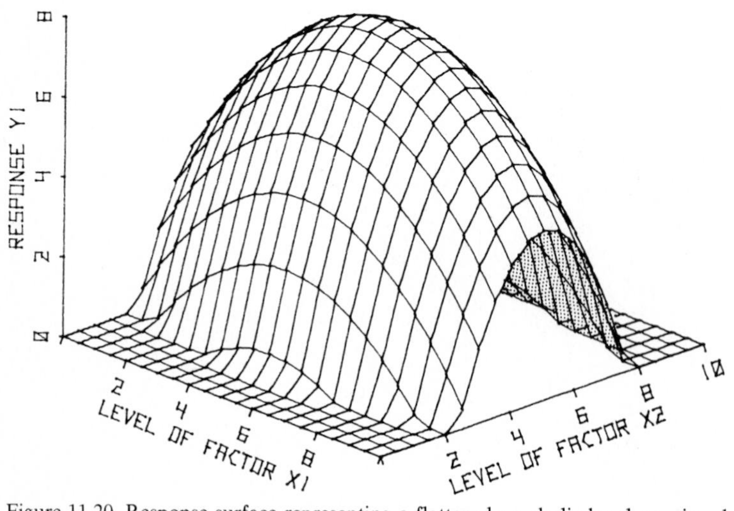

Figure 11.20. Response surface representing a flattened parabolic bowl opening downward.

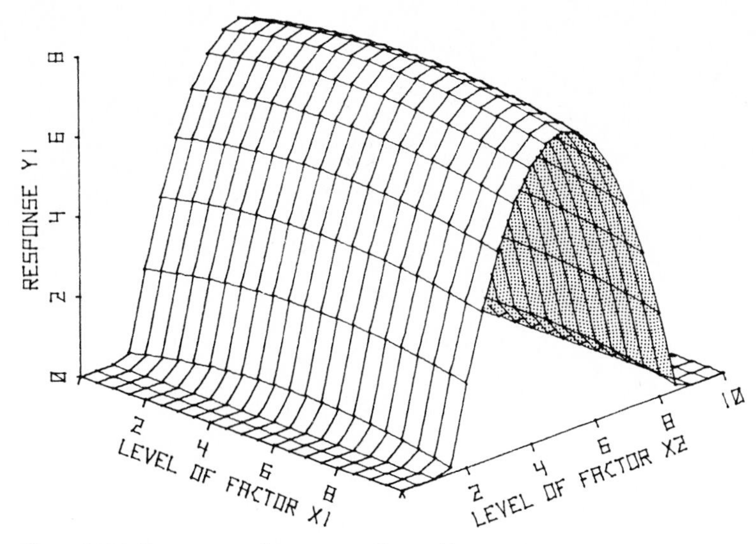

Figure 11.21. Response surface representing a ridge.

the e's are the elements of the rotation matrix $\boldsymbol{E}$ that results from the solution of the eigenvector problem. The corresponding rotated parameter estimates become $\tilde{b}_{11}^{t}$ and $\tilde{b}_{22}^{t}$ ($\tilde{b}_{12}^{t} = 0$). Thus, all non-degenerate two-factor full second-order polynomial models can be reduced to the form

$$y_1 - b_0^t = \tilde{b}_{11}^t x_{21}^{t2} + \tilde{b}_{22}^t x_{22}^{t2} \tag{11.78}$$

The signs and magnitudes of $\tilde{b}_{11}^{t}$ and $\tilde{b}_{22}^{t}$ reveal the essential features of the

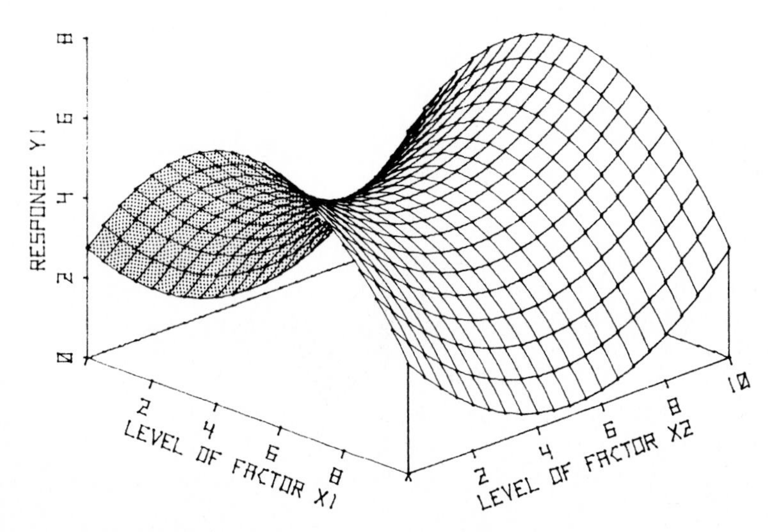

Figure 11.22. Response surface representing a saddle region.

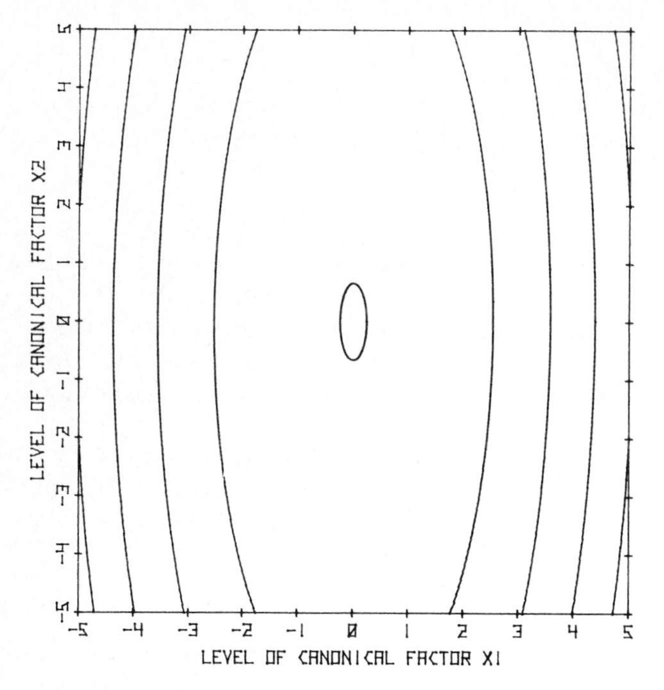

Figure 11.23. Contours of constant response as functions of the canonical axes $\tilde{x}_1^t$ and $\tilde{x}_2^t$ for the response surface of Figure 11.14.

response surface. Table 11.3 gives some of the possibilities; these possibilities are illustrated in Figures 11.18–11.22. For the parameter estimates of Equation 11.64, rotation of the translated factor axes gives

$$\boldsymbol{E} = \begin{bmatrix} e_{11} & e_{12} \\ e_{21} & e_{22} \end{bmatrix} = \begin{bmatrix} 0.9753 & 0.2207 \\ -0.2207 & 0.9753 \end{bmatrix} \tag{11.79}$$

$$\tilde{\boldsymbol{S}} = \begin{bmatrix} -0.1571 & 0 \\ 0 & -0.02069 \end{bmatrix} \tag{11.80}$$

Thus, the canonical factor axis $\tilde{x}_1^t$ is primarily x_1 in character ($0.9753x_1$) but with a slight amount of x_2 character ($0.2207x_2$). Similarly, $\tilde{x}_2^t$ is primarily x_2 ($0.9753x_2$) with a small amount of x_1 ($-0.2207x_1$). Not unexpectedly, in this example the rotated parameter estimates $\tilde{b}_{11}^t$ and $\tilde{b}_{22}^t$ are not very different from the corresponding unrotated values. Because $\tilde{b}_{11}^t$ and $\tilde{b}_{22}^t$ are both negative and because they differ by an order of magnitude, we expect the response surface to be a flattened parabolic bowl opening downward (see Table 11.3). Figure 11.23 graphs the response surface of Equation 11.64 in canonical form.

11.8. Confidence intervals

Confidence intervals for single-factor response surfaces were discussed in Section 10.6. The equations developed for estimating different types of confidence intervals (Equations 10.76, 10.79, 10.80, and 10.81) are entirely general and can be used for multi-factor response surfaces as well.

For example, plotting the standard uncertainty for the estimation of a single value of response (the square root of Equation 10.71) as a function of both x_1 and x_2 for the model expressed by equation 11.54 and the data of Section 11.6 gives the uncertainty surface shown in Figure 11.24 (compare with Figure 10.7). The uncertainty is smallest in the region of greatest experimentation, especially at the center point. The uncertainty is greatest in those regions farthest away from the experimental points, as expected.

The data in Figure 11.24 can be used with Equation 10.76 to generate confidence bands for the fitted model of Equation 11.64. The upper and lower 95% confidence bands are shown as transparent surfaces in Figure 11.25. Other confidence bands can be generated by Equations 10.79–10.81.

11.9. Rotatable designs

The uncertainty contour in Figure 11.24 is "lumpy" because the uncertainty depends not only on the distance from the center of the design, but also on the distance from points at which the other experiments have been carried out. For

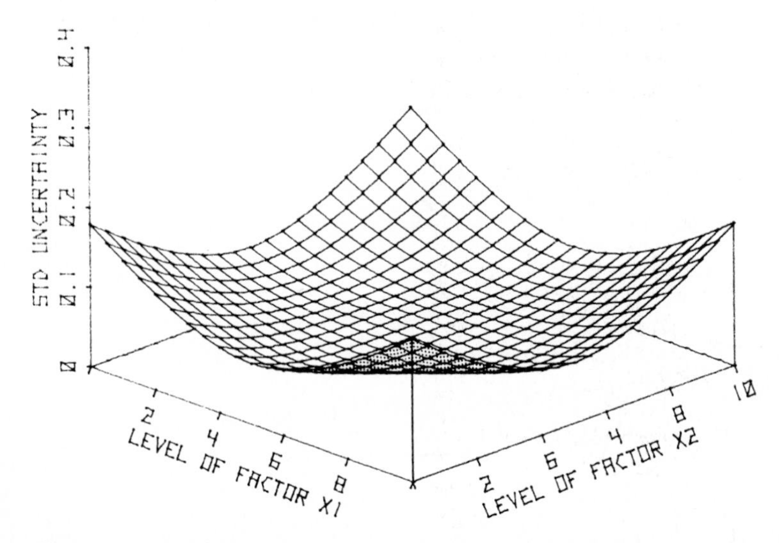

Figure 11.24. Standard uncertainty for estimating one new response as a function of the factors x_1 and x_2. See text for discussion.

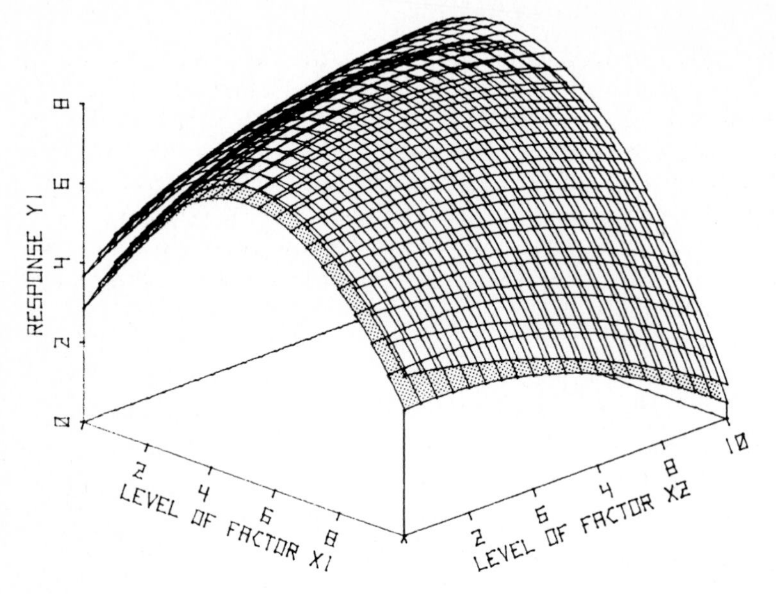

Figure 11.25. Upper and lower 95% confidence bands for the response surface of Figure 11.14.

some applications, it might be desirable that the uncertainty predicted by a full second-order polynomial model be dependent only on the distance from the center of the design and be independent of the location of the experimental points. Such a condition can be achieved by a class of central composite experimental designs known as *rotatable designs*. In effect, the star points are located the same distance from the center of the design as the factorial points. For a two-factor central composite design, the star points are located $\pm 2^{1/2}$ coded units from the center. Thus, the design of Equation 11.57 could be made rotatable if it had specific elements of

$$\boldsymbol{D}^* = \begin{bmatrix} -1 & -1 \\ -1 & 1 \\ 1 & -1 \\ 1 & 1 \\ -2^{1/2} & 0 \\ 2^{1/2} & 0 \\ 0 & -2^{1/2} \\ 0 & 2^{1/2} \\ 0 & 0 \\ 0 & 0 \\ 0 & 0 \\ 0 & 0 \end{bmatrix} \tag{11.81}$$

Assuming the same s_r^2 as was used to draw Figure 11.24, the uncertainty surface for the rotatable design of Equation 11.81 is shown in Figure 11.26. The uncertainty depends only on the distance from the center of the design; i.e., the contours of constant uncertainty are circular about the center of the design.

The use of rotatable designs usually makes sense only in normalized factor spaces (each factor divided by d_{x_1}) because it is difficult to define distance if the factors are measured in different units. For example, if x_1 is measured in °C and x_2 is measured in minutes, the distance of a point (x_{1i}, x_{2i}) from the center of the design (c_{x_1}, c_{x_2}) would be calculated as

$$\text{distance} = \left\{\left[(x_{1i} - c_{x_1})°\text{C}\right]^2 + \left[(x_{2i} - c_{x_2})\ \text{min}\right]^2\right\}^{1/2} \qquad (11.82)$$

However, it is not possible to add $°\text{C}^2$ and min^2. In a normalized factor space the factors are unitless and there is no difficulty with calculating distances. Coded rotatable designs do produce contours of constant response in the uncoded factor space, but in the uncoded factor space the contours are usually elliptical, not circular.

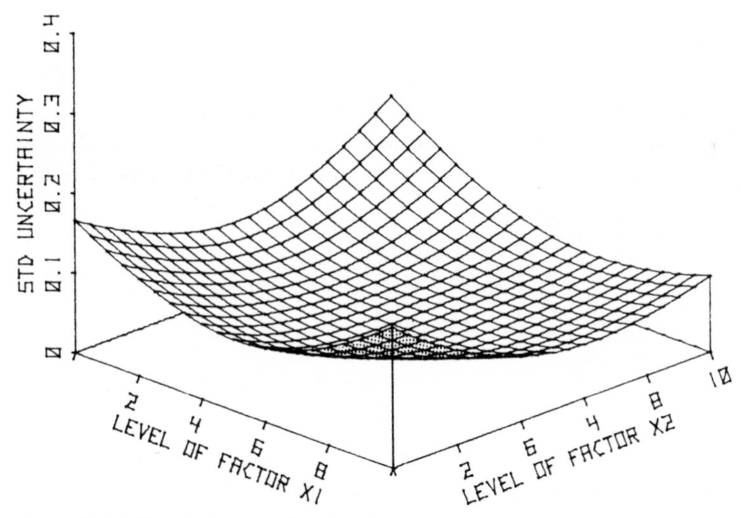

Figure 11.26. Standard uncertainty for estimating one new response as a function of the factors x_1 and x_2 for a rotatable design. Compare with Figure 11.24. See text for discussion.

11.10. Orthogonal designs

An interesting phenomenon occurs if we add four more center points to the design of Equation 11.81.

$$
\boldsymbol{D}^* = \begin{bmatrix} -1 & -1 \\ -1 & 1 \\ 1 & -1 \\ 1 & 1 \\ -2^{1/2} & 0 \\ 2^{1/2} & 0 \\ 0 & -2^{1/2} \\ 0 & 2^{1/2} \\ 0 & 0 \\ 0 & 0 \\ 0 & 0 \\ 0 & 0 \\ 0 & 0 \\ 0 & 0 \\ 0 & 0 \\ 0 & 0 \end{bmatrix} \tag{11.83}
$$

The corresponding matrix least squares treatment for the full second-order polynomial model proceeds as follows.

$$
\boldsymbol{X}^* = \begin{bmatrix} 1 & -1 & -1 & 1 & 1 & 1 \\ 1 & -1 & 1 & 1 & 1 & -1 \\ 1 & 1 & -1 & 1 & 1 & -1 \\ 1 & 1 & 1 & 1 & 1 & 1 \\ 1 & -2^{1/2} & 0 & 2 & 0 & 0 \\ 1 & 2^{1/2} & 0 & 2 & 0 & 0 \\ 1 & 0 & -2^{1/2} & 0 & 2 & 0 \\ 1 & 0 & 2^{1/2} & 0 & 2 & 0 \\ 1 & 0 & 0 & 0 & 0 & 0 \\ 1 & 0 & 0 & 0 & 0 & 0 \\ 1 & 0 & 0 & 0 & 0 & 0 \\ 1 & 0 & 0 & 0 & 0 & 0 \\ 1 & 0 & 0 & 0 & 0 & 0 \\ 1 & 0 & 0 & 0 & 0 & 0 \\ 1 & 0 & 0 & 0 & 0 & 0 \\ 1 & 0 & 0 & 0 & 0 & 0 \end{bmatrix} \tag{11.84}
$$

$$(\boldsymbol{X}^{*\prime}\boldsymbol{X}^{*}) = \begin{bmatrix} 16 & 0 & 0 & 8 & 8 & 0 \\ 0 & 8 & 0 & 0 & 0 & 0 \\ 0 & 0 & 8 & 0 & 0 & 0 \\ 8 & 0 & 0 & 12 & 4 & 0 \\ 8 & 0 & 0 & 4 & 12 & 0 \\ 0 & 0 & 0 & 0 & 0 & 4 \end{bmatrix} \tag{11.85}$$

$$(\boldsymbol{X}^{*\prime}\boldsymbol{X}^{*})^{-1} = \begin{bmatrix} 1/8 & 0 & 0 & -1/16 & -1/16 & 0 \\ 0 & 1/8 & 0 & 0 & 0 & 0 \\ 0 & 0 & 1/8 & 0 & 0 & 0 \\ -1/16 & 0 & 0 & 1/8 & 0 & 0 \\ -1/16 & 0 & 0 & 0 & 1/8 & 0 \\ 0 & 0 & 0 & 0 & 0 & 1/4 \end{bmatrix} \tag{11.86}$$

From this $(\boldsymbol{X}^{*\prime}\boldsymbol{X}^{*})^{-1}$ matrix it is clear that the estimates of b_1^*, b_2^*, b_{11}^*, b_{22}^*, and b_{12}^* do not depend upon each other; that is, there will be no covariance among the parameter estimates associated with the coded factor effects. (The parameter b_0^* is not a factor effect.) If the parameter estimates associated with any one factor in a multifactor design are uncorrelated with those of another, the experimental design is said to be *orthogonal*.

In this example, orthogonality of all factor effects has been achieved by including additional center points in the coded rotatable design of Equation 11.81. Orthogonality of some experimental designs may be achieved simply by appropriate coding (compare Equation 11.26 with Equation 11.20, for example). Because orthogonality is almost always achieved only in coded factor spaces, transformation of the coded parameter estimates ($\hat{\boldsymbol{B}}^*$) back to the uncoded factor space ($\hat{\boldsymbol{B}}$) usually destroys the condition of orthogonality.

Orthogonal designs have been historically important because they often permit simple mathematical formulas to be used to calculate the factor effects (e.g., the Yates algorithm for treating factorial designs), thus avoiding the tedious manual calculations that were previously required for the matrix least squares fitting of linear models to data. A practical disadvantage of orthogonal designs used solely for the purpose of "easy" calculation is that a missed or incorrectly executed experiment prevents the use of simple calculational formulas. However, such an incomplete data set can usually be treated by matrix least squares; this is a relatively trivial matter using modern computers.

11.11. Scaling

Visual presentation of data treatment results can sometimes be misleading if scaling factors are not taken into account. For example, the parabolic bowl shown in Figure 11.19 could be made to look like the ridge in Figure 11.21 if the scaling for the x_1-axis were extended (e.g., if the domain $4 \leqslant x_1 \leqslant 6$, $0 \leqslant x_2 \leqslant 10$ were plotted

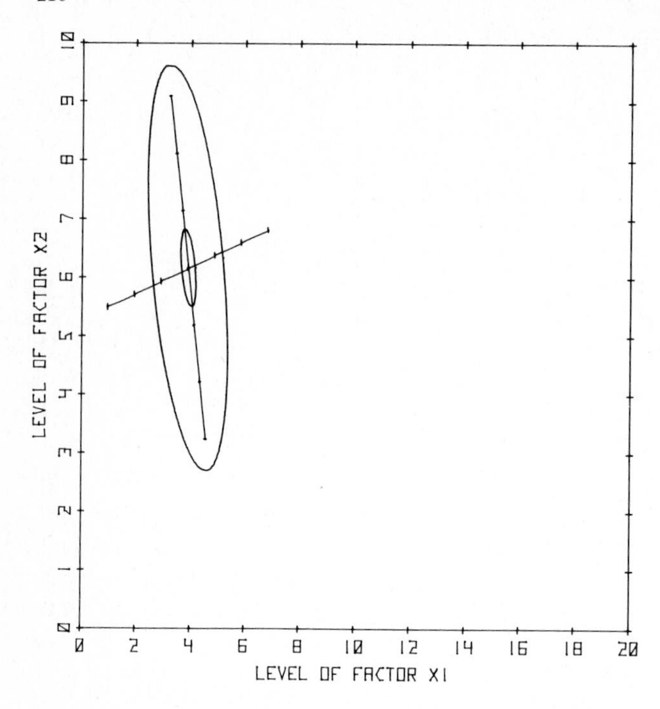

Figure 11.27. Replot of Figure 11.17c, x_1 axis compressed.

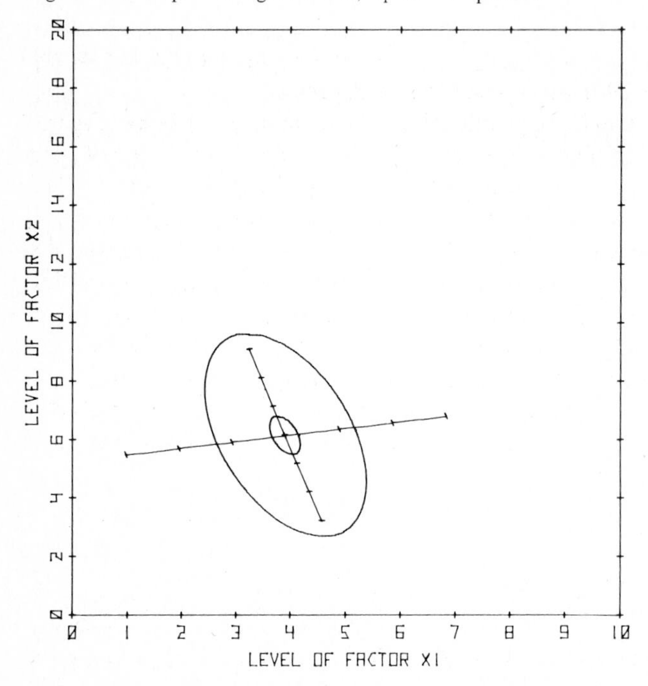

Figure 11.28. Replot of Figure 11.17c, x_2 axis compressed.

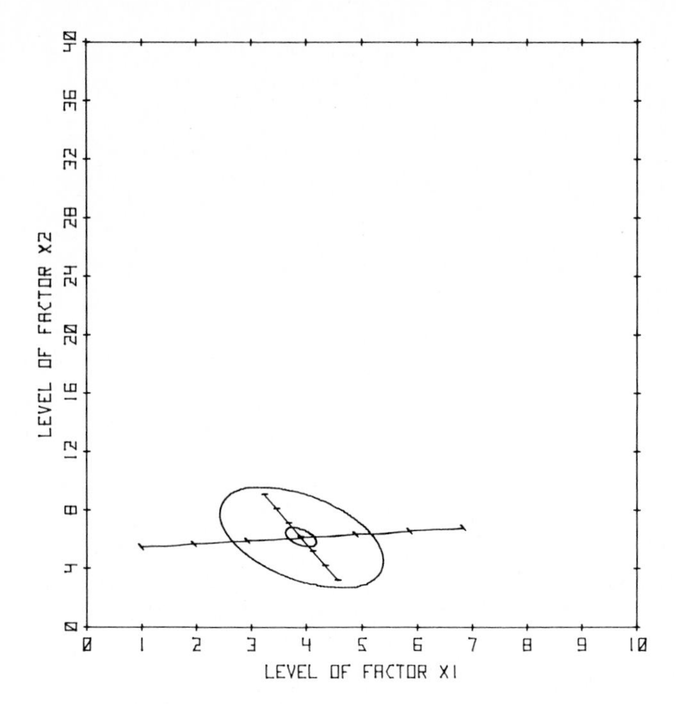

Figure 11.29. Replot of Figure 11.17c, x_2 axis highly compressed.

full scale). Similarly, the ridge of Figure 11.21 could be made to look like the parabolic bowl of Figure 11.19 if the x_1-axis were compressed.

Scaling can also affect apparent relationships in canonical analysis. Figures 11.27–11.29 show different scalings of Figure 11.17c. The canonical axes are at right angles to each other at all times in the coded factor space; it is the scaling of the graphical presentations that distorts them in the visual presentations.

The effects of visual scaling and numerical coding are very similar. However, the "distortions" caused by numerical coding are not always as readily apparent as some distortions caused by visual scaling. For example, rotatable designs in coded factor spaces might not produce rotatable designs in uncoded factor spaces (see Section 11.9). We simply warn the reader that concepts such as "rotatable", "circular", "orthogonal", "perpendicular", and "elliptical" are dependent upon both numerical coding and visual scaling; caution should be exercised in the use and understanding of these concepts.

References

Box, G.E.P., and Draper, N.R. (1969). *Evolutionary Operation. A Statistical Method for Process Improvement.* Wiley, New York.

Box, G.E.P., Hunter, W.G., and Hunter, J.S. (1978). *Statistics for Experimenters. An Introduction to Design, Data Analysis, and Model Building.* Wiley, New York.

Cochran, W.G., and Cox, G.M. (1957). *Experimental Designs*, 2nd ed. Wiley, New York.
Davies, O.L., Ed. (1956). *The Design and Analysis of Industrial Experiments.* Hafner, New York.
Draper, N.R., and Smith, H. (1966). *Applied Regression Analysis.* Wiley, New York.
Mendenhall, W. (1968). *Introduction to Linear Models and the Design and Analysis of Experiments.* Duxbury Press, Belmont, California.
Natrella, M.G. (1963). *Experimental Statistics* (Nat. Bur. of Stand. Handbook 91). US Govt. Printing Office, Washington, DC.
Neter, J., and Wasserman, W. (1974). *Applied Linear Statistical Models. Regression, Analysis of Variance, and Experimental Designs.* Irwin, Homewood, Illinois.
Wilson, E.B., Jr. (1952). *An Introduction to Scientific Research.* McGraw-Hill, New York.

Exercises

11.1. Multifactor systems

Draw a system diagram for a process with which you are familiar. Choose two inputs for investigation. Over what domain might these two inputs be varied and controlled? What linear mathematical model might be used to approximate the effects of these two factors on one of the responses from the system? Design a set of experiments to test the validity of this model.

11.2. Response surfaces

Sketch the response surface that might be expected for the system you chose in Exercise 11.1.

11.3. Factor interaction

Assume the two inputs you chose in Exercise 11.1 will exhibit factor interaction. If they would be expected to interact, what would be the mechanistic basis of that interaction? What might be the approximate mathematical form of that interaction? Would you use a mechanistic or empirical model to approximate the response surface? Why? Sketch the response surface predicted by this model.

11.4. Factor interaction

Eliminate the interaction terms from the model you chose in Exercise 11.3 and sketch the response surface predicted by this simpler model. How dissimilar is this sketch from the sketch made in Exercise 11.3? How important does the interaction term seem to be?

11.5. Homoscedasticity and heteroscedasticity

Is the "response surface" shown in Figure 11.6 homoscedastic or heteroscedastic?

11.6. Canonical analysis

What is the final canonical equation in Section 11.7?

11.7. Factorial designs

Sketch the location of the factor combinations in a 4^2 factorial design. In a 3^3 factorial design. In a 2^4 factorial design. What is f in each of these designs?

11.8. Matrix inversion

Verify that the $(\boldsymbol{X}'\boldsymbol{X})$ matrix obtained for the model of Equation 11.24 and the experimental design of Figure 11.9 cannot be inverted.

11.9. Coding

What is the relationship between Equations 11.40–11.43 and $\boldsymbol{A}$ of Equation 11.31? Could $\boldsymbol{A}$ be derived algebraically using c_{x_1} and d_{x_1} rather than using $\boldsymbol{X}$ and $\boldsymbol{X}^*$?

11.10. Corner designs

Corner designs are located by a starting point from which other factor combinations are generated by moving a positive distance in each factor dimension, one factor dimension at a time. How many factor combinations are there in corner designs? Sketch the location of the factor combinations in 2-, 3-, and 4-factor corner designs. How many parameters can be estimated by corner designs? What linear models are exactly fit by corner designs? Look up the definition of a "simplex". Are corner designs a class of simplex designs?

11.11. Experimental designs

Can the following design be used to fit the model expressed by Equation 11.44? Can it be used to fit the model expressed by Equation 11.15? How is this design related to a star design? To a corner design?

$$D = \begin{bmatrix} 0 & 0 \\ 0 & 1 \\ 0 & 2 \\ 1 & 0 \\ 2 & 0 \end{bmatrix}$$

11.12. Polynomial models

Write full second-order polynomial models for 1, 2, 3, 4, and 5 factors.

11.13. Composite designs

Sketch a 3-factor non-central composite design for which the center of the star coincides with one of the factorial points.

11.14. Canonical analysis

Perform a canonical analysis on the fitted equation $y_{1i} = 5.13 + 0.161x_{1i} - 0.373x_{2i} + 0.517x_{1i}^2 - 1.33x_{1i}^2 - 0.758x_{1i}x_{2i}$. What are the coordinates of the stationary point? What are the characteristics of the response surface in the region of the stationary point (see Table 11.3)?

11.15. Canonical analysis

Write a table similar to Table 11.3 for three-factor systems. What do the possible isoresponse contours look like in three-dimensional factor space? [See, for example, Box, G.E.P. (1954). "The Exploration and Exploitation of Response Surfaces: Some General Considerations and Examples," *Biometrics*, 10, 16.]

11.16. Efficiency of factorial designs

Derive an expression that describes the number of parameters p as a function of the number of factors k for the general model of Equation 11.15 (first-order with interaction and offset). Prepare a table of efficiency ($E = p/n$) of two-level factorial designs for this model (see Table 11.2).

11.17. Response surfaces

Sketch the response surface for Equation 11.10 over the factor space $-10 \leqslant x_1 \leqslant 0$, $0 \leqslant x_2 \leqslant 10$.

11.18. Canonical analysis

Canonical analysis of Equation 11.63 gives $\hat{y}_1 = 8.008 - 0.6285x_1^{*} - 0.08280x_2^{*}$. Compare this with Equation 11.80 which suggests that $\hat{y}_1 = 8.009 - 0.1571\tilde{x}_1^{t} - 0.02069\tilde{x}_2^{t}$, keeping in mind that in translation of axes, $d_{x_1} = d_{x_2} = 1$ whereas in Equation 11.63, $d_{x_1} = d_{x_2} = 2$. Comment.

11.19. Matrix representations

Show that the full two-factor second-order polynomial model may be written $y_{1i} = \beta_0 + \boldsymbol{D}_0 + \boldsymbol{D}_0\boldsymbol{S}\boldsymbol{D}_0'$, where $\boldsymbol{D}_0 = [x_{1i} \quad x_{2i}]$. Show that this may be extended to full three-factor second-order polynomial models.

11.20. Experimental optimization

One important use of experimental designs is to achieve optimum operating conditions of industrial processes. For a discussion of this application, see Box, G.E.P., and Wilson, K.B. (1951). "On the Experimental Attainment of Optimum Conditions," *J. Royal Statist. Soc.*, *B*, 13, 1. This paper is extraordinarily rich in response surface concepts. What is the "steepest ascent technique" discussed in this paper? What models are assumed, and what experimental designs are used?

11.21. Empirical and mechanistic models

A paper by Box, G.E.P., and Youle, P.V. (1955). "The Exploration and Exploitation of Response Surfaces: An Example of the Link Between the Fitted Surface and the Basic Mechanism of the System," *Biometrics*, 11, 287, explores the relationships between an empirical model and a fundamental mechanism. In their Section 9, they discuss some aspects of the process of scientific investigation. What do they perceive as the relationships among experiment, theory, and knowledge?

11.22. Wording

After Equation 12.12, the statement is made, "...it is unlikely that the sodium ion concentration has a statistically significant effect..." Does this statement have content, or is it meaningless? What would be a better way of stating the results?

11.23. Central composite designs

It has been remarked that a three-level two-factor factorial design can be the same as a two-factor central composite design. Comment.

CHAPTER 12

Additional Multifactor Concepts and Experimental Designs

In this final chapter we discuss the multifactor concepts of confounding and randomization. The ideas underlying these concepts are then used to develop experimental designs for discrete or qualitative variables.

12.1. Confounding

Consider the situation of a researcher who believes that the rate of an enzyme catalyzed reaction is affected not only by factors such as temperature, substrate concentration, and pH (see Section 10.1), but also by the concentration of sodium ion ($[Na^+]$) in solution with the enzyme. To investigate this hypothesis, the researcher designs a set of experiments in which all factors are kept constant but one: the concentration of sodium ion is varied from 0 to 10 millimolar (mM) according to the design matrix

$$\boldsymbol{D} = \begin{bmatrix} 0 \\ 1 \\ 2 \\ 3 \\ 4 \\ 5 \\ 6 \\ 7 \\ 8 \\ 9 \\ 10 \end{bmatrix} \tag{12.1}$$

That is, in the first experiment there is no added sodium ion; in the second experiment, $[Na^+] = 1$ mM; in the third experiment, $[Na^+] = 2$ mM; and so on, until in the eleventh and last experiment, $[Na^+] = 10$ mM. Carrying out these experiments (one every ten minutes), the researcher obtains the response matrix

$$\boldsymbol{Y} = \begin{bmatrix} 96 \\ 89 \\ 84 \\ 81 \\ 75 \\ 70 \\ 65 \\ 61 \\ 54 \\ 49 \\ 46 \end{bmatrix} \tag{12.2}$$

The fitted two-parameter straight line model is

$$y_{1i} = 95 - 5x_{1i} + r_{1i} \tag{12.3}$$

or

$$\text{Rate} = 95 - 5[\text{Na}^+]\text{m}M + r_{1i} \tag{12.4}$$

The data and fitted model are shown in Figure 12.1. The parameter estimate b_1 is highly significant.

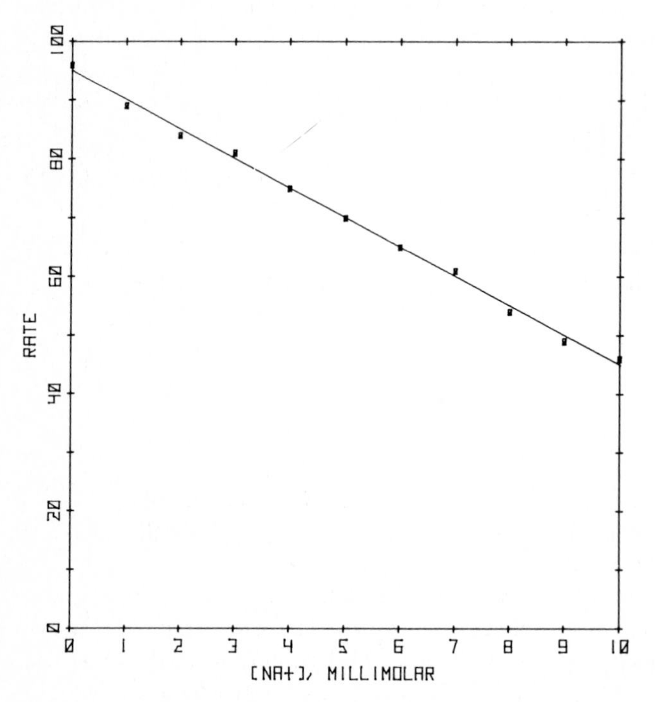

Figure 12.1. Graph of the fitted model $y_{1i} = 95 - 5x_{1i}$ with experimental data.

Based upon this information, the researcher would probably conclude that the concentration of sodium ion does have an effect on the rate of the enzyme catalyzed reaction. However, this conclusion might be wrong because of the lurking presence of a highly correlated, masquerading factor, *time* (see Section 1.2).

A better description of the design matrix would involve not only the concentration of sodium ion (x_1, mM), but also the time at which each experiment was carried out (x_2, min). If we begin our measurement of time with the first experiment, then the design matrix for the previous set of experiments would be

$$\boldsymbol{D} = \begin{bmatrix} 0 & 0 \\ 1 & 10 \\ 2 & 20 \\ 3 & 30 \\ 4 & 40 \\ 5 & 50 \\ 6 & 60 \\ 7 & 70 \\ 8 & 80 \\ 9 & 90 \\ 10 & 100 \end{bmatrix} \tag{12.5}$$

The experimental design in two-dimensional factor space is shown in Figure 12.2.

A simple model that would account for both $[Na^+]$ and time is

$$y_{1i} = \beta_0 + \beta_1 x_{1i} + \beta_2 x_{2i} + r_{1i} \tag{12.6}$$

The $(\boldsymbol{X}'\boldsymbol{X})$ matrix is given by

$$(\boldsymbol{X}'\boldsymbol{X}) = \begin{bmatrix} 11 & 55 & 550 \\ 55 & 385 & 3850 \\ 550 & 3850 & 38\,500 \end{bmatrix} \tag{12.7}$$

Calculation of $(\boldsymbol{X}'\boldsymbol{X})^{-1}$ is not possible, however, because the determinant of Equation 12.7 is zero. Thus, there are an infinite number of solutions to this problem, two of which are

$$y_{1i} = 95 - 5x_{1i} + 0x_{2i} + r_{1i} \tag{12.8}$$

and

$$y_{1i} = 95 + 0x_{1i} - 0.5x_{2i} + r_{1i} \tag{12.9}$$

There is no *unique* combination of b_1 and b_2 that satisfies the condition of least squares; all combinations of b_1 and b_2 such that $b_1 = (0.5 + b_2)/0.1$ will produce a minimum sum of squares of residuals.

The reason for this difficulty is that we are trying to fit a *planar* response surface (Equation 12.6) to data that have been obtained in a *line* (see Figure 12.2). An infinite number of planes can be made to pass through this line; therefore, an infinite number of combinations of b_1 and b_2 satisfy Equation 12.6. (See Sections 4.4 and 5.6 where the analogous difficulty of fitting a straight *line* to a *point* was discussed.) Figure 12.3 shows the graph of Equation 12.8; Figure 12.4 is the graph of Equation 12.9. Both give equally good fits to the data.

Our researcher viewed the system as revealed by Equation 12.8 and Figure 12.3 ($[Na^+]$ responsible for the change in rate). However, Equation 12.9 and Figure 12.4 might be correct instead; the enzyme could be denaturing with time – that is, changing its structure and therefore losing its ability to catalyze the reaction. It might also be true that *both* effects are taking place ($[Na^+]$ could have an effect and the enzyme could be denaturing). We are hopelessly confused about which factor is causing the observed effect. We are *confounded*. The factors x_1 and x_2 are said to be *confounded with each other*. The experimental design is responsible for this confusion because the two factors, $[Na^+]$ and time, are so highly correlated.

12.2. Randomization

If we could somehow destroy the high correlation between $[Na^+]$ and time (see Figure 12.2), then we might be able to unravel the individual effects of these two factors. One way to avoid high correlations among factors is to use uncorrelated designs such as factorial designs (Section 11.3), star designs (Sections 11.5), or central composite designs (Section 11.6). However, when time is a factor it is often

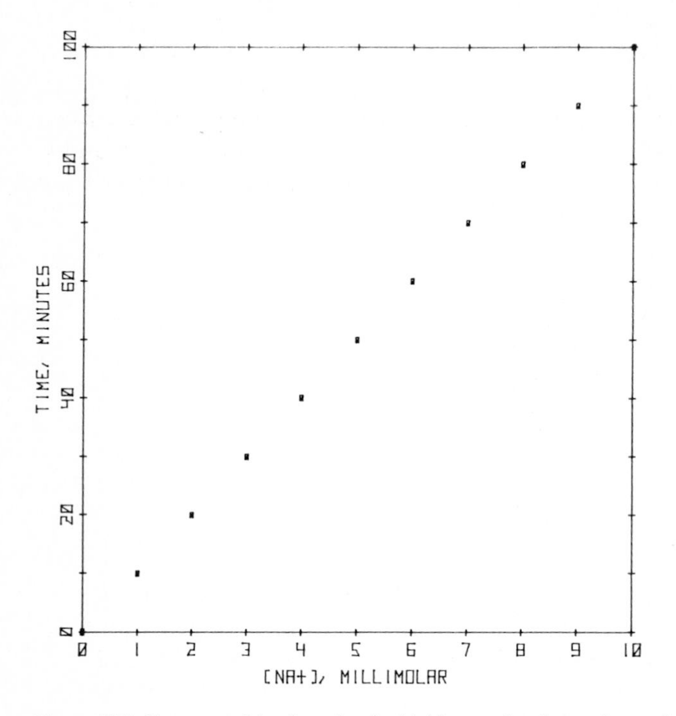

Figure 12.2. Factor combinations for the highly correlated experimental design of Equation 12.5.

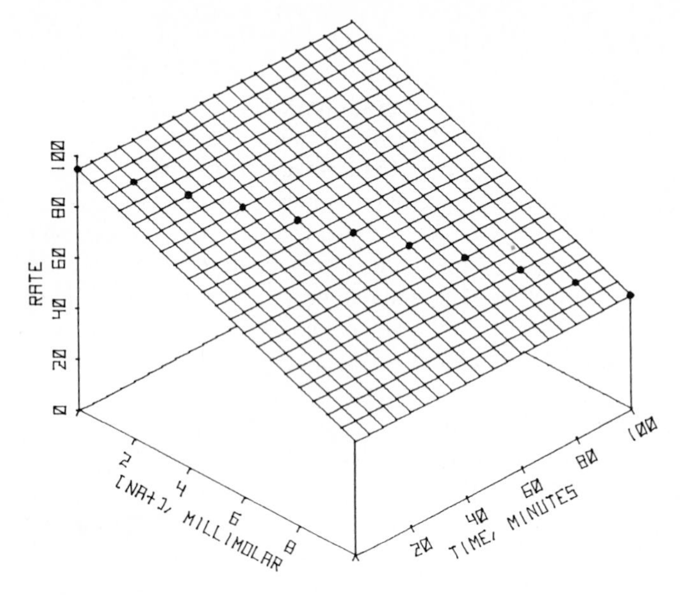

Figure 12.3 Response surface for the fitted model $y_{1i} = 95 - 5x_{1i} + 0x_{2i}$.

difficult to use these highly structured designs because they require running several experiments simultaneously. (E.g., replicate center points demand that several experiments be run at the same fixed levels of all factors; when time is a factor, this means they must all be run at the same fixed level of time – that is, simultaneously.)

Randomization is another approach to avoiding high correlations among factors. For the example of Section 12.1, randomization is accomplished by mixing up the order in which the experiments are carried out. Any of a number of methods might

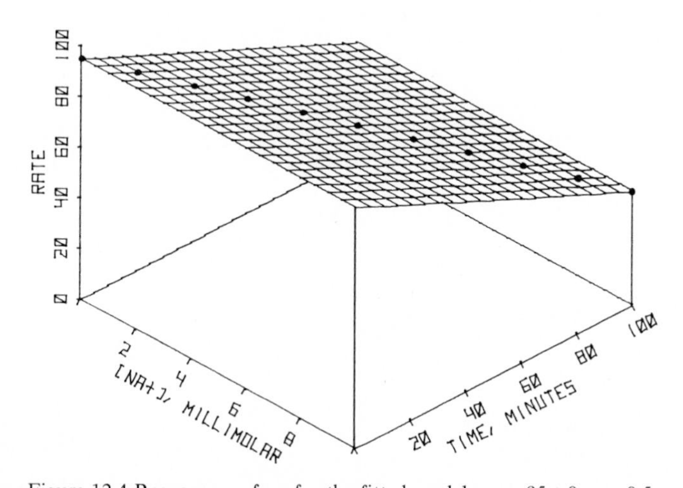

Figure 12.4 Response surface for the fitted model $y_{1i} = 95 + 0x_{1i} - 0.5x_{2i}$.

be used to randomly assign times to concentrations of sodium ion (or, equivalently, randomly assign concentrations of sodium ion to times). One method would be to put eleven slips of paper on which are written the times 0, 10, 20, ... in one bowl, and eleven slips of paper on which are written the sodium ion concentrations 0, 1, 2, ... in another bowl; mix the contents of each bowl; and blindly draw pairs of slips of paper, one slip from each bowl. The resulting pairs would be a random assignment of times and sodium ion concentrations. Other methods of randomization include the use of random number tables, and random number functions in computers. (Random number functions within computers should be used with caution, however; some of them are not as random as one might be led to believe.)

Figure 12.5 illustrates one completely random pairing of times and sodium ion concentration. Very little correlation is evident in this figure ($r^2 = 0.056$).

It must be remembered that randomization *does not guarantee* that factor combinations will be uncorrelated; after all, the pairing shown in Figure 12.2 has exactly the same chance of occurring randomly as the pairing shown in Figure 12.5 does. Randomization is useful because the number of patterns that are very highly correlated is usually so small compared with the total number of possible patterns that there is only a small probability of randomly obtaining a highly correlated pattern. Nevertheless, it is wise to examine the randomized design to fully protect against the unlikely possibility of having obtained a correlated design.

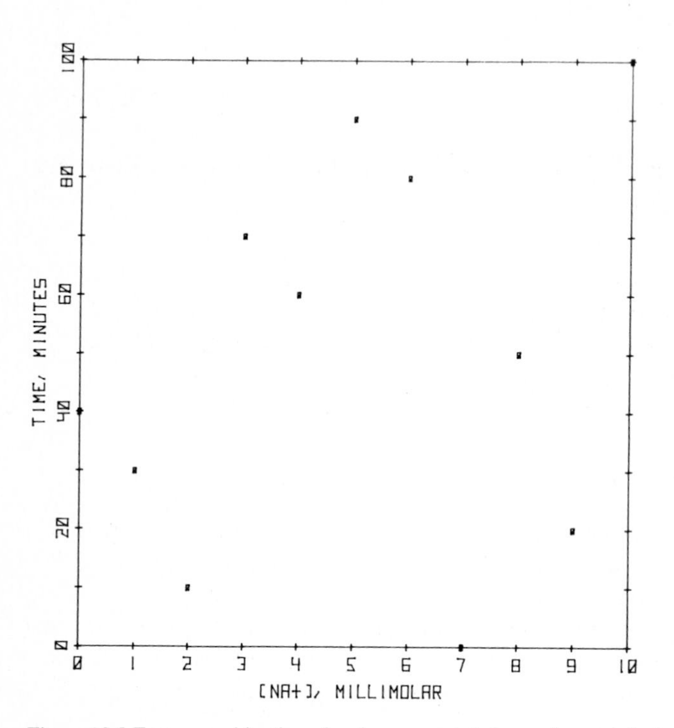

Figure 12.5 Factor combinations for the uncorrelated experimental design of Equation 12.10.

The experimental design matrix for the randomized pattern shown in Figure 12.5 is

$$\boldsymbol{D} = \begin{bmatrix} 7 & 0 \\ 2 & 10 \\ 9 & 20 \\ 1 & 30 \\ 0 & 40 \\ 8 & 50 \\ 4 & 60 \\ 3 & 70 \\ 6 & 80 \\ 5 & 90 \\ 10 & 100 \end{bmatrix} \tag{12.10}$$

Let us assume that the corresponding response matrix is

$$\boldsymbol{Y} = \begin{bmatrix} 96 \\ 89 \\ 84 \\ 81 \\ 75 \\ 70 \\ 65 \\ 61 \\ 54 \\ 49 \\ 46 \end{bmatrix} \tag{12.11}$$

Matrix least squares fitting of the model of Equation 12.6 gives an $(\boldsymbol{X}'\boldsymbol{X})$ matrix that can be inverted. The fitted model is

$$\text{Rate} = 95.04 - 0.009629[\text{Na}^+]\text{m}M - 0.4998 \text{ min} + r_{1i} \tag{12.12}$$

Confidence intervals (Equation 10.66) suggest that it is unlikely the sodium ion concentration has a statistically significant effect upon the rate of the enzyme catalyzed reaction; some other factor, one that is correlated with time, is probably responsible for the observed effect (see Section 1.2 on masquerading factors).

12.3. Completely randomized designs

Suppose we are interested in investigating the effect of fermentation temperature on the percent alcohol response of the wine-making system shown in Figure 1.6. We will assume that ambient pressure has very little effect on the system and that the small variations in response caused by this uncontrolled factor can be included in the residuals. Further, we can use the same type and quantity of yeast in all of our experiments so there will be no (or very little) variation in our results caused by the factor "yeast".

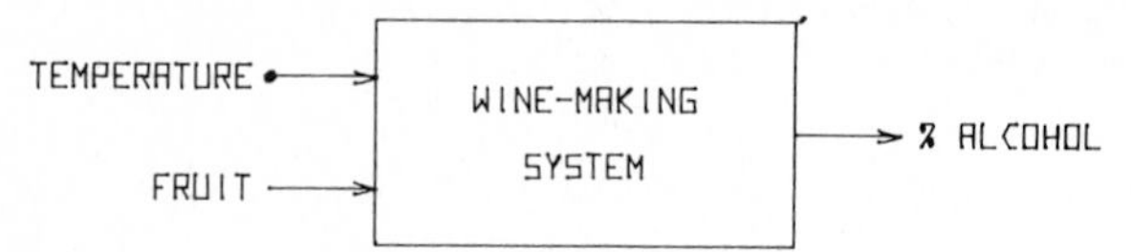

Figure 12.6 General system theory view of a wine-making process.

Because we intend to carry out these experiments in the leisure winter months, our inventory of frozen fruit will be low and we will not have enough of any one type of fruit to be used in all of the experiments. However, we will have available modest amounts of each of twenty types of fruit, so we can randomly assign these fruit types to each experiment and expect to "average out" any effect caused by variability of the qualitative factor, "fruit". A systems view of the experimental arrangement is shown in Figure 12.6.

Let us carry out a so-called "screening experiment" in which we attempt to discover if the fermentation temperature (factor x_1) has a "significant" effect on the response y_1 (% alcohol content). We will choose two levels of temperature, 23°C and 27°C. This is the minimal number of factor levels required to fit the two-parameter model

$$y_{1i} = \beta_0 + \beta_1 x_{1i} + r_{1i} \tag{12.13}$$

where y_{1i} is the % alcohol content and x_{1i} is the fermentation temperature. Table 12.1 contains the experimental design and the results of the investigation. Note that the time order of the experiments has been randomized, and the different fruits have been randomly assigned to the temperatures (with the restriction that ten fruits are assigned to each of the two temperatures). The sugar content of each fruit is also listed. Figure 12.7 shows the factor combinations of fruit number and temperature represented in Table 12.1; the number beside each combination indicates the time order in which the experiments are run.

Least squares fitting of the model expressed by Equation 12.13 to the data in Table 12.1 gives the fitted model

$$y_{1i} = -1.746 + 0.4540 x_{1i} \tag{12.14}$$

Figure 12.8 gives the sum of squares and degrees of freedom tree. The data and fitted model are shown in Figure 12.9. Using Equation 10.66 as a basis, the parameter estimate $b_1 = 0.4540$ is significant at the 84.22% level of confidence:

$$F_{1,18} = b_1^2/s_{b_1}^2 = (0.4540)^2/(7.590 \times 0.0125) = 2.1726 \tag{12.15}$$

Note that the F-ratio in Equation 12.15 would be larger (and therefore more significant) if the effect of temperature were stronger ($b_1 > 0.4540$); however, we have no control over the value of b_1. The F-ratio would also be larger if $s_{b_1}^2$ were smaller; we can have control over $s_{b_1}^2$.

The value of $s_{b_1}^2$ depends upon two quantities: The appropriate element of the $(\boldsymbol{X'X})^{-1}$ matrix and the quantity s_r^2 (see Equation 10.65). The first of these, the

TABLE 12.1
Completely randomized experimental design for determining the effect of temperatture on a wine-making system.

Experiment	Fruit	Sugar, %	Temperature	% Alcohol
1	18	3.38	23	5.10
2	14	10.35	23	11.08
3	3	6.53	27	9.62
4	19	4.28	23	6.62
5	16	1.48	27	5.98
6	4	9.66	27	12.92
7	5	8.26	23	10.20
8	13	3.73	27	8.58
9	9	1.65	27	7.32
10	17	5.30	27	10.64
11	2	7.02	23	9.60
12	6	5.58	23	8.86
13	20	0.82	23	5.44
14	12	2.16	23	6.92
15	7	7.81	27	13.04
16	15	2.60	27	9.28
17	11	5.40	23	9.92
18	8	9.45	27	15.16
19	1	5.73	27	12.58
20	10	9.04	23	13.22

element of the $(\boldsymbol{X}'\boldsymbol{X})^{-1}$ matrix associated with b_1, can be made smaller by increasing the number of experiments (which will *also* increase the number of degrees of freedom in the denominator of the F-ratio and improve the confidence of the estimate) and/or by using a broader region of experimentation in the factor space (e.g., by carrying out experiments at 21°C and 29°C instead of 23°C and 27°C). Decreasing $s_{b_1}^2$ by decreasing the value of s_r^2 might be accomplished by using a model that takes other factors into account, *or by removing the effect of other factors by appropriate experimental design*. We will discuss this latter technique in subsequent sections; in the remainder of this section, we show an example of decreasing $s_{b_1}^2$ by using an expanded model.

Careful inspection of the information in Table 12.1 suggests that sugar content might be a significant factor in determining the percent alcohol content of wine. The plot of percent alcohol vs. percent sugar in Figure 12.10 seems to confirm this. Let us fit the data of Table 12.1 to the model

$$y_{1i} = \beta_0 + \beta_1 x_{1i} + \beta_2 x_{2i} + r_{1i} \qquad (12.16)$$

where y_{1i} is again the percent alcohol, x_{1i} is the temperature, and x_{2i} is the percent sugar in the fruit. Figure 12.11 gives the sums of squares and degrees of freedom tree. The fitted model is

$$y_{1i} = -7.423 + 0.5018x_{1i} + 0.8133x_{2i} \qquad (12.17)$$

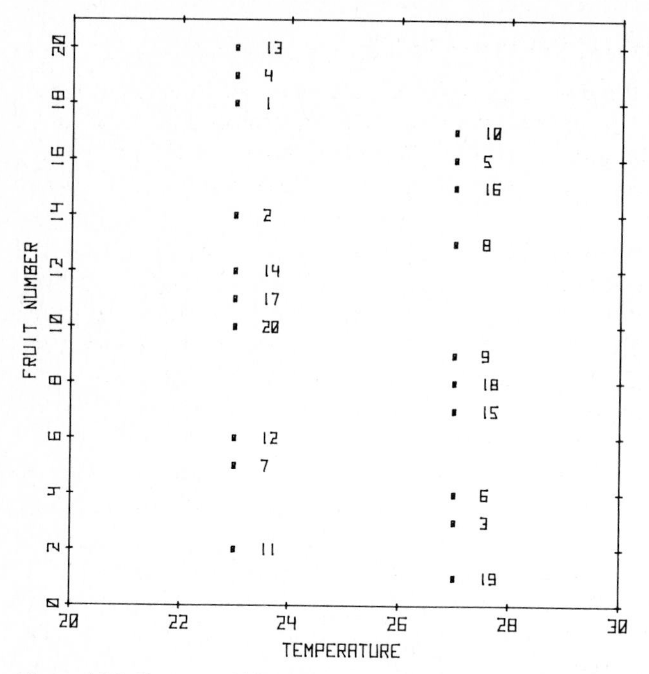

Figure 12.7 Factor combinations for a completely randomized design investigating the effect of temperature. "Fruit number" is an arbitrarily assigned, qualitative factor. Numbers beside factor combinations indicate the time order in which experiments were run.

and is plotted in Figure 12.12. The variance of residuals has been greatly reduced by taking into account the effect of sugar content; s_r^2 is now 1.550 compared to 7.590 when sugar was not included as a factor. The parameter estimate $b_1 = 0.5018$ is now significant at the 99.78% level of confidence:

$$F_{1,17} = b_1^2/s_{b_1}^2 = (0.5018)^2/(1.550 \times 0.0125) = 12.98 \tag{12.18}$$

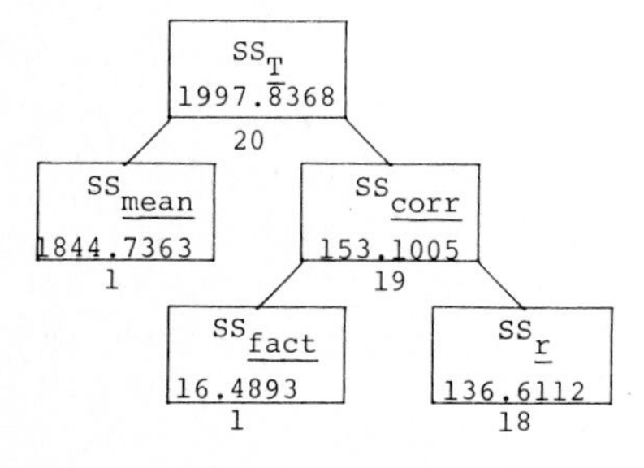

Figure 12.8 Sums of squares and degrees of freedom tree for the completely randomized design.

12.4. Randomized paired comparison designs

In the previous section, the discussion following Equation 12.15 suggested that certain experimental designs could be used to decrease the uncertainty associated with a parameter estimate by minimizing the effect of uncontrolled factors. In this section we discuss one of these experimental designs, the *randomized paired comparison design*.

We have seen that the percent alcohol content of a wine depends not only on the temperature of fermentation, but also on the type of fruit used (in part because of the masquerading factor, sugar content). When different fruit types were randomly assigned to temperature levels, the plot of percent alcohol vs. temperature was very noisy (Figure 12.9), and the effect of temperature was estimated with poor confidence (Equation 12.15). Because we knew that the lurking factor sugar content was important, and because we had measured the level of sugar content in each of the fruits, we were able to account for the effect of sugar in our expanded model (Equations 12.16 and 12.17) and thereby improve the confidence in our estimate of the temperature effect.

But what if we did not know that sugar content was an important factor, or what if we were unable to measure the level of sugar content in each fruit? We would then not have been able to include this factor in our model and the confidence in the temperature effect could not have been improved.

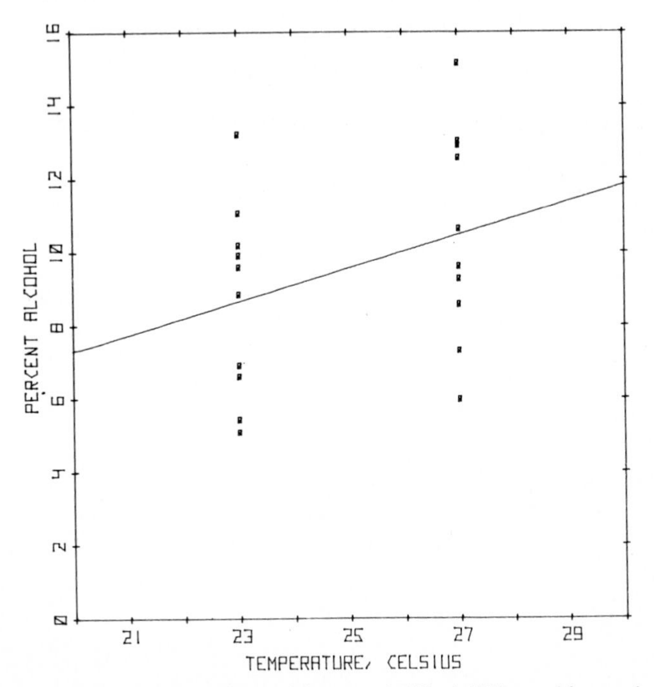

Figure 12.9 Graph of the model $y_{1i} = -1.746 + 0.4540x_{1i}$ with experimental data.

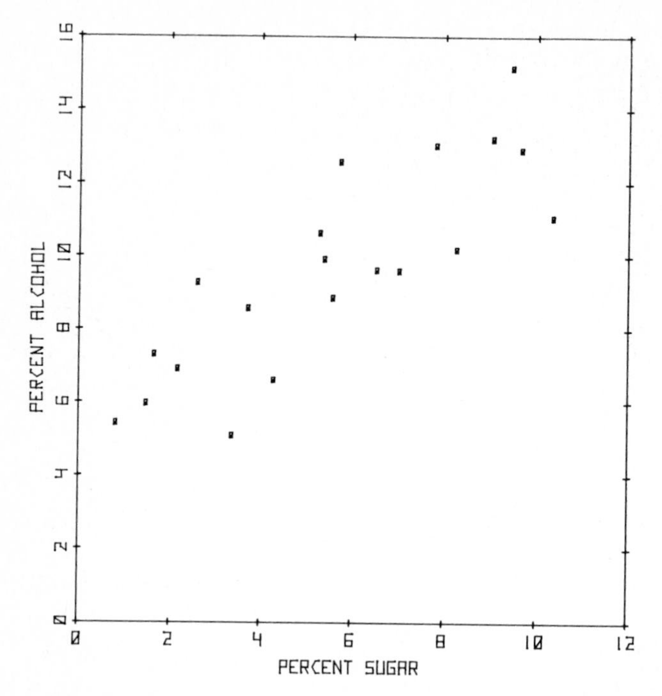

Figure 12.10 Plot of percent alcohol response as a function of sugar content for the 20 fruits listed in Table 12.1

In the completely randomized design, a different fruit was randomly assigned a temperature, either 23°C or 27°C. Let us consider now a different experimental design. We will still employ the same number of experiments (20), but we will use only half as many fruit types, assigning each fruit type to *both* temperatures. Thus, each fruit will be involved in a *pair* of experiments, one experiment at 23°C and the other experiment at 27°C. The experimental design is shown in Table 12.2 and in Figure 12.13. Fitting the model of Equation 12.13 to this data gives

$$y_{1i} = -1.233 + 0.4908x_{1i} \qquad (12.19)$$

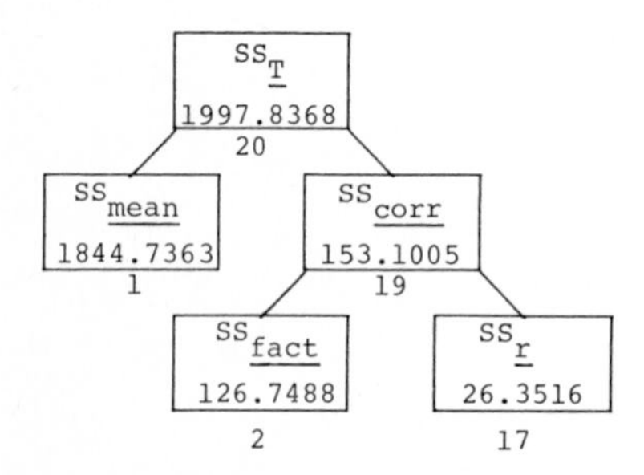

Figure 12.11 Sums of squares and degrees of freedom tree for the two-factor model that includes both temperature and sugar content.

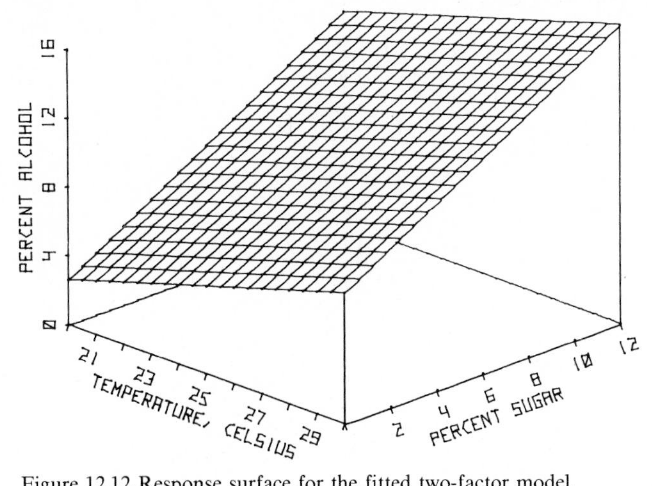

Figure 12.12 Response surface for the fitted two-factor model.

The data and fitted model are shown in Figure 12.14 (compare with Figure 12.9). Based on

$$F_{1,18} = (0.4908)^2/(5.6953 \times 0.0125) = 3.38 \tag{12.20}$$

the temperature effect is significant at the 91.76% level of confidence, only slightly

TABLE 12.2
Randomized paired comparison experimental design for determining the effect of temperature on a wine-making system.

Experiment	Fruit	Temperature	% Alcohol
1	4	27	12.99
2	8	23	13.20
3	7	27	12.98
4	10	23	13.28
5	1	23	10.59
6	5	27	12.13
7	3	23	7.58
8	6	27	10.82
9	2	27	11.56
10	9	23	5.40
11	7	23	11.04
12	9	27	7.28
13	1	27	12.49
14	6	23	8.90
15	4	23	10.83
16	2	23	9.51
17	10	27	15.18
18	5	23	10.21
19	8	27	15.06
20	3	27	9.68

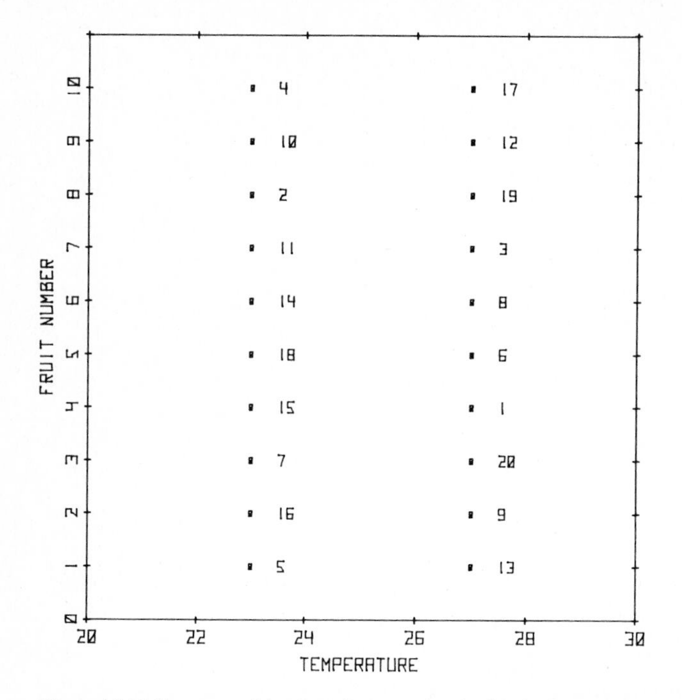

Figure 12.13 Factor combinations for a randomized paired comparison design investigating the effect of temperature. "Fruit number" is an arbitrarily assigned, qualitative factor. Numbers beside factor combinations indicate the time order in which experiments are run.

more significant (by chance) than that of Equation 12.14 (84.22%) for the same model fit to data from the completely random design.

Up to this point, the randomized paired comparison design has not offered any great improvement over the completely randomized design. However, the fact that each fruit has been investigated at a pair of temperatures allows us to carry out a different type of data treatment based upon a series of *paired comparisons*.

We realize that there are a number of factors in addition to temperature that influence the % alcohol response of the wine: sugar content, pressure, magnesium concentration in the fruit, phosphate concentration in the fruit, presence of natural bacteria, etc. Although we strive to keep as many of these factors as controlled and therefore as constant as we can (e.g., pressure), we have no control over many of the other factors, especially those associated with the fruit (see Section 1.2). However, even though we do not have control over these factors, it is nonetheless reasonable to expect that whatever the % alcohol response is at 23° C, the % alcohol response at 27° C *should increase for each of the fruits in our study if temperature has a significant effect*. That is, if we are willing to make the assumption that there are no interactions between the factor of interest to us (temperature) and the other factors that influence the system, the *differences* in responses at 27° C and 23° C should be about the same for each pair of experiments carried out on the same fruit.

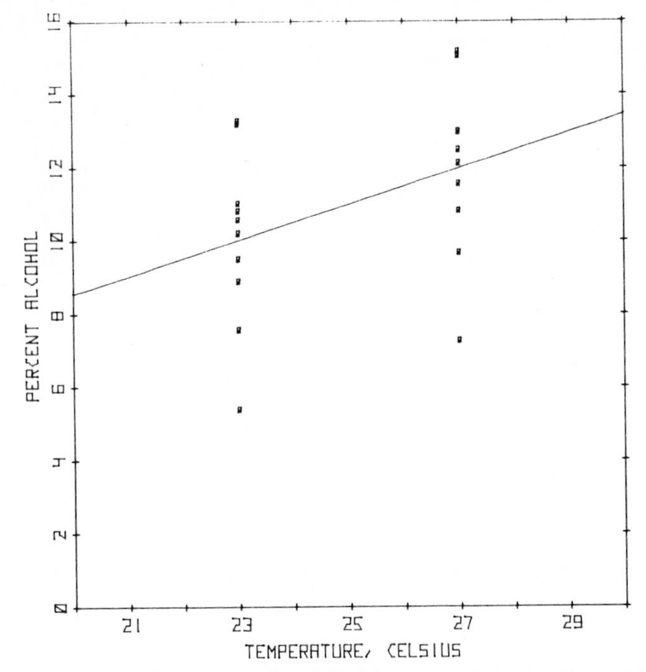

Figure 12.14. Graph of the fitted model $y_{1i} = -1.233 + 0.4908x_{1i}$ with experimental data.

Examination of Table 12.2 tends to confirm this idea. In experiments 10 and 12, for example, fruit number nine had responses of 5.40% and 7.28% at 23°C and 27°C, respectively. In experiments 4 and 17, fruit number ten had responses of 13.28% and 15.18% at 23°C and 27°C, respectively. Even though these two fruits give quite different individual responses at a given temperature, the temperature effect for each fruit is about the same: +1.88% for fruit number nine, and +1.90% for fruit number ten as the temperature is increased from 23°C to 27°C. Similar trends are observed for the other fruits.

There are a number of equivalent ways this paired comparison data can be evaluated; we base our treatment here on a linear model that can be used to estimate the effect of temperature and its significance. Of the twenty original pieces of experimental data, we form the ten pairwise differences listed in Table 12.3. Note that we lose half of our original 20 degrees of freedom in forming these differences, so our resulting data set has only ten degrees of freedom. Thus, we can arbitrarily set the "response" of each fruit equal to zero at 23°C; the "response" of each fruit at 27°C is then simply the calculated difference. Further, we can shift (code) the temperature axis so that 23°C becomes 0 and 27°C becomes +4. An appropriate model for assessing the temperature effect in this coded data system is

$$y_{1i} = \beta_1^* x_{1i}^* + r_{1i} \tag{12.21}$$

TABLE 12.3
Paired differences for determining the effect of temperature on a wine-making system.

Fruit alcohol	ΔT, °C	Δ %
1	4	1.90
2	4	2.05
3	4	2.10
4	4	2.16
5	4	1.92
6	4	1.92
7	4	1.94
8	4	1.86
9	4	1.88
10	4	1.90

This is the straight line model constrained to pass through the origin (see Section 5.3). The fitted model is

$$y_{1i} = 0.4908x_{1i}^* \tag{12.22}$$

and is shown with the data in Figure 12.15. The sums of squares and degrees of freedom tree is given in Figure 12.16. The significance of b_1^* is estimated from

$$F_{1,9} = b_1^2/s_{b_1}^2 = (0.4908)^2/(0.0105 \times 0.00625) = 3670.62 \tag{12.23}$$

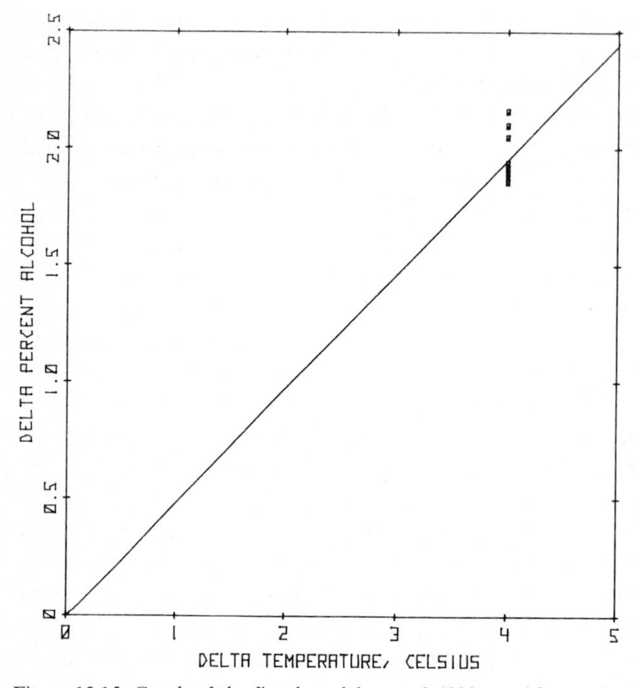

Figure 12.15. Graph of the fitted model $y_{1i} = 0.4908x_{1i}$ with experimental data.

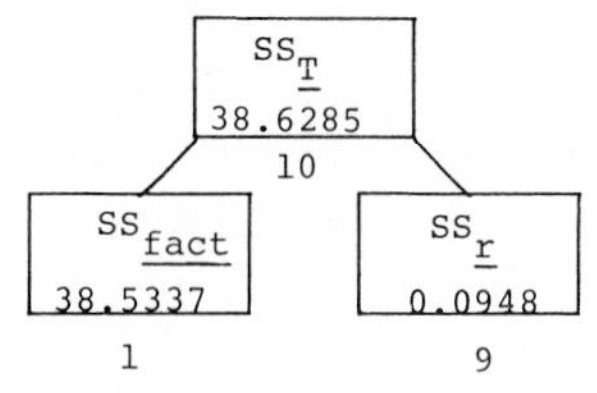

Figure 12.16. Sums of squares and degrees of freedom tree for the randomized paired comparison design.

which in this case is entirely equivalent to the F-test for the significance of regression. F is significant at only slightly less than the 100.000% level of confidence.

Thus, the randomized paired comparison design has allowed *a more sensitive way of viewing our data*, a view that ignores much of the variation caused by the use of different fruits and focuses on pairwise differences associated with the single factor of interest, temperature.

12.5. Randomized complete block designs

The randomized paired comparison design discussed in the previous section separates the effect of a *qualitative* factor, fruit, from the effect of a *quantitative* factor, temperature (see Section 1.2). The *randomized complete block design* discussed in this section allows us to investigate more than one purely qualitative variable and to estimate their quantitative effects.

Suppose the researcher involved with the sodium ion concentration study of Sections 12.1 and 12.2 becomes interested in the wine-making process we have been discussing in Sections 12.3 and 12.4. In particular, let us assume the researcher is interested in determining the effects on the percent alcohol response of adding 10 milligrams of three different univalent cations (Li^+, Na^+, and K^+) and 10 milligrams of four different divalent cations (Mg^{2+}, Ca^{2+}, Sr^{2+}, and Ba^{2+}) in pairs – one univalent and one divalent cation per pair – to one-gallon batches of fermentation mixture. In planning the experiments, the researcher holds as many variables constant as possible, and randomizes to protect against confounding uncontrolled factors with the two factor effects of interest. If experiments are carried out at all combinations of both factors ($3 \times 4 = 12$ factor combinations) with replicates at each factor combination, the resulting 24-experiment randomized design is shown in Table 12.4 and Figure 12.17.

In the classical statistical literature, one of the two qualitative factors is referred to as the "treatments" and the other qualitative factor is referred to as the "blocks". Hence, the term "block designs". In some studies, one of the qualitative factors might be correlated with time, or might even be the factor "time" itself; by carrying out the complete set of experiments in groups (or "blocks") based on this factor, estimated time effects can be removed and the "treatment" effects can be revealed

TABLE 12.4
Randomized complete block design for determining the effects of univalent and divalent cations on a wine-making system.

Experiment	Univalent cation	Divalent cation	% Alcohol
1	K	Sr	3.99
2	K	Ba	1.17
3	Na	Sr	8.25
4	Li	Sr	4.90
5	Li	Ca	6.43
6	K	Ba	1.20
7	Li	Mg	7.61
8	K	Ca	5.76
9	Na	Mg	11.27
10	Na	Ca	10.22
11	Li	Ba	1.97
12	Li	Mg	7.37
13	Na	Ca	10.42
14	Na	Ba	5.60
15	Na	Sr	8.41
16	Li	Sr	4.65
17	Li	Ca	6.56
18	Na	Mg	11.07
19	K	Mg	7.01
20	K	Mg	6.71
21	K	Sr	4.16
22	Na	Ba	5.96
23	K	Ca	6.02
24	Li	Ba	2.39

more clearly. However, it is entirely equivalent to state that by carrying out the complete set of experiments in "treatments", the "block" effects can be revealed more clearly. In situations where the effect of time is not removed by blocking but is instead minimized by randomization, the assignment of the terms "blocks" and "treatments" to the remaining factors of interest is not always straightforward, and the terms are often interchangeable. For example, in the present study, one could view the experiments as involving univalent cation "treatments" and divalent cation "blocks". But the experiments could also be viewed as consisting of univalent cation "blocks" and divalent cation "treatments". In general, we will avoid the use of the terms "treatments" and "blocks" to refer to the levels of one or the other of the qualitative factors, but we will point out that the useful concept of *blocking* allows us to separate the effects of one factor from the effects of another factor.

Because qualitative factors are not continuous, we cannot use a linear model such as $y_{1i} = \beta_0 + \beta_1 x_{1i} + \beta_2 x_{2i} + r_{1i}$ to describe the behavior of this system. For example, if x_1 were to represent the factor "univalent cation", what value would x_1 take when Li was used? Or Na? Or K? There is no rational basis for assigning numerical values to x_1, so we must abandon the familiar linear models containing continuous (quantitative) factors.

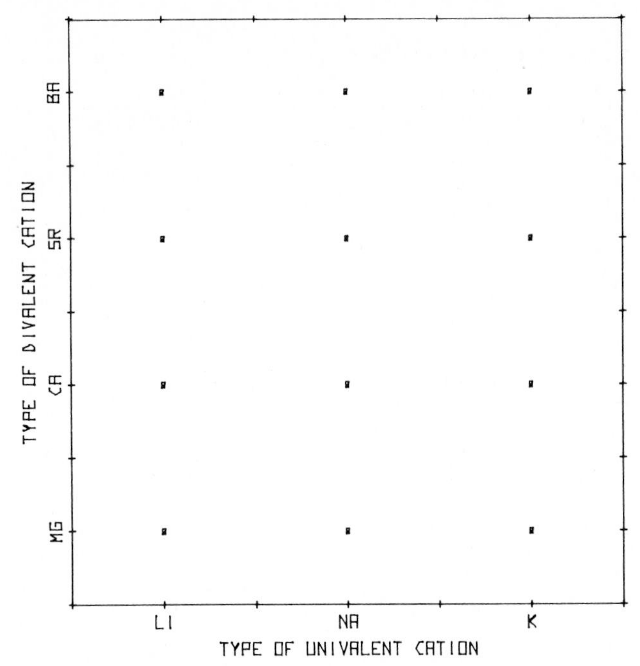

Figure 12.17. Factor combinations for the randomized complete block design investigating two qualitative factors, type of univalent cation and type of divalent cation. Each factor combination is replicated.

Instead, we will view the percent alcohol response from the wine-making process as follows. Let us first pick a reference factor combination. Any of the twelve factor combinations could be used; we will choose the combination Li-Mg (the lower left corner of the design in Figure 12.17) as our reference. For that particular reference combination, we could write the linear model

$$y_{1i} = \beta_{\text{LiMg}} + r_{1i} \tag{12.24}$$

which is analogous to the zero-factor model $y_{1i} = \beta_0 + r_{1i}$. The subscript on β_{LiMg} indicates which factor combination we are using as our reference.

Let us now assume that no matter what level of divalent cation we are using, changing the univalent cation from Li to Na will give the same change in response. This is very similar to the reasoning used in treating the data from the randomized paired comparison design of Section 12.4. Thus, if Na were used instead of Li, we might write the model

$$y_{1i} = \beta_{\text{LiMg}} + \beta_{\text{Na}} + r_{1i} \tag{12.25}$$

where the new term β_{Na} is a measure of the *difference* in response caused by using Na instead of Li. Likewise, if K were used instead of Li, we might write

$$y_{1i} = \beta_{\text{LiMg}} + \beta_{\text{K}} + r_{1i} \tag{12.26}$$

where β_{K} is a measure of the *difference* in response caused by using K instead of Li.

In a similar way, we might substitute Ca, Sr, or Ba for Mg and write the models

$$y_{1i} = \beta_{LiMg} + \beta_{Ca} + r_{1i} \tag{12.27}$$

$$y_{1i} = \beta_{LiMg} + \beta_{Sr} + r_{1i} \tag{12.28}$$

and

$$y_{1i} = \beta_{LiMg} + \beta_{Ba} + r_{1i} \tag{12.29}$$

where the terms β_{Ca}, β_{Sr}, and β_{Ba} are measures of the *differences* in response caused by using Ca, Sr, or Ba instead of Mg.

If we used Ca instead of Mg and used Na instead of Li, a valid model would be

$$y_{1i} = \beta_{LiMg} + \beta_{Na} + \beta_{Ca} + r_{1i} \tag{12.30}$$

If we were to continue this scheme of two-factor substitutions, we would accumulate a total of twelve different models, one for each factor combination in the study. However, it is possible to combine all twelve of these separate models in a single model through the use of "dummy variables". These dummy variables can be used to "turn on" or "turn off" various terms in the combined model in such a way that the twelve separate models result.

Let us create a dummy variable, x_{Na}. For a given experiment i, we will assign to x_{Na} the value 0 if sodium ion *is not* included in the study, or the value 1 if sodium ion *is* included in the study. We might then write

$$y_{1i} = \beta_{LiMg} + \beta_{Na}x_{Nai} + r_{1i} \tag{12.31}$$

If Li were included in a particular experiment, $x_{Nai} = 0$ and Equation 12.31 would reduce to Equation 12.24; if Na were included in the experiment, $x_{Nai} = 1$ and Equation 12.31 would be the same as Equation 12.25.

We can also create the dummy variables x_{K}, x_{Ca}, x_{Sr}, and x_{Ba} for use with the model

$$y_{1i} = \beta_{LiMg} + \beta_{Na}x_{Nai} + \beta_{K}x_{Ki} + \beta_{Ca}x_{Cai} + \beta_{Sr}x_{Sri} + \beta_{Ba}x_{Bai} + r_{1i} \tag{12.32}$$

to designate any one of the factor combinations shown in Figure 12.17. Note that there is one reference parameter, β_{LiMg}, analogous to β_0 in the other linear models; *two* difference parameters (β_{Na} and β_{K}) for the *three* univalent cations; and *three* difference parameters (β_{Ca}, β_{Sr}, and β_{Ba}) for the *four* divalent cations, giving a total of six parameters. The matrix of parameter coefficients $\boldsymbol{X}$ for the experiments listed in Table 12.4 and the model of Equation 12.32 is

$$\boldsymbol{X} = \begin{bmatrix} & Na & K & Ca & Sr & Ba \\ 1 & 0 & 1 & 0 & 1 & 0 \\ 1 & 0 & 1 & 0 & 0 & 1 \\ 1 & 1 & 0 & 0 & 1 & 0 \\ 1 & 0 & 0 & 0 & 1 & 0 \\ 1 & 0 & 0 & 1 & 0 & 0 \\ 1 & 0 & 1 & 0 & 0 & 1 \\ 1 & 0 & 0 & 0 & 0 & 0 \\ 1 & 0 & 1 & 1 & 0 & 0 \\ 1 & 1 & 0 & 0 & 0 & 0 \\ 1 & 1 & 0 & 1 & 0 & 0 \\ 1 & 0 & 0 & 0 & 0 & 1 \\ 1 & 0 & 0 & 0 & 0 & 0 \\ 1 & 1 & 0 & 1 & 0 & 0 \\ 1 & 1 & 0 & 0 & 0 & 1 \\ 1 & 1 & 0 & 0 & 1 & 0 \\ 1 & 0 & 0 & 0 & 1 & 0 \\ 1 & 0 & 0 & 1 & 0 & 0 \\ 1 & 1 & 0 & 0 & 0 & 0 \\ 1 & 0 & 1 & 0 & 0 & 0 \\ 1 & 0 & 1 & 0 & 0 & 0 \\ 1 & 0 & 1 & 0 & 1 & 0 \\ 1 & 1 & 0 & 0 & 0 & 1 \\ 1 & 0 & 1 & 1 & 0 & 0 \\ 1 & 0 & 0 & 0 & 0 & 1 \end{bmatrix} \tag{12.33}$$

Treating the data by conventional matrix least squares techniques gives

$$\hat{\boldsymbol{B}} = \begin{bmatrix} b_{LiMg} \\ b_{Na} \\ b_K \\ b_{Ca} \\ b_{Sr} \\ b_{Ba} \end{bmatrix} = \begin{bmatrix} 7.53 \\ 3.67 \\ -0.73 \\ -0.94 \\ -2.78 \\ -5.46 \end{bmatrix} \tag{12.34}$$

The interpretation of these parameter estimates is straightforward. The parameter b_{LiMg} estimates the response for the reference factor combination containing Li and Mg. The estimated value of 7.53 compares favorably with the results of experiments 7 and 12 in Table 12.4, replicates for Li and Mg with responses of 7.61 and 7.37, respectively. The parameter estimate $b_{Na} = 3.67$ suggests that replacing Li with Na causes an increase in response to 11.20; experiments 9 and 18 in Table 12.4 have responses of 11.27 and 11.07, respectively. Replacing Li with K, however, causes a

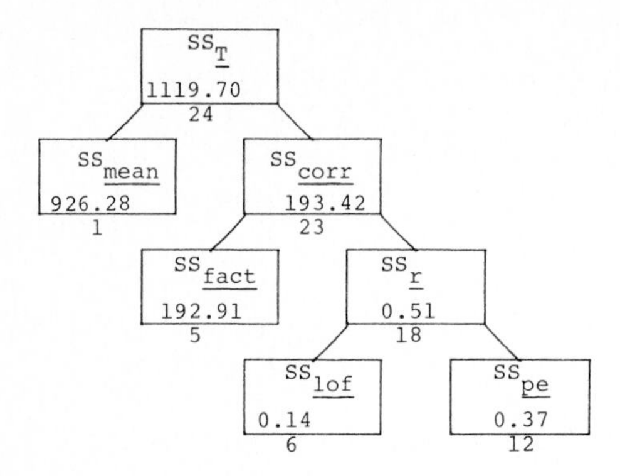

Figure 12.18. Sums of squares and degrees of freedom tree for the randomized complete block design.

decrease in response of -0.73 (b_K). Similarly, replacing Mg with Ca, Sr, or Ba causes decreases in response of -0.94 (b_{Ca}), -2.78 (b_{Sr}), and -5.46 (b_{Ba}).

The response for experiment 6 in Table 12.4 would be estimated as 7.53 (b_{LiMg}) minus 0.73 (b_K, the difference caused by replacing Li with K) minus 5.46 (b_{Ba}, the difference caused by replacing Mg with Ba) = 1.34; the measured response was 1.20.

Figure 12.18 shows a sums of squares and degrees of freedom tree for the data of Table 12.4 and the model of Equation 12.32. The significance of the parameter estimates may be obtained from Equation 10.66 using s_r^2 and $(X'X)^{-1}$ to obtain the variance-covariance matrix. The $(X'X)^{-1}$ matrix for the present example is

$$(X'X)^{-1} = \begin{bmatrix} 0.250 & -0.125 & -0.125 & -0.167 & -0.167 & -0.167 \\ -0.125 & 0.250 & 0.125 & 0 & 0 & 0 \\ -0.125 & 0.125 & 0.250 & 0 & 0 & 0 \\ -0.167 & 0 & 0 & 0.333 & 0.167 & 0.167 \\ -0.167 & 0 & 0 & 0.167 & 0.333 & 0.167 \\ -0.167 & 0 & 0 & 0.167 & 0.167 & 0.333 \end{bmatrix} \tag{12.35}$$

Note that for the randomized complete block design, there is no covariance between the effects of the univalent cations and the effects of the divalent cations.

The randomized complete block design has provided a sensitive way of viewing the data from this set of experiments involving two qualitative factors. The linear model using dummy variables ignores much of the variation in the data by again focusing on pairwise differences associated with the different discrete levels of the factors of interest.

12.6. Coding of randomized complete block designs

Because of tradition, the model of Equation 12.32 is seldom used for the randomized complete block design. Instead, a somewhat different but essentially equivalent model is used:

$$y_{ij} = \mu + \gamma_i + \tau_j + \epsilon_{ij} \tag{12.36}$$

where y_{ij} is the response in the ith "block" receiving the jth "treatment" (note the different meaning attached to i), μ is the average response (equivalent to $\bar{y}_1$), γ_i is a "block" effect, τ_j is a "treatment" effect, and the ϵ_{ij} are "errors" (residuals) between what is observed and what the model predicts. Two equality constraints are placed on the γ_i's and τ_j's (see Section 2.3):

$$\sum_i \gamma_i = 0 \tag{12.37}$$

$$\sum_j \tau_j = 0 \tag{12.38}$$

For the example used in Section 12.5, there would be three γ's and four τ's (or, perhaps, four γ's and three τ's). If γ is associated with the qualitative factor "univalent cation", then γ_{Li} would be the average "block" difference in response between the experiments involving Li *and the overall mean*; γ_{Na} and γ_K would be the corresponding differences for experiments involving Na and K. Similarly, τ_{Mg}, τ_{Ca}, τ_{Sr}, and τ_{Ba} would be the average "treatment" differences for experiments involving the divalent cations Mg, Ca, Sr, and Ba. Thus, the full model would be

$$y_{ij} = \mu + \gamma_{Li} + \gamma_{Na} + \gamma_K + \tau_{Mg} + \tau_{Ca} + \tau_{Sr} + \tau_{Ba} + \epsilon_{ij} \tag{12.39}$$

Again, certain terms are "turned on" or "turned off" to correspond to a particular factor combination ij. Notice that Equation 12.39 has eight parameters; Equation 12.32 has only six parameters. It would appear that the two models are not equivalent because they contain different numbers of parameters; however, the equality constraints of Equations 12.37 and 12.38 take away two degrees of freedom from Equation 12.39 and it actually has only six *independent* parameters (see Section 2.3). The equality constraints require that

$$\gamma_{Li} = -\gamma_{Na} - \gamma_K \tag{12.40}$$

and

$$\tau_{Mg} = -\tau_{Ca} - \tau_{Sr} - \tau_{Ba} \tag{12.41}$$

The relationships between the parameters of Equations 12.32 and 12.39 are easily discovered.

$$\beta_{LiMg} = \mu + \gamma_{Li} + \tau_{Mg} = \mu - \gamma_{Na} - \gamma_K - \tau_{Ca} - \tau_{Sr} - \tau_{Ba} \tag{12.42}$$

$$\beta_{Na} = \gamma_{Na} - \gamma_{Li} = \gamma_{Na} - (-\gamma_{Na} - \gamma_K) = 2\gamma_{Na} + \gamma_K \tag{12.43}$$

$$\beta_K = \gamma_K - \gamma_{Li} = \gamma_{Na} + 2\gamma_K \tag{12.44}$$

$$\beta_{Ca} = \tau_{Ca} - \tau_{Mg} = \tau_{Ca} - (-\tau_{Ca} - \tau_{Sr} - \tau_{Ba}) = 2\tau_{Ca} + \tau_{Sr} + \tau_{Ba} \tag{12.45}$$

$$\beta_{Sr} = \tau_{Sr} - \tau_{Mg} = \tau_{Ca} + 2\tau_{Sr} + \tau_{Ba} \tag{12.46}$$

$$\beta_{Ba} = \tau_{Ba} - \tau_{Mg} = \tau_{Ca} + \tau_{Sr} + 2\tau_{Ba} \tag{12.47}$$

If

$$\boldsymbol{B} = \begin{bmatrix} \beta_{\text{LiMg}} \\ \beta_{\text{Na}} \\ \beta_{\text{K}} \\ \beta_{\text{Ca}} \\ \beta_{\text{Sr}} \\ \beta_{\text{Ba}} \end{bmatrix} \tag{12.48}$$

and

$$\boldsymbol{B}^* = \begin{bmatrix} \mu \\ \gamma_{\text{Na}} \\ \gamma_{\text{K}} \\ \tau_{\text{Ca}} \\ \tau_{\text{Sr}} \\ \tau_{\text{Ba}} \end{bmatrix} \tag{12.49}$$

then by Equation 10.48, $\boldsymbol{B} = \boldsymbol{A}^{-1}\boldsymbol{B}^*$ and it is evident that Equations 12.42–12.47 are equivalent to

$$\begin{bmatrix} \beta_{\text{LiMg}} \\ \beta_{\text{Na}} \\ \beta_{\text{K}} \\ \beta_{\text{Ca}} \\ \beta_{\text{Sr}} \\ \beta_{\text{Ba}} \end{bmatrix} = \begin{bmatrix} 1 & -1 & -1 & -1 & -1 & -1 \\ 0 & 2 & 1 & 0 & 0 & 0 \\ 0 & 1 & 2 & 0 & 0 & 0 \\ 0 & 0 & 0 & 2 & 1 & 1 \\ 0 & 0 & 0 & 1 & 2 & 1 \\ 0 & 0 & 0 & 1 & 1 & 2 \end{bmatrix} \begin{bmatrix} \mu \\ \gamma_{\text{Na}} \\ \gamma_{\text{K}} \\ \tau_{\text{Ca}} \\ \tau_{\text{Sr}} \\ \tau_{\text{Ba}} \end{bmatrix} \tag{12.50}$$

Similarly, $\boldsymbol{B}^* = \boldsymbol{A}\boldsymbol{B}$ and

$$\begin{bmatrix} \mu \\ \gamma_{\text{Na}} \\ \gamma_{\text{K}} \\ \tau_{\text{Ca}} \\ \tau_{\text{Sr}} \\ \tau_{\text{Ba}} \end{bmatrix} \begin{bmatrix} 1 & 1/3 & 1/3 & 1/4 & 1/4 & 1/4 \\ 0 & 2/3 & -1/3 & 0 & 0 & 0 \\ 0 & -1/3 & 2/3 & 0 & 0 & 0 \\ 0 & 0 & 0 & 3/4 & -1/4 & -1/4 \\ 0 & 0 & 0 & -1/4 & 3/4 & -1/4 \\ 0 & 0 & 0 & -1/4 & -1/4 & 3/4 \end{bmatrix} \begin{bmatrix} \beta_{\text{LiMg}} \\ \beta_{\text{Na}} \\ \beta_{\text{K}} \\ \beta_{\text{Ca}} \\ \beta_{\text{Sr}} \\ \beta_{\text{Ba}} \end{bmatrix} \tag{12.51}$$

The remaining parameters of Equation 12.36 (γ_{Li} and τ_{Mg}) are calculated from Equations 12.40 and 12.41. Thus, if the model of Equation 12.32 has been used to treat the data from a randomized complete block design, the results may be readily converted to the form of Equation 12.36.

References

Box, G.E.P., Hunter, W.G., and Hunter, J.S. (1978). *Statistics for Experimenters. An Introduction to Design, Data Analysis, and Model Building*. Wiley, New York.
Cochran, W.G., and Cox, G.M. (1957). *Experimental Designs*, 2nd ed. Wiley, New York.
Davies, O.L., Ed. (1956). *The Design and Analysis of Industrial Experiments*. Hafner, New York.
Draper, N.R., and Smith, H. (1966). *Applied Regression Analysis*. Wiley, New York.
Hicks, C.R. (1973). *Fundamental Concepts in the Design of Experiments*, 2nd. ed. Holt, Rinehart, and Winston, New York.
Kempthorne, O. (1952). *The Design and Analysis of Experiments*. Wiley, New York.
Mendenhall, W. (1968). *Introduction to Linear Models and the Design and Analysis of Experiments*. Duxbury Press, Belmont, California.
Natrella, M.G. (1963). *Experimental Statistics* (Nat. Bur. of Stand. Handbook 91). US Govt. Printing Office, Washington, DC.
Neter, J., and Wasserman, W. (1974). *Applied Linear Statistical Models. Regression, Analysis of Variance, and Experimental Designs*. Irwin, Homewood, Illinois.
Wilson, E.B., Jr. (1952). *An Introduction to Scientific Research*. McGraw-Hill, New York.

Exercises

12.1. Confounding

Historical data from production facilities often give rise to highly confounded results. Consider a process in which temperature and pressure are considered to be inputs, and yield is an output. If the data are based on *observations* rather than *experiments*, why might temperature and pressure be confounded? [See, for example, Box, G.E.P. (1954). "The Exploration and Exploitation of Response Surfaces: Some General Considerations and Examples", *Biometrics*, 10, 16.]

12.2. Confounding

Use matrix least squares to fit the model $y_{1i} = \beta_0 + \beta_1 x_{1i} + \beta_2 x_{2i} + r_{1i}$ to the following data:

i	x_{1i}	x_{2i}	y_{1i}
1	17	38.5	93.6
2	14	34.0	12.3
3	20	43.0	33.5
4	13	32.5	87.4
5	19	41.5	3.0
6	16	37.0	105.0
7	22	46.0	47.3
8	15	35.5	88.6
9	18	40.0	88.2
10	21	44.5	51.7

12.3. Randomization

The following ten random numbers were obtained in the following sequence from a computer: 1, 2, 10, 3, 4, 9, 5, 6, 7, 8. Comment.

12.4. Randomization

Find a table of random numbers and use it to reorder the experimental design of Equation 12.1. [See, for example, Cohran, W.G., and Cox, G.M. (1957). *Experimental Designs*, 2nd ed., p. 569. Wiley, New York.]

12.5. Randomization

"Randomization may be thought of as insurance, and, like insurance, may sometimes be too expensive". [Natrella, M.G. (1963). *Experimental Statistics* (Nat. Bur. of Stand. Handbook 91), p. 11-4. US Govt. Printing Office, Washington, DC.] Comment. Give a set of circumstances in which it might be too expensive to randomize.

12.6. Randomization

"Randomization affords insurance against uncontrollable disturbances in the sense that such disturbances have the same chance of affecting each of the factors under study, and will be balanced out in the long run". [Natrella, M.G. (1963). *Experimental Statistics* (Nat. Bur. of Stand. Handbook 91), p. 13-1. US Govt. Printing Office, Washington, DC.] Will the uncontrollable disturbances be balanced out "in the short run"?

12.7. Blocking

One of the advantages often put forth for grouping experiments in blocks is that it allows the "treatment effects" to be obtained with (usually) better precision. The "block effects" are usually ignored. Is information being wasted when this is done? Give an example for which the block effects might provide useful information.

12.8. Completely randomized designs

"One fact compensates to some extent for the higher experimental errors [uncertainties of the completely randomized design] as compared with other designs. For a given number of treatments and a given number of experimental units, complete randomization provides the maximum number of degrees of freedom for the estimation of error". [Cochran, W.G., and Cox, G.M. (1957). *Experimental Designs*, 2nd ed., p. 96. Wiley, New York.] Why is this an advantage? Is it more important for large n or small n?

12.9. Completely randomized designs

"This plan is simple, and is the best choice when the experimental material is homogeneous and background conditions can be well controlled during the experiment". [Natrella, M.G. (1963). *Experimental Statistics* (Nat. Bur. of Stand. Handbook 91), p. 13-1. US Govt. Printing Office, Washington, DC.] Comment.

12.10. Randomized paired comparison designs

Suppose someone comes to you with the hypothesis that shoes worn on the right foot receive more wear than shoes worn on the left foot. Design a randomized paired comparison experiment to test this hypothesis. How might it differ from a completely randomized design?

12.11. Randomized complete block design

Suppose a sports enthusiast wants to see which of the following has the greatest effect on minimizing bicyclists' time to complete a one-mile ride: resting 10 minutes immediately before the race, doing 50 pushups 20 minutes before the race, running in place 10 minutes immediately before the race, drinking 0.5 liters of orange juice 5 minutes before the race, or drinking 0.5 liters of water 15 minutes before the race. Design a randomized complete block design to help answer the question raised by the sports enthusiast. What are the "blocks" and "treatments" in your design?

12.12. Randomized complete block designs

Randomized paired comparison designs can be thought of as randomized complete block designs in which the blocking is done in pairs to eliminate unwanted sources of variability. If the data in Section 12.4 are treated as a randomized complete block design, what is the model? How many degrees of freedom are available for estimating the temperature effect? How many "block effects" must be estimated? How many degrees of freedom are there for s_{pe}^2?

12.13. Randomized complete block designs

It has been suggested that a set of factor combinations such as that in Section 12.5 is simply a 3×4 factorial design. Comment.

12.14. Balanced incomplete block designs

When the "block size" is smaller than the number of "treatments" to be evaluated, *incomplete block designs* may be used. [See Yates, F. (1936). "A New Method of Arranging Variety Trials Involving a Large Number of Varieties", *Jour.*

Agr. Sci., 26, 424.] *Balanced incomplete block designs* give approximately the same precision to all pairs of treatments.

The following is a *symmetrical* (i.e., the number of blocks equals the number of treatments) balanced incomplete block design:

Block	Treatments
1	A B D
2	B C E
3	C D F
4	D E G
5	E F A
6	F G B
7	G A C

How many times does a given treatment appear with each of the other treatments? Graph the design with blocks on the vertical axis and treatments on the horizontal axis. If the usual model for this design is $y_{ij} = \mu + \gamma_i + \tau_j + \epsilon_{ij}$, write an equivalent linear model using β's, and write the corresponding $\boldsymbol{X}$ matrix.

12.15. Latin square designs

Randomized block designs are useful when there is a single type of inhomogeneity (the "block effects") detracting from precise estimates of the factor effect of interest (the "treatment effects"). When there are two types of inhomogeneity, a *Latin square* design is often useful. The model is $y_{ijk} = \mu + \gamma_i + \sigma_j + \tau_k + \epsilon_{ijk}$. The following is a 4×4 Latin square design:

A B C D
C D A B
B C D A
D A B C

If the "columns" represent the γ_i and the "rows" represent the σ_j, what do the letters *A*, *B*, *C*, and *D* represent? If the corresponding numerical results are

46 48 46 40
27 28 36 27
40 49 51 44
40 55 48 43

what are the three estimated differences (β's) between *A* and: *B*, *C*, and *D*?

12.16. Youden square designs

Another useful experimental design for minimizing the effects of two types of inhomogeneity is the *Youden square design*. Latin squares must have the same number of levels for both of the blocking factors and the treatment factor; Youden squares must have the same number of levels for the treatment factor and one of the blocking factors, but the number of levels for the other blocking factor can be smaller. Thus, Youden squares are more efficient than Latin squares, especially as the number of "treatment" levels gets large.

How is the following Youden square design related to the Latin square design of Problem 12.15?

A	*B*	*C*	*D*
C	*D*	*A*	*B*
D	*A*	*B*	*C*

If the results are

46	48	46	40
27	28	36	27
40	55	48	43

what are the three differences (β's) between *A* and: *B*, *C*, and *D*?

What is the relationship – Youden square designs : Latin square designs :: balanced incomplete block designs : randomized complete block designs?

12.17. Graeco-Latin square designs

If there are *three* types of blocking factors, *Graeco-Latin square designs* can be used to minimize their effects. The following is a 4×4 Graeco-Latin square. What do α, β, γ, and σ represent?

$A\alpha$	$B\beta$	$C\gamma$	$D\sigma$
$B\sigma$	$A\gamma$	$D\beta$	$C\alpha$
$C\beta$	$D\alpha$	$A\sigma$	$B\gamma$
$D\gamma$	$C\sigma$	$B\alpha$	$A\beta$

12.18. Other designs

Look up one of the following experimental designs and explain the basis of the design, its strengths, its weaknesses, and its major areas of application: Plackett-Burman designs, Box-Behnken designs, cross-over designs, hyper-Graeco-Latin square designs, fractional factorial designs, split-plot designs, quasi-Latin square designs, partially balanced incomplete block designs, lattice designs, rectangular lattice designs, cubic lattice designs, chain block designs.

12.19. Experimental design

"In general, we should try to think of all variables that could possibly affect the results, select as factors as many variables as can reasonably be studied, and use planned grouping where possible". [Natrella, M.G. (1963). *Experimental Statistics* (Nat. Bur. of Stand. Handbook 91), p. 11-4. US Govt. Printing Office, Washington, DC.] Comment.

APPENDIX A

Matrix Algebra

Matrix algebra provides a concise and practical method for carrying out the mathematical operations involved in the design of experiments and in the treatment of the resulting experimental data.

A.1. Definitions

A *matrix* is a rectangular array of numbers. Many types of data are tabulated in arrays. For example, baseball fans are familiar with a tabulation of data similar to the following array:

	Won	Lost	Pct
Houston	41	22	0.651
Cincinnati	40	25	0.615
Los Angeles	36	28	0.563
San Francisco	28	32	0.467
San Diego	29	35	0.453
Atlanta	25	36	0.410

Not only is the value of each element in the matrix important, but the location of each element is also significant. Fans of the Atlanta team would be dismayed to see the sixth *row* of the array,

Atlanta 25 36 0.410

ranking Atlanta in last place. The baseball fans might also be interested in the winning percentages given by the third *column* of the array,

0.651
0.615
0.563
0.467
0.453
0.410

If we omit the row and column headings and focus our attention on the arrays of numbers in this example, we are dealing with the matrices

$$A = \begin{bmatrix} 41 & 22 & 0.651 \\ 40 & 25 & 0.615 \\ 36 & 28 & 0.563 \\ 28 & 32 & 0.467 \\ 29 & 35 & 0.453 \\ 25 & 36 & 0.410 \end{bmatrix}, \qquad B = [25 \quad 36 \quad 0.410], \qquad C = \begin{bmatrix} 0.651 \\ 0.615 \\ 0.563 \\ 0.467 \\ 0.453 \\ 0.410 \end{bmatrix}$$

The *dimensions* of a matrix are given by stating first the number of rows and then the number of columns that it has. Thus, matrix $\boldsymbol{A}$ shown above has six rows and three columns, and is said to be a 6×3 (read "six by three") matrix. Matrix $\boldsymbol{B}$ shown above has one row and three columns and is a 1×3 matrix. Matrix $\boldsymbol{C}$ is a 6×1 matrix. Generally, a matrix $\boldsymbol{M}$ that has r rows and c columns is called an $r \times c$ matrix and can be identified as such by the notation $\boldsymbol{M}_{rc}$.

If the number of rows is equal to the number of columns in the matrix, the matrix is said to be a *square matrix*. For example, given the two simultaneous equations

$$2x_1 + 4x_2 = 5, \qquad x_1 - 2x_2 = 1$$

the coefficients of the two unknowns x_1 and x_2 constitute a 2×2 square matrix

$$\begin{bmatrix} 2 & 4 \\ 1 & -2 \end{bmatrix}$$

If a matrix contains only one row, it is called a *row matrix* or a *row vector*. The matrix $\boldsymbol{B}$ shown above is an example of a 1×3 row vector. Similarly, a matrix containing only one column is known as a *column matrix* or *column vector*. The matrix $\boldsymbol{C}$ shown above is a 6×1 column vector. One use of vectors is to represent the location of a point in an orthogonal coordinate system. For example, a particular point in a three-dimensional space can be represented by the 1×3 row vector

$$[7 \quad 4 \quad 9]_{1 \times 3}$$

where the first element (7) represents the x_1-coordinate, the second element (4) represents the x_2-coordinate, and the third element (9) represents the x_3-coordinate.

Capital italic letters in bold-face are used by typesetters to represent matrices. The values in an array, or the *elements* of the array, are denoted using the corresponding small italic letters with appropriate subscripts. Thus, $\boldsymbol{A}_{ij}$ denotes the element in the ith row and jth column of the matrix $\boldsymbol{A}$. The individual elements of the previously defined $\boldsymbol{A}$ matrix are

$$\begin{array}{lll} a_{11} = 41, & a_{12} = 22, & a_{13} = 0.651 \\ a_{21} = 40, & a_{22} = 25, & a_{23} = 0.615 \\ a_{31} = 36, & a_{32} = 28, & a_{33} = 0.563 \\ a_{41} = 28, & a_{42} = 32, & a_{43} = 0.467 \\ a_{51} = 29, & a_{52} = 35, & a_{53} = 0.453 \\ a_{61} = 25, & a_{62} = 36, & a_{63} = 0.410 \end{array}$$

Two matrices are *equal* ($\boldsymbol{A} = \boldsymbol{B}$) if and only if their dimensions are identical and their corresponding elements are equal ($a_{ij} = b_{ij}$ for all i and j).

The *transpose* $\boldsymbol{X}'$ of a matrix $\boldsymbol{X}$ is formed by interchanging its rows and columns; that is, the element x'_{ij} in row i and column j of the transpose matrix is equal to the element x_{ji} in row j and column i of the original matrix. For example, if

$$\boldsymbol{X} = \begin{bmatrix} 1 & 3 \\ 1 & -1 \\ 1 & 0 \end{bmatrix}$$

then

$$\boldsymbol{X}' = \begin{bmatrix} 1 & 1 & 1 \\ 3 & -1 & 0 \end{bmatrix}$$

Note that the first *row* of $\boldsymbol{X}$ becomes the first *column* of $\boldsymbol{X}'$, the second row of $\boldsymbol{X}$ becomes the second column of $\boldsymbol{X}'$, and so on. If $\boldsymbol{X}$ is a $p \times q$ matrix, then $\boldsymbol{X}'$ is a $q \times p$ matrix.

If the transpose of a matrix is identical in every element to the original matrix (that is, if $\boldsymbol{A}' = \boldsymbol{A}$), then the matrix is called a *symmetric matrix*. Thus, a symmetric matrix has all elements $\boldsymbol{A}_{ij}$ equal to all elements $\boldsymbol{A}_{ji}$; it is symmetric with respect to its principal diagonal from upper left to lower right. A symmetric matrix is necessarily a square matrix, because otherwise its transpose would have different dimensions and could not be identical to it.

A special case of the symmetric matrix is the *diagonal matrix*, in which all the off-diagonal elements are zero. For example,

$$\boldsymbol{M} = \begin{bmatrix} 2 & 0 & 0 \\ 0 & 3 & 0 \\ 0 & 0 & 1 \end{bmatrix}$$

is a 3×3 diagonal matrix where $m_{11} = 2$, $m_{22} = 3$, $m_{33} = 1$, and $m_{ij} = m_{ji} = 0$ for all $i \neq j$.

The *identity matrix* $\boldsymbol{I}$ is a diagonal matrix which has all 1's on the diagonal; for example, the 3×3 identity matrix is

$$\boldsymbol{I} = \begin{bmatrix} 1 & 0 & 0 \\ 0 & 1 & 0 \\ 0 & 0 & 1 \end{bmatrix}$$

A.2. Matrix addition and subtraction

The *sum* of two matrices is obtained by adding the corresponding elements of the two matrices. For example, given

$$\boldsymbol{A} = \begin{bmatrix} 2 & 3 & 1 \\ -1 & 0 & 5 \end{bmatrix}$$

and

$$B = \begin{bmatrix} -4 & 2 & 1 \\ 1 & 3 & -2 \end{bmatrix}$$

then the sum S is

$$S = A + B = \begin{bmatrix} 2+(-4) & 3+2 & 1+1 \\ (-1)+1 & 0+3 & 5+(-2) \end{bmatrix} = \begin{bmatrix} -2 & 5 & 2 \\ 0 & 3 & 3 \end{bmatrix}$$

Note that the resulting matrix has the same dimensions as the original matrices; two matrices may be added together if and only if they have identical dimensions. For example, if

$$T = \begin{bmatrix} 1 & 1 \\ 2 & 4 \\ 0 & -1 \end{bmatrix}$$

then S and T cannot be added together. When the dimensions of two matrices are the same, they are said to be *conformable for addition*. Matrix addition is *commutative* and *associative*:

$$A + B = B + A \qquad \text{(commutative)}$$

$$A + (B + C) = (A + B) + C \qquad \text{(associative)}$$

The *negative* $-A$ of a matrix A is simply the matrix whose elements are the negatives of the corresponding elements of A.

The *difference* between two matrices is obtained by subtracting the corresponding elements of the second matrix from the elements of the first. For example, given the previously defined matrices A and B, their difference is

$$D = A - B = \begin{bmatrix} 6 & 1 & 0 \\ -2 & -3 & 7 \end{bmatrix}$$

A.3. Matrix multiplication

The product of two matrices AB exists if and only if the number of rows in the second matrix B is the same as the number of columns in the first matrix A. If this is the case, the two matrices are said to be *conformable for multiplication*. If A is an $m \times p$ matrix and B is a $p \times n$ matrix, then the product C is an $m \times n$ matrix:

$$C_{m \times n} = A_{m \times p} B_{p \times n}$$

The number of *rows* (m) in the product matrix C is given by the number of rows in the first matrix A, and the number of *columns* (n) in the product matrix is given by the number of columns in the second matrix B.

Each of the elements of the product matrix, $\boldsymbol{C}=\boldsymbol{AB}$, is found by multiplying each of the p elements in a column of $\boldsymbol{B}$ by the corresponding p elements in a row of $\boldsymbol{A}$ and taking the sum of the intermediate products. Algebraically, an element c_{ij} is calculated

$$c_{ij}=\sum_{k=1}^{p} a_{ik}b_{kj}$$

For example, given the matrices

$$\boldsymbol{A}=\begin{bmatrix}1 & 2 & 3\\ 4 & 5 & 6\\ 7 & 8 & 9\end{bmatrix}$$

$$\boldsymbol{B}=\begin{bmatrix}1 & 0\\ 2 & 3\\ 4 & 1\end{bmatrix}$$

To calculate the element c_{11} in the product matrix $\boldsymbol{AB}$:

$$\begin{bmatrix}1 & 2 & 3\\ - & - & -\\ - & - & -\end{bmatrix}\begin{bmatrix}1 & -\\ 2 & -\\ 4 & -\end{bmatrix}=\begin{bmatrix}1\times 1+2\times 2+3\times 4 & -\\ - & -\\ - & -\end{bmatrix}=\begin{bmatrix}17 & -\\ - & -\\ - & -\end{bmatrix}$$

To form the element c_{32} in the product matrix, we find

$$\begin{bmatrix}- & - & -\\ - & - & -\\ 7 & 8 & 9\end{bmatrix}\begin{bmatrix}- & 0\\ - & 3\\ - & 1\end{bmatrix}=\begin{bmatrix}- & -\\ - & -\\ 7\times 0+8\times 3+9\times 1 & -\end{bmatrix}=\begin{bmatrix}- & -\\ - & -\\ - & 33\end{bmatrix}$$

The entire product matrix may be calculated similarly.

$$\begin{bmatrix}1 & 2 & 3\\ 4 & 5 & 6\\ 7 & 8 & 9\end{bmatrix}\begin{bmatrix}1 & 0\\ 2 & 3\\ 4 & 1\end{bmatrix}=\begin{bmatrix}17 & 9\\ 38 & 21\\ 59 & 33\end{bmatrix}$$

An alternative layout of the matrices is often useful, especially when the matrice are large. The right matrix is raised, and the product matrix is moved to the left into the space that has been made available. The rows of the left matrix and the columns of the right matrix now "point" to the location of the corresponding product element.

$$
\begin{array}{c} \\ \begin{bmatrix} 3 & 15 & 7 & 9 & 12 \\ 11 & 3 & -1 & 5 & 3 \\ 2 & 1 & -2 & 3 & 13 \end{bmatrix} \end{array}
\begin{array}{c} \begin{bmatrix} 1 & 2 & -1 & 1 & 2 \\ -1 & -1 & 2 & 1 & -1 \\ -2 & 1 & 1 & 1 & -1 \\ 2 & 1 & 3 & 1 & 1 \\ 1 & -1 & -3 & -1 & 1 \end{bmatrix} \\ \begin{bmatrix} - & - & - & - & - \\ - & - & 0 & - & - \\ - & - & - & - & - \end{bmatrix} \end{array}
$$

$$
\begin{array}{c} \\ \begin{bmatrix} 3 & 15 & 7 & 9 & 12 \\ 11 & 3 & -1 & 5 & 3 \\ 2 & 1 & -2 & 3 & 13 \end{bmatrix} \end{array}
\begin{array}{c} \begin{bmatrix} 1 & 2 & -1 & 1 & 2 \\ -1 & -1 & 2 & 1 & -1 \\ -2 & 1 & 1 & 1 & -1 \\ 2 & 1 & 3 & 1 & 1 \\ 1 & -1 & -3 & -1 & 1 \end{bmatrix} \\ \begin{bmatrix} 4 & -5 & 25 & 22 & 5 \\ 23 & 20 & 0 & 15 & 28 \\ 24 & -9 & -32 & -9 & 21 \end{bmatrix} \end{array}
$$

Another example involves an identity matrix:

$$\begin{bmatrix} 2 & 3 \\ -1 & 7 \end{bmatrix}\begin{bmatrix} 1 & 0 \\ 0 & 1 \end{bmatrix} = \begin{bmatrix} 2 & 3 \\ -1 & 7 \end{bmatrix}$$

This example illustrates why the identity matrix $\boldsymbol{I}$ is so named: it serves the same role as the number 1 does in the multiplication of ordinary real numbers.

$$\boldsymbol{AI} = \boldsymbol{IA} = \boldsymbol{A}$$

Note that the multiplication of matrices is distributive

$$\boldsymbol{A}(\boldsymbol{B} + \boldsymbol{C}) = \boldsymbol{AB} + \boldsymbol{AC}$$

and associative

$$(\boldsymbol{AB})\boldsymbol{C} = \boldsymbol{A}(\boldsymbol{BC})$$

but that it is not, in general, commutative,

$$\boldsymbol{AB} \neq \boldsymbol{BA}$$

This general non-commutative property of matrix multiplication is in contrast with ordinary algebra.

The product of a number and a matrix is another matrix obtained by multiplying each of the elements of the matrix by the number. For example,

$$2\begin{bmatrix} 2 & 3 \\ 1 & 6 \end{bmatrix} = \begin{bmatrix} 4 & 6 \\ 2 & 12 \end{bmatrix}$$

A.4. Matrix inversion

The inverse $\boldsymbol{A}^{-1}$ of a matrix $\boldsymbol{A}$ serves the same role in matrix algebra that the reciprocal of a number serves in ordinary algebra. That is, for a nonzero number $\boldsymbol{A}$ in ordinary algebra,

$$a(1/a) = (1/a)a = 1$$

whereas in matrix algebra

$$\boldsymbol{A}\boldsymbol{A}^{-1} = \boldsymbol{A}^{-1}\boldsymbol{A} = \boldsymbol{I}$$

where $\boldsymbol{I}$ is an identity matrix. Multiplying a matrix by an inverse matrix is analogous to division with numbers, an operation not defined in matrix algebra. The inverse of a matrix exists only for square matrices, and for any square matrix there can exist at most only one inverse. The inverse of an $n \times n$ square matrix is another $n \times n$ square matrix. As will be seen, not all square matrices have inverses.

Finding the inverse of a large matrix is a tedious process, usually requiring a computer. The inverse matrices for 2×2 and 3×3 matrices, however, can be easily calculated by hand using the following formulas.

Given the 2×2 matrix

$$\boldsymbol{A} = \begin{bmatrix} a & b \\ c & d \end{bmatrix}$$

the inverse matrix may be found by calculating

$$\boldsymbol{A}^{-1} = \begin{bmatrix} d/D & -b/D \\ -c/D & a/D \end{bmatrix}$$

where

$$D = ad - cb$$

is the *determinant* of the 2×2 matrix $\boldsymbol{A}$.

Given the 3×3 matrix

$$\boldsymbol{B} = \begin{bmatrix} a & b & c \\ d & e & f \\ g & h & k \end{bmatrix}$$

then the inverse matrix is

$$\boldsymbol{B}^{-1} = \begin{bmatrix} p & q & r \\ s & t & u \\ v & w & x \end{bmatrix}$$

where

$$\begin{aligned} p &= (ek - hf)/D, & q &= -(bk - hc)/D, & r &= (bf - ec)/D \\ s &= -(dk - gf)/D, & t &= (ak - gc)/D, & u &= -(af - dc)/D \\ v &= (dh - ge)/D, & w &= -(ah - gb)/D, & x &= (ae - db)/D \end{aligned}$$

and the determinant D is calculated as

$$\begin{aligned} D &= a(ek - hf) - b(dk - gf) + c(dh - ge) \\ &= aek + cgf + cdh - ahf - bdk - cge \end{aligned}$$

If the determinant of the matrix to be inverted is zero, the calculations to be performed are undefined. This suggests a general rule: *a square matrix has an inverse if and only if its determinant is not equal to zero.* A matrix having a zero determinant is said to be *singular* and has no inverse.

As an example consider the 2×2 matrix

$$\boldsymbol{A} = \begin{bmatrix} 1 & 3 \\ 2 & 4 \end{bmatrix}$$

The determinant is

$$D = 1 \times 4 - 2 \times 3 = -2$$

Thus, the inverse matrix is

$$\boldsymbol{A}^{-1} = \begin{bmatrix} 4/-2 & -3/-2 \\ -2/-2 & 1/-2 \end{bmatrix} = \begin{bmatrix} -2 & 1.5 \\ 1 & -0.5 \end{bmatrix}$$

This result can be verified by multiplying the original matrix by the inverse

$$\boldsymbol{A}\boldsymbol{A}^{-1} = \begin{bmatrix} 1 & 3 \\ 2 & 4 \end{bmatrix} \begin{bmatrix} -2 & 1.5 \\ 1 & -0.5 \end{bmatrix}$$

$$= \begin{bmatrix} 1 \times (-2) + 3 \times 1 & 1 \times 1.5 + 3 \times (-0.5) \\ 2 \times (-2) + 4 \times 1 & 2 \times 1.5 + 4 \times (-0.5) \end{bmatrix} = \begin{bmatrix} 1 & 0 \\ 0 & 1 \end{bmatrix}$$

which was to be expected.

As an example of a 3×3 matrix inversion, consider the matrix

$$\boldsymbol{B} = \begin{bmatrix} 4 & 3 & 2 \\ 6 & 5 & 8 \\ 10 & 1 & 6 \end{bmatrix}$$

The determinant D is

$$D = 4 \times (5 \times 6 - 1 \times 8) - 3 \times (6 \times 6 - 10 \times 8) + 2 \times (6 \times 1 - 10 \times 5) = 132$$

$$\boldsymbol{B}^{-1} = \begin{bmatrix} \dfrac{(5 \times 6 - 1 \times 8)}{132} & \dfrac{-(3 \times 6 - 1 \times 2)}{132} & \dfrac{(3 \times 8 - 5 \times 2)}{132} \\ \dfrac{-(6 \times 6 - 10 \times 8)}{132} & \dfrac{(4 \times 6 - 10 \times 2)}{132} & \dfrac{-(4 \times 8 - 6 \times 2)}{132} \\ \dfrac{(6 \times 1 - 10 \times 5)}{132} & \dfrac{-(4 \times 1 - 10 \times 3)}{132} & \dfrac{(4 \times 5 - 6 \times 3)}{132} \end{bmatrix}$$

$$= \begin{bmatrix} 0.167 & -0.121 & 0.106 \\ 0.333 & 0.030 & -0.152 \\ -0.333 & 0.197 & 0.015 \end{bmatrix}$$

The verification that this inverse matrix is correct is left as an exercise

A special case for matrix inversion is that of a diagonal matrix. The inverse of the diagonal matrix

$$
\boldsymbol{C} = \begin{bmatrix} c_{11} & 0 & 0 & \dots & 0 \\ 0 & c_{22} & 0 & \dots & 0 \\ 0 & 0 & c_{33} & \dots & 0 \\ \vdots & \vdots & \vdots & \vdots & \vdots \\ 0 & 0 & 0 & \dots c_{nn} & \end{bmatrix}
$$

is another diagonal matrix of the form

$$
\boldsymbol{C}^{-1} = \begin{bmatrix} 1/c_{11} & 0 & 0 & \dots & 0 \\ 0 & 1/c_{22} & 0 & \dots & 0 \\ 0 & 0 & 1/c_{33} & \dots & 0 \\ \vdots & \vdots & \vdots & \vdots & \vdots \\ 0 & 0 & 0 & \dots & 1/c_{nn} \end{bmatrix}
$$

APPENDIX B

Critical Values of t

	Probability.												
n	·9	·8	·7	·6	·5	·4	·3	·2	·1	·05	·02	·01	·001
1	·158	·325	·510	·727	1·000	1·376	1·963	3·078	6·314	12·706	31·821	63·657	636·619
2	·142	·289	·445	·617	·816	1·061	1·386	1·886	2·920	4·303	6·965	9·925	31·598
3	·137	·277	·424	·584	·765	·978	1·250	1·638	2·353	3·182	4·541	5·841	12·924
4	·134	·271	·414	·569	·741	·941	1·190	1·533	2·132	2·776	3·747	4·604	8·610
5	·132	·267	·408	·559	·727	·920	1·156	1·476	2·015	2·571	3·365	4·032	6·869
6	·131	·265	·404	·553	·718	·906	1·134	1·440	1·943	2·447	3·143	3·707	5·959
7	·130	·263	·402	·549	·711	·896	1·119	1·415	1·895	2·365	2·998	3·499	5·408
8	·130	·262	·399	·546	·706	·889	1·108	1·397	1·860	2·306	2·896	3·355	5·041
9	·129	·261	·398	·543	·703	·883	1·100	1·383	1·833	2·262	2·821	3·250	4·781
10	·129	·260	·397	·542	·700	·879	1·093	1·372	1·812	2·228	2·764	3·169	4·587
11	·129	·260	·396	·540	·697	·876	1·088	1·363	1·796	2·201	2·718	3·106	4·437
12	·128	·259	·395	·539	·695	·873	1·083	1·356	1·782	2·179	2·681	3·055	4·318
13	·128	·259	·394	·538	·694	·870	1·079	1·350	1·771	2·160	2·650	3·012	4·221
14	·128	·258	·393	·537	·692	·868	1·076	1·345	1·761	2·145	2·624	2·977	4·140
15	·128	·258	·393	·536	·691	·866	1·074	1·341	1·753	2·131	2·602	2·947	4·073
16	·128	·258	·392	·535	·690	·865	1·071	1·337	1·746	2·120	2·583	2·921	4·015
17	·128	·257	·392	·534	·689	·863	1·069	1·333	1·740	2·110	2·567	2·898	3·965
18	·127	·257	·392	·534	·688	·862	1·067	1·330	1·734	2·101	2·552	2·878	3·922
19	·127	·257	·391	·533	·688	·861	1·066	1·328	1·729	2·093	2·539	2·861	3·883
20	·127	·257	·391	·533	·687	·860	1·064	1·325	1·725	2·086	2·528	2·845	3·850
21	·127	·257	·391	·532	·686	·859	1·063	1·323	1·721	2·080	2·518	2·831	3·819
22	·127	·256	·390	·532	·686	·858	1·061	1·321	1·717	2·074	2·508	2·819	3·792
23	·127	·256	·390	·532	·685	·858	1·060	1·319	1·714	2·069	2·500	2·807	3·767
24	·127	·256	·390	·531	·685	·857	1·059	1·318	1·711	2·064	2·492	2·797	3·745
25	·127	·256	·390	·531	·684	·856	1·058	1·316	1·708	2·060	2·485	2·787	3·725
26	·127	·256	·390	·531	·684	·856	1·058	1·315	1·706	2·056	2·479	2·779	3·707
27	·127	·256	·389	·531	·684	·855	1·057	1·314	1·703	2·052	2·473	2·771	3·690
28	·127	·256	·389	·530	·683	·855	1·056	1·313	1·701	2·048	2·467	2·763	3·674
29	·127	·256	·389	·530	·683	·854	1·055	1·311	1·699	2·045	2·462	2·756	3·659
30	·127	·256	·389	·530	·683	·854	1·055	1·310	1·697	2·042	2·457	2·750	3·646
40	·126	·255	·388	·529	·681	·851	1·050	1·303	1·684	2·021	2·423	2·704	3·551
60	·126	·254	·387	·527	·679	·848	1·046	1·296	1·671	2·000	2·390	2·660	3·460
120	·126	·254	·386	·526	·677	·845	1·041	1·289	1·658	1·980	2·358	2·617	3·373
∞	·126	·253	·385	·524	·674	·842	1·036	1·282	1·645	1·960	2·326	2·576	3·291

Taken from Table III of Fisher and Yates: *Statistical Tables for Biological, Agricultural and Medical Research*, published by Longman Group Ltd., London (previously published by Oliver and Boyd, Edinburgh), and by permission of the authors and publishers.

APPENDIX C

Critical Values of F, α = 0.05

5 Per Cent. Points of e^{2z}

n_2 \ n_1	1	2	3	4	5	6	8	12	24	∞
1	161·4	199·5	215·7	224·6	230·2	234·0	238·9	243·9	249·0	254·3
2	18·51	19·00	19·16	19·25	19·30	19·33	19·37	19·41	19·45	19·50
3	10·13	9·55	9·28	9·12	9·01	8·94	8·84	8·74	8·64	8·53
4	7·71	6·94	6·59	6·39	6·26	6·16	6·04	5·91	5·77	5·63
5	6·61	5·79	5·41	5·19	5·05	4·95	4·82	4·68	4·53	4·36
6	5·99	5·14	4·76	4·53	4·39	4·28	4·15	4·00	3·84	3·67
7	5·59	4·74	4·35	4·12	3·97	3·87	3·73	3·57	3·41	3·23
8	5·32	4·46	4·07	3·84	3·69	3·58	3·44	3·28	3·12	2·93
9	5·12	4·26	3·86	3·63	3·48	3·37	3·23	3·07	2·90	2·71
10	4·96	4·10	3·71	3·48	3·33	3·22	3·07	2·91	2·74	2·54
11	4·84	3·98	3·59	3·36	3·20	3·09	2·95	2·79	2·61	2·40
12	4·75	3·88	3·49	3·26	3·11	3·00	2·85	2·69	2·50	2·30
13	4·67	3·80	3·41	3·18	3·02	2·92	2·77	2·60	2·42	2·21
14	4·60	3·74	3·34	3·11	2·96	2·85	2·70	2·53	2·35	2·13
15	4·54	3·68	3·29	3·06	2·90	2·79	2·64	2·48	2·29	2·07
16	4·49	3·63	3·24	3·01	2·85	2·74	2·59	2·42	2·24	2·01
17	4·45	3·59	3·20	2·96	2·81	2·70	2·55	2·38	2·19	1·96
18	4·41	3·55	3·16	2·93	2·77	2·66	2·51	2·34	2·15	1·92
19	4·38	3·52	3·13	2·90	2·74	2·63	2·48	2·31	2·11	1·88
20	4·35	3·49	3·10	2·87	2·71	2·60	2·45	2·28	2·08	1·84
21	4·32	3·47	3·07	2·84	2·68	2·57	2·42	2·25	2·05	1·81
22	4·30	3·44	3·05	2·82	2·66	2·55	2·40	2·23	2·03	1·78
23	4·28	3·42	3·03	2·80	2·64	2·53	2·38	2·20	2·00	1·76
24	4·26	3·40	3·01	2·78	2·62	2·51	2·36	2·18	1·98	1·73
25	4·24	3·38	2·99	2·76	2·60	2·49	2·34	2·16	1·96	1·71
26	4·22	3·37	2·98	2·74	2·59	2·47	2·32	2·15	1·95	1·69
27	4·21	3·35	2·96	2·73	2·57	2·46	2·30	2·13	1·93	1·67
28	4·20	3·34	2·95	2·71	2·56	2·44	2·29	2·12	1·91	1·65
29	4·18	3·33	2·93	2·70	2·54	2·43	2·28	2·10	1·90	1·64
30	4·17	3·32	2·92	2·69	2·53	2·42	2·27	2·09	1·89	1·62
40	4·08	3·23	2·84	2·61	2·45	2·34	2·18	2·00	1·79	1·51
60	4·00	3·15	2·76	2·52	2·37	2·25	2·10	1·92	1·70	1·39
120	3·92	3·07	2·68	2·45	2·29	2·17	2·02	1·83	1·61	1·25
∞	3·84	2·99	2·60	2·37	2·21	2·10	1·94	1·75	1·52	1·00

Taken from Table V of Fisher and Yates: *Statistical Tables for Biological, Agricultural and Medical Research*, published by Longman Group Ltd., London (previously published by Oliver and Boyd, Edinburgh), and by permission of the authors and publishers.

Subject Index

F

N

O

P

Q

R

S

T

V

W